VOLUME ONE HUNDRED AND ONE

SEMICONDUCTORS AND SEMIMETALS

Future Directions in Silicon Photonics

SERIES EDITORS

CHENNUPATI JAGADISH

Distinguished Professor
Department of Electronic Materials Engineering
Research School of Physics and Engineering
Australian National University
Canberra, ACT2601, Australia

ZETIAN MI

Professor
Department of Electrical Engineering and Computer Science
University of Michigan
1310 Beal Avenue
Ann Arbor, MI 48109
United States of America

VOLUME ONE HUNDRED AND ONE

SEMICONDUCTORS AND SEMIMETALS

Future Directions in Silicon Photonics

Edited by

SEBASTIAN LOURDUDOSS
Department of Applied Physics,
KTH-Royal Institute of Technology,
Stockholm, Sweden

JOHN E. BOWERS
Institute for Energy Efficiency,
University of California Santa Barbara,
Santa Barbara, CA, United States

CHENNUPATI JAGADISH
Department of Electronic Materials Engineering
Research School of Physics and Engineering
Australian National University
Canberra, ACT2601, Australia

ELSEVIER

ACADEMIC PRESS
An imprint of Elsevier

Academic Press is an imprint of Elsevier
50 Hampshire Street, 5th Floor, Cambridge, MA 02139, United States
525 B Street, Suite 1650, San Diego, CA 92101, United States
The Boulevard, Langford Lane, Kidlington, Oxford OX5 1GB, United Kingdom
125 London Wall, London, EC2Y 5AS, United Kingdom

First edition 2019

ISBN: 978-0-12-818857-6
ISSN: 0080-8784

For information on all Academic Press publications
visit our website at https://www.elsevier.com/books-and-journals

Publisher: Zoe Kruze
Acquisition Editor: Jason Mitchell
Editorial Project Manager: Peter Llewellyn
Production Project Manager: Abdulla Sait
Cover Designer: Greg Harris

Typeset by SPi Global, India

Contents

Contributors

Carlos Alonso-Ramos
Centre for Nanoscience and Nanotechnology (C2N), CNRS, Université Paris Sud, Université Paris Saclay, Palaiseau, France. (ch1)

Yasuhiko Arakawa
Institute for Nano Quantum Information Electronics, The University of Tokyo, Tokyo, Japan. (ch3)

Marina Baryshnikova
Imec, Leuven, Belgium. (ch8)

Charles Baudot
Technology R&D, STMicroelectronics SAS, Crolles, France. (ch1)

Badhise Ben Bakir
University Grenoble Alpes and CEA, LETI, Grenoble, France. (ch1)

Daniel Benedikovic
Centre for Nanoscience and Nanotechnology (C2N), CNRS, Université Paris Sud, Université Paris Saclay, Palaiseau, France. (ch1)

Mathias Berciano
Centre for Nanoscience and Nanotechnology (C2N), CNRS, Université Paris Sud, Université Paris Saclay, Palaiseau, France. (ch1)

Frédéric Bœuf
Technology R&D, STMicroelectronics SAS, Crolles, France. (ch1)

John Bowers
Institute for Energy Efficiency, University of California Santa Barbara, Santa Barbara, CA, United States. (ch9)

Pierre Brianceau
University Grenoble Alpes and CEA, LETI, Grenoble, France. (ch1)

Eric Cassan
Centre for Nanoscience and Nanotechnology (C2N), CNRS, Université Paris Sud, Université Paris Saclay, Palaiseau, France. (ch1)

Ismael Charlet
Technology R&D, STMicroelectronics SAS, Crolles, France. (ch1)

Sébastien Cremer
Technology R&D, STMicroelectronics SAS, Crolles, France. (ch1)

Pedro Damas
Centre for Nanoscience and Nanotechnology (C2N), CNRS, Université Paris Sud, Université Paris Saclay, Palaiseau, France. (ch1)

Yannick De Koninck
Imec, Leuven, Belgium. (ch8)

Lucas Deniel
Centre for Nanoscience and Nanotechnology (C2N), CNRS, Université Paris Sud, Université Paris Saclay, Palaiseau, France. (ch1)

Phuong T. Do
Centre for Nanoscience and Nanotechnology (C2N), CNRS, Université Paris Sud, Université Paris Saclay, Palaiseau, France. (ch1)

Dorian Doser
Centre for Nanoscience and Nanotechnology (C2N), CNRS, Université Paris Sud, Université Paris Saclay, Palaiseau, France. (ch1)

Maurin Douix
Centre for Nanoscience and Nanotechnology (C2N), CNRS, Université Paris Sud, Université Paris Saclay, Palaiseau; Technology R&D, STMicroelectronics SAS, Crolles, France. (ch1)

Elena Duran Valdeiglesias
Centre for Nanoscience and Nanotechnology (C2N), CNRS, Université Paris Sud, Université Paris Saclay, Palaiseau, France. (ch1)

Jocelyn Durel
Technology R&D, STMicroelectronics SAS, Crolles, France. (ch1)

Franck Fournel
University Grenoble Alpes and CEA, LETI, Grenoble, France. (ch1)

Elodie Ghegin
Technology R&D, STMicroelectronics SAS, Crolles, France. (ch1)

Sylvain Guerber
Centre for Nanoscience and Nanotechnology (C2N), CNRS, Université Paris Sud, Université Paris Saclay, Palaiseau; Technology R&D, STMicroelectronics SAS, Crolles, France. (ch1)

Karim Hassan
University Grenoble Alpes and CEA, LETI, Grenoble, France. (ch1)

Christophe Jany
University Grenoble Alpes and CEA, LETI, Grenoble, France. (ch1)

Bernardette Kunert
Imec, Leuven, Belgium. (ch8)

Nadezda Kuznetsova
Imec, Leuven, Belgium. (ch8)

Jinkwan Kwoen
Institute for Nano Quantum Information Electronics, The University of Tokyo, Tokyo, Japan. (ch3)

Christian Lafforgue
Centre for Nanoscience and Nanotechnology (C2N), CNRS, Université Paris Sud, Université Paris Saclay, Palaiseau, France. (ch1)

Robert Langer
Imec, Leuven, Belgium. (ch8)

Kei May Lau
Department of Electronic and Computer Engineering, Hong Kong University of Science and Technology, Kowloon, Hong Kong. (ch7)

Xavier Le Roux
Centre for Nanoscience and Nanotechnology (C2N), CNRS, Université Paris Sud, Université Paris Saclay, Palaiseau, France. (ch1)

Sebastian Lourdudoss
Department of Applied Physics, KTH-Royal Institute of Technology, Stockholm, Sweden. (ch5)

Peter Ludewig
NAsP III/V GmbH, Marburg, Germany. (ch6)

Guillaume Marcaud
Centre for Nanoscience and Nanotechnology (C2N), CNRS, Université Paris Sud, Université Paris Saclay, Palaiseau, France. (ch1)

Delphine Marris-Morini
Centre for Nanoscience and Nanotechnology (C2N), CNRS, Université Paris Sud, Université Paris Saclay, Palaiseau, France. (ch1)

Shinji Matsuo
NTT Device Technology Laboratories, NTT Corporation, Atsugi, Kanagawa, Japan. (ch2)

Yves Mols
Imec, Leuven, Belgium. (ch8)

Stéphane Monfray
Technology R&D, STMicroelectronics SAS, Crolles, France. (ch1)

Takahiro Nakamura
Photonics Electronics Technology Research Association, The University of Tokyo, Tokyo, Japan. (ch3)

Shigeya Naritsuka
Department of Materials Science and Engineering, Faculty of Science and Technology, Meijo University, Nagoya, Japan. (ch4)

Fabrice Nemouchi
University Grenoble Alpes and CEA, LETI, Grenoble, France. (ch1)

Justin Norman
Institute for Energy Efficiency, University of California Santa Barbara, Santa Barbara, CA, United States. (ch9)

Marianna Pantouvaki
Imec, Leuven, Belgium. (ch8)

Diego Perez-Galacho
Centre for Nanoscience and Nanotechnology (C2N), CNRS, Université Paris Sud, Université Paris Saclay, Palaiseau, France. (ch1)

Joan-Manel Ramirez
Centre for Nanoscience and Nanotechnology (C2N), CNRS, Université Paris Sud, Université Paris Saclay, Palaiseau, France. (ch1)

Philippe Rodriguez
University Grenoble Alpes and CEA, LETI, Grenoble, France. (ch1)

Loic Sanchez
University Grenoble Alpes and CEA, LETI, Grenoble, France. (ch1)

Samuel Serna
Centre for Nanoscience and Nanotechnology (C2N), CNRS, Université Paris Sud, Université Paris Saclay, Palaiseau, France. (ch1)

Bei Shi
Department of Electronic and Computer Engineering, Hong Kong University of Science and Technology, Kowloon, Hong Kong. (ch7)

Yuting Shi
Photonics Research Group, Ghent University-imec, Ghent, Belgium. (ch8)

Wolfgang Stolz
Structure and Technology Research Laboratory, Material Sciences Center and Department of Physics, Philipps-University; NAsP III/V GmbH, Marburg, Germany. (ch6)

Yan-Ting Sun
Department of Applied Physics, KTH-Royal Institute of Technology, Stockholm, Sweden. (ch5)

Bertrand Szelag
University Grenoble Alpes and CEA, LETI, Grenoble, France. (ch1)

Vladyslav Vakarin
Centre for Nanoscience and Nanotechnology (C2N), CNRS, Université Paris Sud, Université Paris Saclay, Palaiseau, France. (ch1)

Joris Van Campenhout
Imec, Leuven, Belgium. (ch8)

Dries Van Thourhout
Photonics Research Group, Ghent University-imec, Ghent, Belgium. (ch8)

Léopold Virot
Centre for Nanoscience and Nanotechnology (C2N), CNRS, Université Paris Sud, Université Paris Saclay, Palaiseau; Technology R&D, STMicroelectronics SAS, Crolles; University Grenoble Alpes and CEA, LETI, Grenoble, France. (ch1)

Laurent Vivien
Centre for Nanoscience and Nanotechnology (C2N), CNRS, Université Paris Sud, Université Paris Saclay, Palaiseau, France. (ch1)

Kerstin Volz
Structure and Technology Research Laboratory, Material Sciences Center and Department of Physics, Philipps-University, Marburg, Germany. (ch6)

Yating Wan
Institute for Energy Efficiency, University of California Santa Barbara, Santa Barbara, CA, United States. (ch9)

JianHao Zhang
Centre for Nanoscience and Nanotechnology (C2N), CNRS, Université Paris Sud, Université Paris Saclay, Palaiseau, France. (ch1)

Preface

It is with pleasure we are bringing out this volume entitled *Future Directions in Silicon Photonics* to provide future directions in this emerging field and to expand on research covered in the previous volume on *Silicon Photonics*. The motivation has become even stronger now than when we published the first volume as more and more industries have actively joined to implement silicon photonics in their portfolio as its anticipated market should reach 1 billion USD by 2023 with the compound annual growth rate (CAGR) of ~22% between 2017 and 2023.[a]

This volume collects the state-of-the-art results on achieving silicon photonic components and circuits and future directions toward fulfilling the promise of silicon photonics.

In Chapter 1, *Building blocks of silicon photonics*, Laurent Vivien and coworkers present a useful introduction to silicon photonics by considering the functionalities of the individual components that make up silicon photonic integrated circuits. After an adequate description of the physical phenomena used by the cohabiting components on silicon, a hybrid integration strategy is exemplified via demonstration of hybrid optical modulations and hybrid III–V/Si transmitter.

Chapters 2 and 3 deal with heterogeneously integrated devices by means of a bonding approach.

In Chapter 2, *Heterogeneously integrated III–V photonic devices on Si*, Shinji Matsuo presents several integrated devices of III–V/Si achieved by wafer/die bonding. With adequate prior physical modeling, a laterally injected membrane laser containing DFB and DBR sections and a λ-scale embedded active region photonic crystal (LEAP) lasers both integrated on Si/SiO_2 waveguide and emitting at 1.5 μm have been demonstrated. The latter exhibits record-low 4.4 fj/bit operating transmission energy. Mach–Zehnder modulators incorporating InGaAsP/Si MOS capacitor with $V_\pi L$ performance of 0.09 Vcm are also demonstrated. Schemes for chip-to-chip and on-chip interconnects are also deviced.

In Chapter 3, *Quantum dot lasers for silicon photonics*, Yasuhiko Arakawa and coworkers present the usefulness of quantum dot (QD) lasers for silicon photonics. After summarizing their pioneering work from its

[a] https://www.marketresearchfuture.com/reports/silicon-photonics-market-2809

infancy and the evolution of QD laser, the authors argue for its supremacy for silicon photonics. Flip-chip bonding of 1.3-μm QD lasers on silicon photonic integrated circuits is treated in detail taking into account the alignment issues. Low coupling loss of 3 dB between QD laser and silicon photonic integrated circuits is achieved. Furthermore, the feeble temperature sensitivity inherent to QD laser is demonstrated in the temperature range of 25–125 °C when it is used with a silicon optical interposer for error-free data links up to 12 Gb/s. Wafer bonding of QD laser directly onto Si and via metal and also on a waveguide for an evanescent laser is treated, and impressive results are presented. In such hybrid lasers, DFB and DBR sections are also involved. In addition to bonding approaches, the epitaxial growth of QD lasers on silicon in the 1.3-μm wavelength range is presented.

Chapters 4–9 deal with epitaxial approaches for silicon photonics. Several different methods are undertaken by the authors of these chapters.

In Chapter 4, *Microchannel epitaxy of III–V layers on Si substrates*, Shigeya Naritsuka provides an extensive treatment on microchannel epitaxy (MCE) useful for dislocation reduction in the highly mismatched epitaxial growth. Besides using equilibrium epitaxial techniques such as liquid-phase epitaxy and chloride-transported epitaxy for controlling the directional growth by facet formation, a nonequilibrium technique such as molecular beam epitaxy has been put into use to demonstrate superior directional growth. The principles of various forms of MCE employed are described with ample experimental results. The results demonstrate a dramatic reduction of dislocations in GaAs/Si and InP/Si and dislocation-free areas large enough for fabricating vertical cavity surface-emitting lasers (VCSELs) on silicon.

In Chapter 5, *Epitaxial lateral overgrowth of III–V semiconductors on Si for photonic integration*, Yan-Ting Sun and coworkers treat the extension of MCE by conducting epitaxial lateral overgrowth (ELOG) of InP on Si and demonstrate a very good optical quality of InP. They demonstrate the feasibility of fabricating 1.55-μm microdisk lasers from multiquantum wells on ELOG InP/Si. Besides, a modified form of ELOG called corrugated epitaxial lateral overgrowth (CELOG) has been proposed to obtain direct III–V/Si heterojunctions, which was not feasible with ELOG. CELOG InP/Si exhibits low threading dislocation density at the interface and enhanced carrier lifetime, which resulted in proof-of-concept photoresponse of the processed InP/Si heterojunction photodiode.

In Chapter 6, *Monolithic integration of lattice-matched Ga(NAsP)-based laser structures on CMOS-compatible Si(001) wafers for Si-photonics applications*, K. Volz and coworkers present the current status of understanding for the novel lattice-matched Ga(NAsP)-based laser material as a possible solution for monolithically integrated optoelectronic and photonic functionalities on CMOS-compatible Si(001) substrates. The epitaxial growth by applying metal-organic vapor-phase epitaxy (MOVPE) and the detailed evaluation of the structural, electrical, electronic as well as optoelectronic properties are detailed. Electroluminescence in the range of 850–1200 nm has been demonstrated with careful optimization of structure and growth. In addition, the unique integration scenario into the standard CMOS-production process for the realization of Si-photonics-integrated circuits is outlined.

In Chapter 7, *Growth of III–V semiconductors and lasers on silicon substrates by MOCVD*, Bei Shi and coworkers demonstrate an attractive technology to prepare III–V devices directly on silicon by initiating growth from V-grooves in Si and subsequent planarization. The authors address fundamental materials challenges related to achieving compliant substrates and routes to overcome them. GaAs/Si and InP/Si are the main focus, and they have achieved 1.3 μm as well as 1.55 μm electrically pumped lasers.

In Chapter 8, *Nano-ridge laser monolithically grown on (001) Si*, Dries Van Thourhourt and coworkers present yet another design for growing III–V devices on silicon by initiating growth from silicon trenches. They exploit aspect ratio trapping and nano-ridge engineering at the same time for novel integration approaches. On the GaAs nano-ridges on Si, InGaAs/GaAs-based lasers operating with optical pumping have been demonstrated.

In Chapter 9, *Quantum dot microcavity lasers on silicon substrates*, Yating Wan and coworkers consider microcavity lasers on silicon for small-footprint and low-threshold light sources. The dense, energetically confined, and spatially isolated QD gain medium is utilized for circumventing crystal defects in heteroepitaxy, as well as for scaling to ultra-small dimensions in microcavities. The former allows for laser growth and processing up to 450 mm Si wafer with minimal compromise in light emission efficiency, and the latter enables device miniaturization for ultra-dense photonic integration with complex functionality and economy of scale. They provide an introduction to this technology followed by technological demonstrations of QD microcavity devices. In addition, the outlook of using III–V/Si epitaxy to form a bandwidth scale-up, energy scale-down, and volume manufacturable Si photonics solution is presented.

As we mentioned in Volume 99 of this series on *Silicon Photonics*, this field is truly a multidisciplinary one involving physics, materials science, photonics, and advanced technologies. Already between the publication of Volume 99 and this volume on *Future Directions in Silicon Photonics*, we are witnessing tremendous developments in this field. Breakthroughs in reliability of the devices rapidly progress to a point warranting discussion for near-term commercial viability. Hence to end this preface we repeat what we wrote in the previous volume: "Everyone involved in silicon photonics will find in this volume solutions to solve one or more problems that one may face today or ideas to advance in their endeavour. We hope that this volume will be a useful reference book for the scientists and engineers striving for large-volume integrated silicon photonic devices and circuits with high performance, large integration density, low footprints, low-cost, low energy consumption and low latency."

We would like to thank all the contributing authors for sharing their knowledge and experience by putting in lot of their precious time. We also thank Peter Llewellyn, Jason Mitchell, and the whole production team at Elsevier.

Sebastian Lourdudoss
KTH—Royal Institute of Technology, Stockholm, Sweden
John E. Bowers
University of California Santa Barbara, United States
Chennupati Jagadish
Australian National University, Canberra, Australia

CHAPTER ONE

Building blocks of silicon photonics

Laurent Vivien[a,*], Charles Baudot[b], Frédéric Bœuf[b], Bertrand Szelag[c], Carlos Alonso-Ramos[a], Daniel Benedikovic[a], Delphine Marris-Morini[a], Eric Cassan[a], Sylvain Guerber[a,b], Maurin Douix[a,b], Léopold Virot[a,b,c], Philippe Rodriguez[c], Fabrice Nemouchi[c], Christophe Jany[c], Badhise Ben Bakir[c], Xavier Le Roux[a], Diego Perez-Galacho[a], Mathias Berciano[a], Guillaume Marcaud[a], Ismael Charlet[b], Lucas Deniel[a], Christian Lafforgue[a], JianHao Zhang[a], Samuel Serna[a], Pedro Damas[a], Phuong T. Do[a], Dorian Doser[a], Jocelyn Durel[b], Elodie Ghegin[b], Vladyslav Vakarin[a], Joan-Manel Ramirez[a], Stéphane Monfray[b], Sébastien Cremer[b], Elena Duran Valdeiglesias[a], Loic Sanchez[c], Franck Fournel[c], Pierre Brianceau[c], Karim Hassan[c]

[a]Centre for Nanoscience and Nanotechnology (C2N), CNRS, Université Paris Sud, Université Paris Saclay, Palaiseau, France
[b]Technology R&D, STMicroelectronics SAS, Crolles, France
[c]University Grenoble Alpes and CEA, LETI, Grenoble, France
*Corresponding author: e-mail address: laurent.vivien@u-psud.fr

Contents

Semiconductors and Semimetals, Volume 101
ISSN 0080-8784
https://doi.org/10.1016/bs.semsem.2019.07.006

1. Introduction

Silicon is one of those materials that drastically changed the scientific, technological and societal world from the invention of the transistor by Bell Labs in 1947 (Ross, 1998). The amazing shrinkage in feature size and the increase in transistor numbers (Texas Instruments, n.d.) led the development of the electronics industry toward cheaper, smaller and faster devices. This integrated circuits (IC) evolution has primarily been driven by the so-called Moore's law, announced by Gordon Moore in 1965, predicting an exponential increase of the number of transistors per chip each 18 months (Moore, 1965).

Silicon (Si) is unbeatably the preferred material for the semiconductor industry because it has very interesting characteristics that may be advantageously exploited. In particular, (i) it is the second most abundant element on earth after oxygen, making it a very cheap material; (ii) its energy bandgap of 1.1 eV allows for efficient devices at room temperature; (iii) it presents higher processing temperature than other common semiconductors; and (iv) it is also ease of growing silicon dioxide and related materials on Si, allowing for fabrication techniques such as parallel and planar circuit processing (Grimmeiss and Kasper, 2009). Year after year, highly efficient silicon-based transistors and ICs have been developed. However, recently used metal-based electrical interconnects (EIs) are reaching their performance limits and are facing difficulties to keep up with the surge in the data rate demand over the last decade. The communications bottleneck is identified as one of the grand challenges in the progress of silicon computation and can be stated as: while individual logic elements have become significantly faster, computational speed is limited by the communication between different parts of a processor (Friedman et al., 2005). The main limitations of EI's are, among others, the constrained interconnect shrinkage, high-energy consumption, and signal delay, as well as limited bandwidth and data rate (Friedman et al., 2005; Miller, 2009; Ron et al., 2001). Particularly, power consumption in data processing is now so large that it is starting to be environmentally significant. To overcome this limitation, optical interconnects (OIs), where an optical signal is used instead of an electrical signal, have been considered, for many years, to provide a solution to the communication bottleneck and to energy consumption by replacing electrical wires with faster and low power optical devices in integrated circuits (Dangel et al., 2015). This visionary objective in the 1990s has matured, and year by year, silicon has become the revolutionary platform for photonics, especially through the emergence of versatile silicon-on-insulator (SOI) substrates. Indeed, additionally to extraordinary electrical properties, silicon also offers a range of excellent optical properties, including its large wavelength transparency window from 1.1 up to 7 μm. Furthermore, the large refractive index contrast between silicon, Si ($n_{Si} \approx 3.47$) and silicon dioxide, SiO_2 ($n_{SiO_2} \approx 1.45$) enables development of compact devices that are suitable for large-scale and high-density integration on a chip. The strong evolution of silicon photonics has opened new routes for many applications, especially in all-optical communication systems. Indeed, the rapid increase of the digital data volume circulated around the globe, mainly within bandwidth-hungry datacenters, modified the horizon of silicon photonics. This evolution is driven by the change of Internet communications used via mobile devices and the development of the "Internet of Things." Therefore, some new challenges have to be considered in terms of speed, power consumption, flexibility, and reliability for the implementation of optical transceivers in the future datacenters.

For a few years, some industrial solutions combining both photonics and electronics have been available in order to address datacenter requirements (http://www.luxtera.com; https://www.intel.com/content/www/us/en/architecture-and-technology/silicon-photonics/silicon-photonics-overview.html). In parallel, due to the low-cost and high-volume capability of silicon photonics, biological and chemical sensors have also become available as commercial products (https://www.genalyte.com; Densmore et al., 2009). As reported in Fig. 1, nowadays, transceivers for datacenter and packaged resonators for bio-and chemical sensing are the first silicon photonics products available in the market. This fast evolution from fundamental physical concepts in the 1980s toward products has generated a strong interest in others applications. Indeed, additionally to the development of the next generation of devices and circuits for advanced datacom, including wavelength division multiplexing systems (Dong, 2016), all optical signal processing (Koos et al., 2009), and intra-processor links (Atabaki et al., 2018; Sun et al., 2015), new applications, circuits, and devices are now considered: optical gyroscopes (Srinivasan et al., 2014), on-chip amplifiers (Rönn et al., 2019), LIDAR (Martin et al., 2018), optical spectroscopy (Liu et al., 2018), Quantum photonics (Sibson et al., 2017), microwave and opto-RF photonics circuits (Pérez et al., 2017), to name a few.

Efficient on-chip devices to emit, modulate, and detect light have to be developed in the silicon photonics platform (Fig. 2). Thereby, numerous

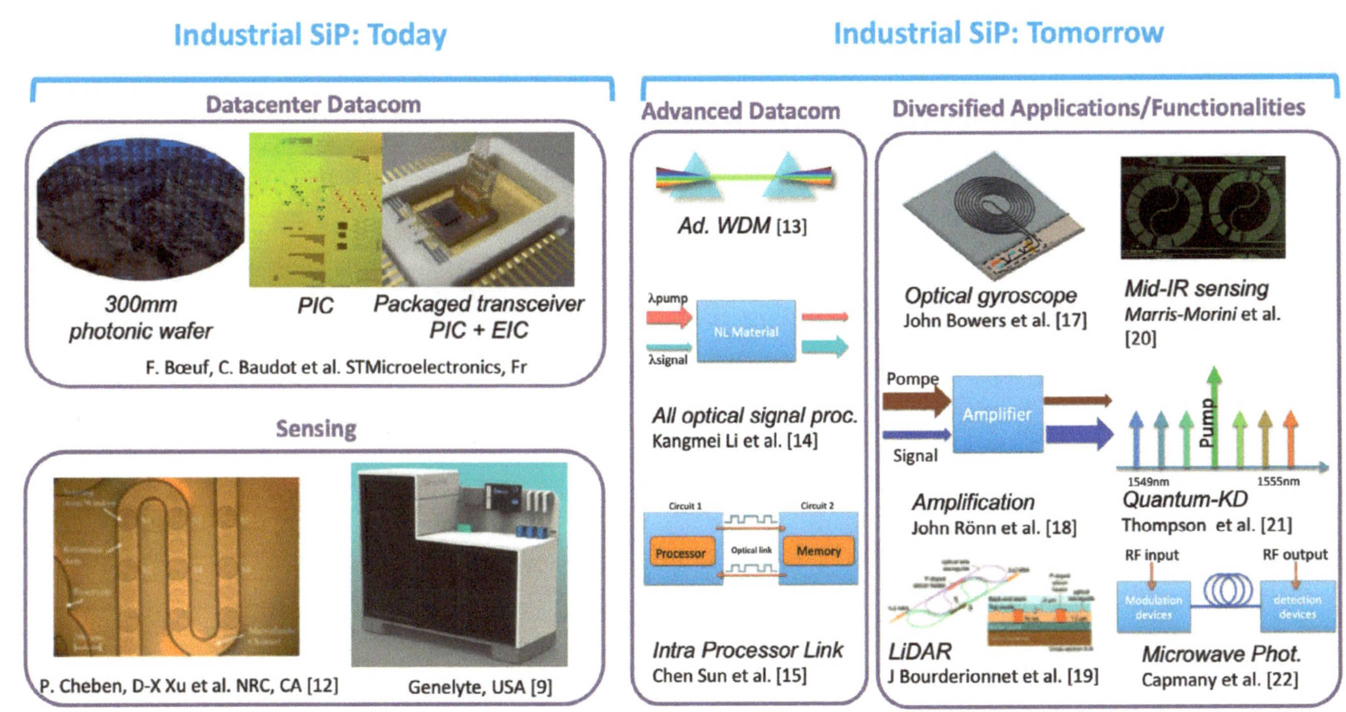

Fig. 1 Evolution scheme of the industrial silicon photonics (Vivien et al., 2018). PIC, photonic integrated circuits; EIC, electronic integrated circuits; Ad. WDM, advanced wavelength division multiplexing; Proc., processors; quantum-KD, quantum key distribution.

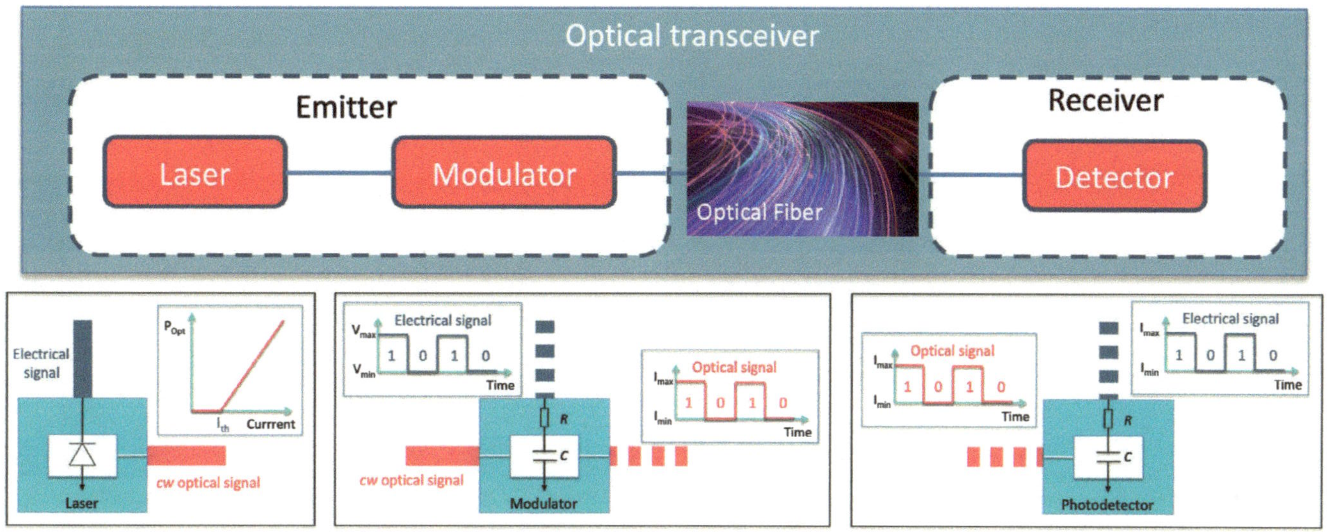

Fig. 2 Optical transceiver: main building blocks including laser, modulator and detector.

devices have then been demonstrated to realize a complete optical link, including efficient input/output (I/O) light couplers (Alonso-Ramos et al., 2014; Baudot et al., 2014; Benedikovic et al., 2015a,b, 2017; Boeuf et al., 2016; Chen et al., 2016a, 2017a,b; Ding et al., 2015; Halir et al., 2009; Kopp et al., 2011; Li et al., 2014; Mekis et al., 2011; Roelkens et al., 2006; Snyder and O'Brien, 2013; Taillaert et al., 2002, 2004; Vermeulen et al., 2010; Wang et al., 2010; Watanabe et al., 2017; Xu et al., 2014; Yang et al., 2014; Zaoui et al., 2014), low-loss passive waveguides, and optical resonators (Bogaerts et al., 2012), high-speed modulators (Ding et al., 2013; Ferrotti et al., 2016a,b; Kupijai et al., 2016; Li et al., 2017; Pérez-Galacho et al., 2015; Perez-Galacho et al., 2017; Romero-García et al., 2017; Shao et al., 2017; Sharif Azadeh et al., 2014; Streshinsky et al., 2013; Thomson et al., 2013; Vermeulen et al., 2016; Wang et al., 2013, 2017a; Xiao et al., 2017; Xiong et al., 2015; Xu et al., 2013; Yong et al., 2017; Zheng et al., 2017), as well as high-speed and high-responsivity photodetectors (Assefa et al., 2010a; Byrd et al., 2017; Chen et al., 2015a, 2016b; DeRose et al., 2011; Going et al., 2015; Li et al., 2012; Liao et al., 2011; Lischke et al., 2015; Michel et al., 2010; Nam et al., 2015; Okyay et al., 2006; Virot et al., 2017; Vivien et al., 2009, 2012; Zhang et al., 2014). Due to the huge impact of silicon photonics in many application domains, most of these components are recently fabricated in accessible industrial and academic foundries via multi-project wafer (MPW) shuttle services (CEA/Leti, IMEC, AIM Photonics, IME, ...) that are widely available to the silicon photonics community. Among all devices, the optical source on silicon remains a daily challenge, because silicon is an indirect band semiconductor material with poor efficiency in light emission. From band and strain engineering, some promising results have been obtained with germanium-on-silicon lasers (Camacho-Aguilera et al., 2012; El Kurdi et al., 2016). However, today, the most common approach is based on hybrid co-integration of III–V materials with silicon (Ben Bakir et al., 2011; de Valicourt et al., 2012; Fang et al., 2006; Jones et al., 2019; Keyvaninia et al., 2013; Lamponi et al., 2012; Le Liepvre et al., 2013; Wada and Kamijoh, 1996; Wang et al., 2017b), well-known for their efficient laser properties in a wide wavelength range.

In this chapter, an overview of the main silicon-based building blocks, including passive devices, optical modulators, photodetectors, and light sources is presented. As pure silicon devices to generate, modulate, and detect light are not realistic yet, hybrid integration of active materials on the silicon photonics platform is also discussed in the last section with the presentation of few examples that highlight benefits of hybrid silicon photonics.

2. Silicon-based building blocks

An integrated optical link, i.e., optical transceiver, is generally formed at least by three main optoelectronic elements, as schematically illustrated in Fig. 2: the emitters, responsible for the electrical-optical signal conversion; the analogic or digital optical modulators to code the information, and the receivers, where the optical signal is converted back to an electrical domain. To complete the link, structures, which guide the light into the different elements of the photonic IC and ensure the in/out light coupling from/to optical fiber are included. Therefore, each building block has to be implemented so that its overall speed, size, the global loss (insertion and propagation losses), and energy consumption satisfies the required limits, driven by the target application as well as its compatibility with silicon technology in microelectronics pilot lines. For optical interconnects, the energy consumption of transceivers should be 50–200 fJ/bit and should meet the global demands in the near future, i.e., this value should be lowered to 10 fJ/bit, according to an analysis presented by Miller (David and Miller, 1997).

2.1 Passive devices

The objective of this section is not to report on an exhaustive overview of all possibilities afforded by silicon photonics for the development of passive devices. The aim is to present the two fundamental blocks necessary to build up a photonic circuit: waveguides as basic component, allowing the implementation of more complex devices such as resonators (Bogaerts et al., 2012), wavelength multiplexers (Cheben et al., 2007), and modulators (Perez-Galacho et al., 2017), to list a few, and the surface grating couplers as a direct link toward the external world accessed through the optical fibers.

Thanks to the high index contrast between silicon and silicon dioxide cladding in the SOI platform, strong modal confinement is readily achievable. This leads to a great flexibility in light guiding on a chip, i.e., compact waveguides, splitters and resonators of sharp bends. Several types of waveguide geometries have been implemented in silicon, each targeting a different optical functionality, while providing specific advantages and drawbacks. Fig. 3 schematically shows the four most commonly used waveguide geometries. Main characteristics of these waveguides are briefly summarized below:

- *Strip waveguide*: This geometry relies on a full-etch step fabrication process, resulting in a high index contrast confinement in both vertical and horizontal directions. Such waveguide configuration allows tight bending radii, of around 5 μm (Bogaerts et al., 2012), which make them

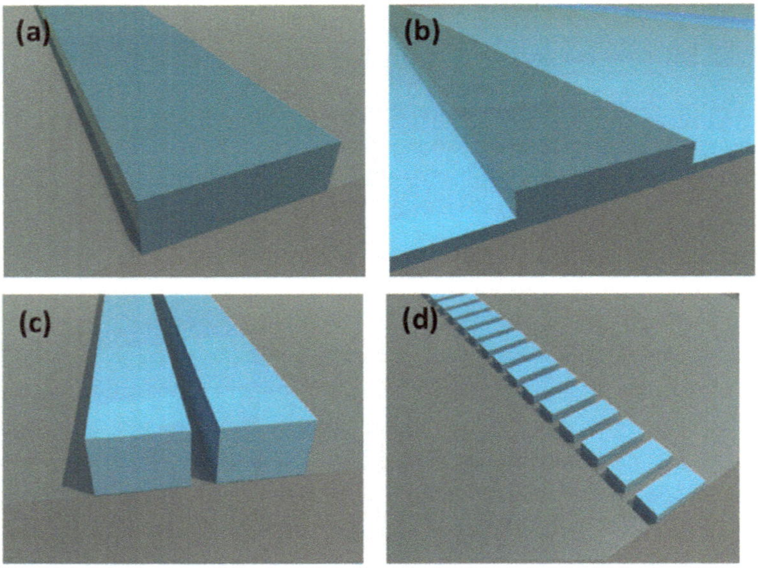

Fig. 3 Schematic representation of most commonly used waveguide geometries in Si photonics: (A) strip, (B) rib, (C) slot, (D) sub-wavelength.

very interesting for the implementation of ultra-compact devices. On the other hand, strip waveguides are very sensitive to sidewall roughness, resulting in comparatively high propagation loss, with typical values of dB/cm (Boeuf et al., 2016).

- *Rib waveguide:* A shallow- or deep-etch waveguide arrangement is used to reduce lateral index contrast, yielding a reduction of the propagation loss (down to sub-dB/cm (Lardenois et al., 2003)) and relaxing conditions for single-mode operation (Soref et al., 1991). Rib waveguides are commonly used for opto-electronic devices, especially modulators and photodetectors, as the lateral Si slab allows putting doping regions apart from the guiding zone.
- *Slot waveguide*: Unlike conventional strip and rib geometries, slot waveguides exploit boundary conditions for light polarized in the chip plane, to funnel the light within the low-index region between the two silicon rails (Almeida et al., 2004). This particular guiding mechanism, with a remarkably strong confined evanescent field, make slot waveguides a very promising approach for the hybrid integration of active materials (Durán-Valdeiglesias et al., 2016) or the implementation of evanescent-wave sensors (Passaro and La Notte, 2012). Still, higher

propagation loss may arise due to the strong interaction of light with sidewall roughness, requiring tight control of fabrication processes (Passaro and La Notte, 2012).

- *Sub-wavelength waveguides*: By periodically etching silicon with a pitch shorter than half of the wavelength, it is possible to suppress diffraction effects, implementing a metamaterial waveguide with engineered optical properties (Halir et al., 2014). Sub-wavelength waveguides exhibit a propagation loss comparable to that of conventional strip counterparts (Halir et al., 2014), yet with flexible confinement and dispersion engineering that allowed the demonstration of low-loss crossings (Bock et al., 2010) and high-efficiency fiber-chip couplers (Cheben et al., 2015). However, trade-offs between the mode confinement engineering and the bending radius are typically required (Wang et al., 2016).

From these waveguide geometries, numerous possibilities are offered by adjusting the height and width of the waveguides. Furthermore, in order to reduce the two photon absorption (TPA) in silicon at high optical power and the phase error due to the strong light coupling into Si waveguides, Silicon Nitride integration can also be considered as an alternative.

As silicon photonics applications require multi-interchip connections via optical fibers, efficient coupling interfaces are recognized as essential functional components in planar lightwave circuits. The light coupling is a non-trivial task, especially for high-index contrast platforms such as silicon-on-insulator, due to the large mode size disparity between modes of sub-micron waveguide cores and optical fibers, since dimensions of such waveguides are around two-orders of magnitude smaller than that of standard optical fibers (SMF-28).

The fiber-chip surface grating couplers (SGCs) have appeared as an elementary, yet recently popular approach to couple the light to or from integrated waveguide circuits, as demonstrated by seminal works of Dakss et al. (1970) and Tamir and Peng (1977). Over the years, the SGCs have shown marked maturity to afford efficient, robust, and scalable coupling functionality for off-chip optical fiber attachments (Alonso-Ramos et al., 2014; Baudot et al., 2014; Benedikovic et al., 2015a,b, 2017; Boeuf et al., 2016; Chen et al., 2016a, 2017a,b; Ding et al., 2015; Halir et al., 2009; Kopp et al., 2011; Li et al., 2014; Mekis et al., 2011; Roelkens et al., 2006; Snyder and O'Brien, 2013; Taillaert et al., 2002, 2004; Vermeulen et al., 2010; Wang et al., 2010; Watanabe et al., 2017; Xu et al., 2014; Yang et al., 2014; Zaoui et al., 2014). The SGCs do not require additional post-manufacturing steps for a chip facet preparation such as dicing, polishing, and cleaving, while enabling flexible positioning on the chip with cost-effective wafer-scale accessories, as well as the compatibility with integration and packaging scenarios (Boeuf et al., 2016; Kopp et al., 2011; Li et al., 2014; Mekis et al., 2011; Snyder and O'Brien, 2013). The SGCs are formed by a periodic diffraction grating in slab waveguide to allow a resonant light coupling between the planar circuit and a typically tilted optical fiber, positioned above the device, as schematically shown in Fig. 4A.

To date, an extensive set of SGC design strategies have been reported, with a steady progress in improving the fiber-chip coupling efficiency. Several decisive aspects *en bloc* contribute to the overall coupling performance, including the back-reflections at the waveguide-to-grating junction, the mode mismatch between the Gaussian-like optical fiber mode and the exponential profile of the grating beam, as well as the parasitic optical power that radiates toward the bottom substrate. The first two factors are typically addressed simultaneously by using the SGC apodization (Benedikovic et al., 2015a; Halir et al., 2009; Kopp et al., 2011; Li et al., 2014; Mekis et al., 2011; Taillaert et al., 2004), i.e., diffraction grating of a variable coupling strength that matches the near-field Gaussian mode profile. This encompasses approaches such as a duty cycle variation (Kopp et al., 2011; Mekis et al., 2011; Taillaert et al., 2004), etch depth optimization (Li et al., 2014), or a sub-wavelength grating (SWG) refractive index engineering (Benedikovic et al., 2015a; Halir et al., 2009). However, over the years, the SGC directionality, typically defined as the ratio of the power coupled toward the optical fiber and the total out-coupled optical power (Benedikovic et al., 2015a), has been identified as a main limiting factor for high-efficiency coupling (Xu et al., 2014).

The state-of-the-art solutions exclusively exploit the effect of thin-film interference by forming partial or full reflectors underneath the coupling device. This way, the light radiated downward (into the substrate) is effectively re-utilized in order to assure in-phase diffraction with the preferential first-order radiation beam that is constructively coupled to the optical fiber mode. This may be realized through the etching depth optimization (Kopp et al., 2011; Li et al., 2014; Taillaert et al., 2004), BOX thinning (Boeuf et al., 2016; Ding et al., 2015; Mekis et al., 2011), or radiation angle adjustment (Benedikovic et al., 2015a), typically combined with additional fabrication steps of backside wafer metallization (Benedikovic et al., 2015a; Ding et al., 2015; Zaoui et al., 2014) or flip-chip wafer-to-wafer bonding (Baudot et al., 2014; Boeuf et al., 2016; Kopp et al., 2011; Mekis et al., 2011; Wang et al., 2010). Experimentally, fiber-chip coupling efficiencies down to

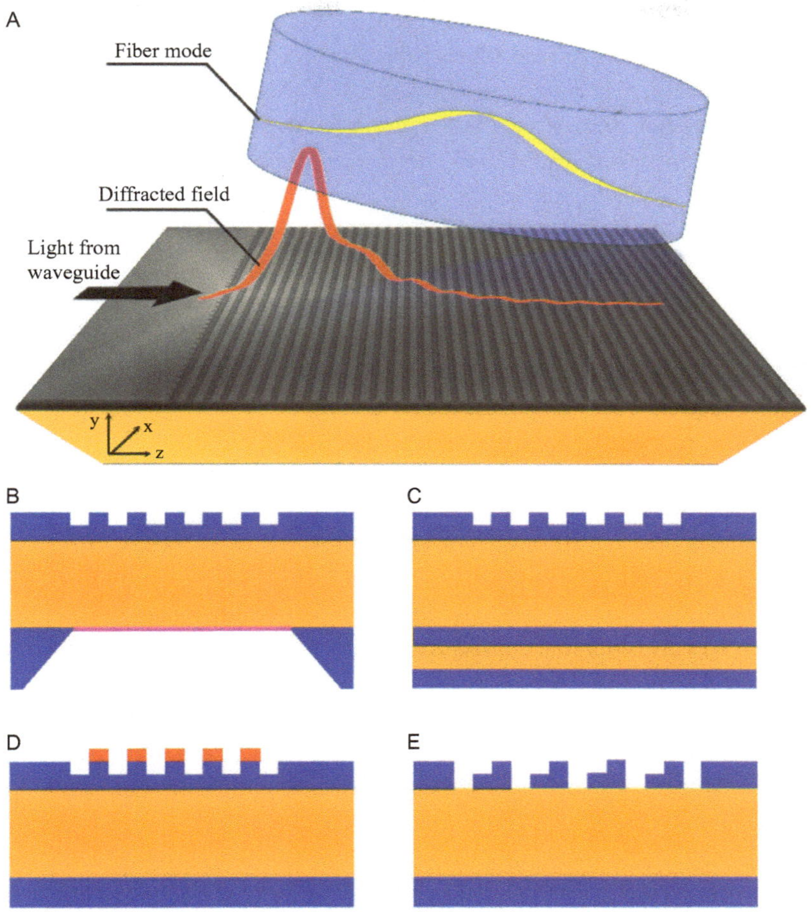

Fig. 4 (A) A fiber-chip surface grating coupler. The diffraction grating resonantly couples the light injected from the waveguide with an exponentially-decaying profile toward a Gaussian-like optical fiber mode. Schematic illustrations of the state-of-the-art approaches to achieve a high-efficiency coupling between optical fibers and integrated circuits. (B) SGC with a metal mirror implemented via backside processing (Benedikovic et al., 2015a; Ding et al., 2015; Zaoui et al., 2014). (C) SGC with a Bragg mirror implemented via wafer-to-wafer bonding or by using a customized SOI wafers (Baudot et al., 2014; Boeuf et al., 2016; Kopp et al., 2011; Mekis et al., 2011; Wang et al., 2010). (D) SGC with an overlayer realized on top of device (Roelkens et al., 2006; Vermeulen et al., 2010; Yang et al., 2014). (E) Blazed SGC with an L-shaped geometry (Baudot et al., 2014; Ding et al., 2015; Wang et al., 2010; Zaoui et al., 2014).

−0.5 dB have been demonstrated hitherto. Schematics of these concepts are illustrated in Fig. 4B and C, respectively.

Alternatively, maximizing the intrinsic SGC directionality, i.e., breaking the vertical device symmetry, has also emerged in recent years as a profound design scenario to vastly improve the overall coupling efficiency (Alonso-Ramos et al., 2014; Benedikovic et al., 2015b, 2017; Chen et al., 2016a, 2017a,b; Roelkens et al., 2006; Vermeulen et al., 2010; Watanabe et al., 2017; Yang et al., 2014). In this concept, the optical power constructively diffracted to the upper cladding is maximized, while at the same time the parasitic power leaked into the substrate is conversely minimized. Efficient SGC designs have been theoretically elaborated and experimentally validated, primarily by using subsidiary low and high refractive index overlayers (Roelkens et al., 2006; Vermeulen et al., 2010; Yang et al., 2014) or by exploiting the blazing effect in interleaved or in *L*-shaped device arrangements (Alonso-Ramos et al., 2014; Benedikovic et al., 2015b, 2017; Chen et al., 2016a, 2017a,b; Watanabe et al., 2017). Those approaches are shown in Fig. 4D and E, respectively, affording a tested coupling performance down to −1 dB up to date. Most recently, the later blazing concept has received a considerable amount of attention, particularly due to its inherent versatility (Alonso-Ramos et al., 2014; Benedikovic et al., 2015b; Chen et al., 2017a), compatibility with processes and tools established in nanophotonics foundries (Benedikovic et al., 2017), and waveguide platform portability (Chen et al., 2016a, 2017b), delivering high-efficiency and robust coupling reliance.

Today, the major challenge in light coupling in silicon photonics circuits resides in the limited spectral bandwidth of SGCs. Some solutions based on edge coupling are studied: inverted taper (Almeida et al., 2003; Shoji et al., 2002), subwavelength taper (Cheben et al., 2015), and silicon nitride edge coupler (Maegami et al., 2016). These solutions provide broadband light coupling and polarization insensitivity; however, they are not suitable for large-scale wafer testing. Light coupling from/to optical fiber is still considered as an open and challenging task, which has to be solved in order to fully develop reliable multi-spectral photonic integrated circuits.

2.2 Optical modulation

Optical modulation is carried out by modifying the optical properties of the material (Si, Ge, III–V, Polymer, etc.) to change the behaviors of the light propagating through it. This modulation is mainly performed by an external electrical signal, which changes the real and/or imaginary parts of the refractive index of the particular material. The real part (n) is associated to the phase velocity and the imaginary part (α) is associated to the absorption. This allows classifying optical modulators in two main groups: (i) Electro-refraction modulators and (ii) Electro-absorption modulators, respectively. The most common physical effects exploited for optical modulation in photonics are Thermo-Optic (TO) (Cocorullo et al., 1999), Pockels (Damas et al., 2016, 2017; Jacobsen et al., 2006), Franz-Keldysh (FK) (Berciano et al., 2018; Feng et al., 2013), Quantum confined Stark Effect (QCSE) (Gao et al., 2017a,b; Kolahdouz et al., 2012; Kuo et al., 2005; Lever et al., 2011; Liu et al., 2008; Marris-Morini et al., 2016; Zang et al., 2017), and Free-Carrier Plasma Dispersion effects (Ding et al., 2013; Ferrotti et al., 2016a,b; Kupijai et al., 2016; Li et al., 2017; Pérez-Galacho et al., 2015; Perez-Galacho et al., 2017; Romero-García et al., 2017; Shao et al., 2017; Sharif Azadeh et al., 2014; Streshinsky et al., 2013; Thomson et al., 2013; Vermeulen et al., 2016; Wang et al., 2013, 2017a; Xiao et al., 2017; Xiong et al., 2015; Xu et al., 2013; Yong et al., 2017; Zheng et al., 2017).

The evaluation of the optical modulator performance is carried out through a typically wide set of parameters, so-called Figures-of-Merit (FoM). The most commonly used are the following:

- *The Insertion Loss (IL)*: It is the total loss of the modulator section including in/out coupling and propagation losses. For a phase modulator, the insertion loss for L_π length (IL_π) is given. L_π is the modulator length to achieve π phase shift.
- *The Electro-Optical Bandwidth (BW)*: This characteristic corresponds to the maximum speed of the modulator (analog signal). For most modulators, this bandwidth could be theoretically analyzed through the equivalent RC circuit of the electrical structures. Speed characteristics are also studied using digital signals (data rate). Then, eye diagram measurements are required as well as Bit-Error-Rate (BER) analysis to define the quality of the data transmission.
- *The extinction ratio (ER)*: It characterizes the ratio of optical intensities in on/off states. ER is often reported at a given data rate, which is often lower than the ER given in DC.
- *The modulation efficiency ($V_\pi L_\pi$)*: Only reported for an electro-refraction modulator inducing a variation of the effective index (Δn_{eff}) at a given wavelength λ. $V_\pi L_\pi$ is the product of the voltage (V_π) and the length (L_π) of the phase shifter needed to achieve π phase shift.

$$V_\pi L_\pi = \frac{V_\pi \lambda}{2.\Delta n_{eff}(V_\pi)} \tag{1}$$

- *The power consumption (J/bit)*: Power consumption depends of the capacitance (C) of the electrical structure, the voltage swing (V), and the photogenerated current (I_{ph}) at a given data rate (B). The latter is mainly considered for an electro-absorption modulator. Power consumptions is driven by two equations:

$$E_1 = \frac{1}{4} C V^2 \tag{2}$$

$$E_2 = \frac{1}{B} I_{ph} V \tag{3}$$

We can notice that a comparison of published results on optical modulators is not often straightforward due to the variety of afore-described optical and electrical characteristics. In the following, an overview of the main physical effects typically elaborated to change the optical intensity is reported.

2.2.1 Thermo-optic effect

The Thermo-Optic effect is a slow effect, which is mainly used to tune the operating point of photonic devices or as an actuator for switches. Indeed, under a temperature change (ΔT), the refractive index variation (Δn) of the material also changes following this expression in silicon: $\Delta n/\Delta T = 1.86 \times 10^{-4}\ K^{-1}$, denoted as the thermo-optic material coefficient (in this case coefficient of silicon). In some cases, when the electric power applied to the device is high, the temperature may rise, leading to a shift of the component behavior. In most of applications, the thermo-optic effect has to be addressed, either to compensate the thermal drift or to tune the operating wavelength.

2.2.2 Pockels effects

The Pockels effect is the dominant effect used in optical modulators based on Lithium Niobate ($LiNbO_3$) and Indium Phosphide (InP). However, silicon is a centro-symmetric crystal. An immediate consequence is the non-existence of the second-order optical susceptibility (χ^2) in the material, the fundamental material property behind the origin of the Pockels effect.

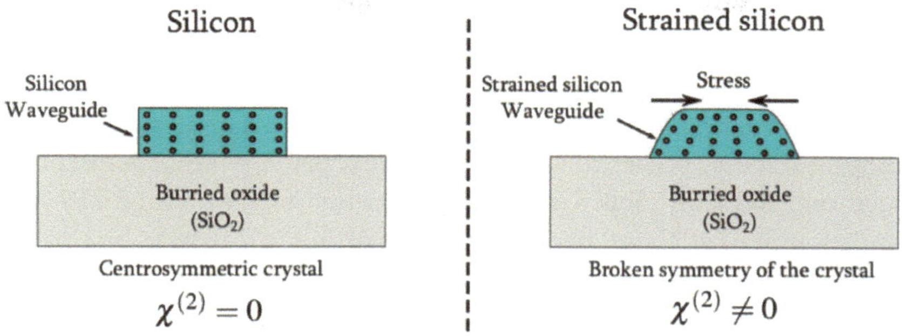

Fig. 5 The process of straining silicon to break the symmetry and induce χ^2 dependent phenomena.

Nevertheless, it has been proposed and evaluated that the strain can be used to break the symmetry of the silicon crystal, as schematically represented in Fig. 5. This process is capable of generating the normally inhibited second-order susceptibility ($\chi^2 \neq 0$) and enables nonlinear phenomena like the Pockels effect (Berciano et al., 2018; Damas et al., 2016, 2017; Jacobsen et al., 2006). The motivation is to develop strained silicon photonic devices for optical modulation: with a prospect for high-speed, low-loss, compact, low-power consumption, large optical bandwidth, and silicon compatible modulators.

A lot of efforts have been made in the recent years toward breaking the centro-symmetry of silicon and achieving the Pockels effect. Even though the results look promising, they are still far from the device point-of-view.

2.2.3 Electro-absorption effects

Electro-absorption (EA) modulators are a class of modulators that employ an electric field to induce a variation of the absorption coefficient (α) of the material, which directly changes the light intensity. Direct modulation of the optical intensity is obtained and the performance metrics depend on the absorption coefficient variation and the working spectral range. There are two main EA mechanisms in silicon photonics: the Franz-Keldysh effect and the (QCSE). Both of these effects have been demonstrated in Germanium and Germanium alloys (Chaisakul et al., 2014; Feng et al., 2013; Gao et al., 2017a,b; Kolahdouz et al., 2012; Kuo et al., 2005; Lever et al., 2011; Liu et al., 2008; Marris-Morini et al., 2016; Zang et al., 2017), providing a platform for silicon-compatible materials for high speed, low power, and compact modulators.

In 2008, a MIT group demonstrated the first EA GeSi modulator based on the Franz-Keldysh effect (FKE), working around the C and L bands of telecommunication wavelengths (Feng et al., 2013; Liu et al., 2008). From this demonstration, several groups developed Ge technology to achieve compact and low-power consumption modulators. A typical EA modulator based on FKE from bulk Ge has been reported to have around 8–10 dB extinction ratio at about 1600 nm wavelength using a reverse bias voltage between 0 and 4 V for a device with an active area of <45 μm^2, with a corresponding insertion loss of 5–10 dB and ~100 fJ/bit dynamic energy consumption. Adding a very small fraction of Si into Ge, 1550 nm operation has been reached (Feng et al., 2013; Liu et al., 2008).

The use of bulk Ge for light modulation based on FKE has shown two main limitations for datacom and low power consumption applications. First, the operating wavelength is limited to material band gap energy, i.e., 1550 nm. Second, swing bias voltages are still higher than 1 V, the maximum voltage delivered by CMOS circuits. These limitations can potentially be overcome by using the quantum-confined Stark effect (QCSE) in Ge multiple Quantum wells (MQWs). The benefits of quantum well structures result from its discrete energy levels thanks to the quantum confinement effect. Some recent demonstrations on high-speed optical modulation have been reported at wavelengths from 1300 to 1550 nm (Chaisakul et al., 2014; Gao et al., 2017a,b; Kolahdouz et al., 2012; Kuo et al., 2005; Lever et al., 2011; Marris-Morini et al., 2016; Zang et al., 2017).

2.2.4 Free-carrier plasma dispersion effect

The Free-Carrier Plasma Dispersion (FCPD) effect relates the changes in free carrier concentration in a semiconductor (electrons and holes) to a change in the complex refractive index. As the free carrier concentration increases, the imaginary part of the refractive index (absorption coefficient) increases, while the real part (phase) decreases. This effect was theoretically described by the Drude-Lorentz model (Schumann and Phillips, 1967; Singh, 2010; Soref and Bennett, 1987):

$$\Delta n = -\frac{e^2 \lambda^2}{8\pi^2 c^2 \varepsilon_0 n}\left(\frac{\Delta N_e}{m_e} + \frac{\Delta N_h}{m_h}\right) \tag{4}$$

$$\Delta \alpha = -\frac{e^3 \lambda^2}{4\pi^2 c^3 \varepsilon_0 n}\left(\frac{\Delta N_e}{m_e^2 \mu_e} + \frac{\Delta N_h}{m_h^2 \mu_h}\right) \tag{5}$$

where e is the charge of the electron, c is the speed of light, λ is the wavelength, n is the refractive index of the unperturbed material, ε_0 is the vacuum permittivity, ΔN_e and ΔN_h are the free carrier concentrations of electrons and holes, respectively, m_e and m_h are the electron and hole respective masses, μ_e and μ_h are the mobilities of electron and holes, respectively. From Eqs. (3) and (4), it is possible to model the effect of free-carriers into the refractive index of any materials.

In silicon photonics, FCPD is still the main effect used to develop high-speed and efficient optical modulators. In the 1990s, visionary works from Soref and Bennett reported on experimental data on the variations of the refractive index as well as absorption of doped silicon wafers and defined a set of empirical equations for wavelengths around 1.3 and 1.55 μm (Soref and Bennett, 1987). As an example, for $\lambda = 1.3\,\mu m$ they take the form:

$$\Delta n = -6.2 \times 10^{-22}\,\Delta N_e - 6 \times 10^{-18}(\Delta N_h)^{0.8} \tag{6}$$

$$\Delta \alpha = 6 \times 10^{-18}\,\Delta N_e + 4 \times 10^{-18}\,\Delta N_h \tag{7}$$

As the electro-absorption generated by the FCPD effect is not strong enough for modulation, optical modulators based on this effect are mainly designed as electro-refraction modulators (i.e., phase shifters), minimizing absorption changes. Three main kinds of FCPD effect can be used to change carrier concentration variation into a waveguide: injection, accumulation, or depletion of carriers. The electronic structures to achieve such carrier variations are illustrated in Fig. 6.

Carrier injection is based on PIN junctions, as reported in Fig. 6A–C. Under a forward bias, a current flows through the junction, and the intrinsic zone (in green) is flooded with carriers (electrons and holes). A proper overlap between this zone and the optical mode introduces a change in the refractive index (i.e., phase change). This FCPD effect is the most efficient effect, but it is power hungry and limited in speed.

Carrier depletion is normally based on a PN junction (see Fig. 6D–F). Under a reverse bias, a depletion (intrinsic) zone appears in the vicinity of the PN junction. The width of the depletion zone increases as a function of the reverse bias, removing carriers from both P and N doped regions. The widening of the depletion zone introduces a change in the refractive index and then a phase variation of the guided mode. As a unipolar effect (only one carrier), this effect is intrinsically high-speed. However, as the depletion zone is limited, the efficiency is lower than the carrier injection effect. In silicon photonics, carrier depletion is still the main effect used

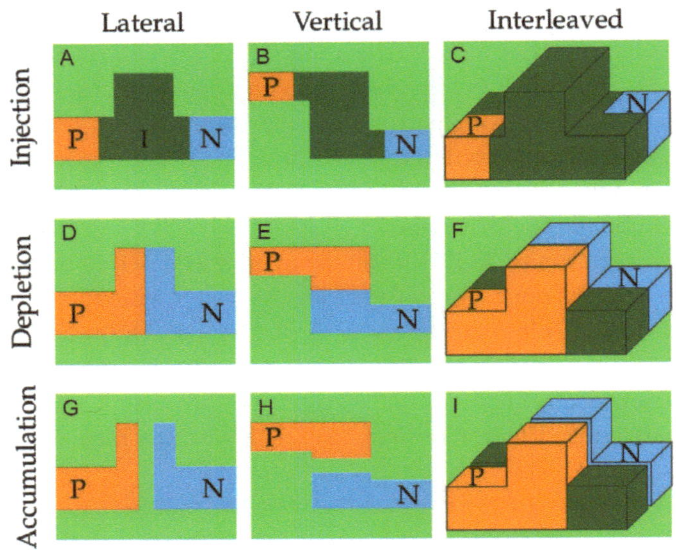

Fig. 6 Waveguide modulators based on the free carrier plasma dispersion effect: Carrier injection: (A) lateral pin diode, (B) vertical pin diode, (C) interdigitated pin diode; Carrier depletion: (D) lateral pn diode, (E) vertical pn diode, (F) interdigitated pn diode; Carrier accumulation: (G) lateral capacitance, (H) vertical capacitance and (I) interdigitated capacitance.

for optical modulation, because it combines high-speed, low insertion loss, and moderate power consumption with a simple technological process.

Carrier accumulation employs p-doped, oxide, n-doped (PON) capacitor structures (see Fig. 6G–I). Under a bias voltage, electric charges accumulate apart the oxide interfaces, i.e., holes in the p-doped part and electrons in the n-doped part. This accumulation of carriers induces a change in the refractive index in the silicon close to the interface with the oxide. This effect combines a judicious mixture between carrier injection and depletion. The main drawback is still the complex processing and the limited bandwidth due to the large capacitance of the structure; however, promising results have been already obtained (Abraham et al., 2014; Douix et al., 2018; Fujikata et al., 2013; Sodagar et al., 2015; Van Campenhout et al., 2012; Webster et al., 2014).

Remarkable OOK (On-Off Keying) modulation characteristics have already been obtained with performance close to state-of-the-art lithium niobate modulators. Indeed, 40 Gbps, and even more, have been demonstrated with extinction ratios larger than 6 dB and insertion loss below

5 dB. The modulation efficiency $V_{\pi}L_{\pi}$ is 1–2 V cm for carrier depletion effects, while it is below 1 V cm for carrier accumulation modulators. From these characteristics, current trends are to first develop a reliable complex circuit that integrates lasers, Si modulators, and detectors, and second, to demonstrate advanced multi-level modulation formats, including pulse amplitude modulation (M-PAM), phase shift keying (M-PSK), and quadrature amplitude modulation (M-QAM) based on FCPD.

Despite amazing characteristics, the requirements in terms of ultra-low power consumption are not reachable to date. Indeed, values of about one pJ/bit were obtained in optimized FCPD structures, which is quite far from the objective: a few fJ/bit. Then, some alternative solutions have to be developed and a promising trend may be the hybrid integration with other materials that are typically considered to be more efficient compared to the silicon-only approaches. This includes materials such as polymers, nanomaterials like graphene, chalcogenide glasses, or carbon nanotubes functional oxides (BTO, BST, etc.), and of course III–V compounds. In Section 3, hybrid integration of III–V materials on a silicon photonics platform will be discussed and some demonstrations will be presented.

2.2.5 Optical receivers

Optical receivers are one of the main building blocks of the full Si photonics chain. Indeed, as it is positioned at the end of the optical link, it governs the whole power budget of the link, i.e., the delivered optical power of the lasers, the insertion loss of the modulators, and the loss of passive structures. A photodetector needs a minimum power to achieve the BER specifications at a given data rate.

In the silicon photonics platform, germanium (Ge) rapidly became the material of choice, because of its strong absorption coefficient in the near-infrared wavelength range (up to 1.6 μm) and its compatibility with CMOS processes, favoring monolithic integration. Despite the large lattice parameter mismatch between Ge and Si (4.2%), the epitaxial growth of high-quality and thick crystalline germanium layers on silicon has been successfully demonstrated with low dislocation concentration in Ge (Hartmann et al., 2004). Today, the Ge photodetector is well identified as the most robust and mature device in the silicon photonics platform. Indeed, it offers cutting-edge performance directly comparable to its III–V counterparts.

Two strategies have been developed and reported for the integration of Ge photodetectors into silicon waveguides: (i) evanescent coupling and (ii) butt coupling (Fig. 7A and B, respectively) (Assefa et al., 2010a;

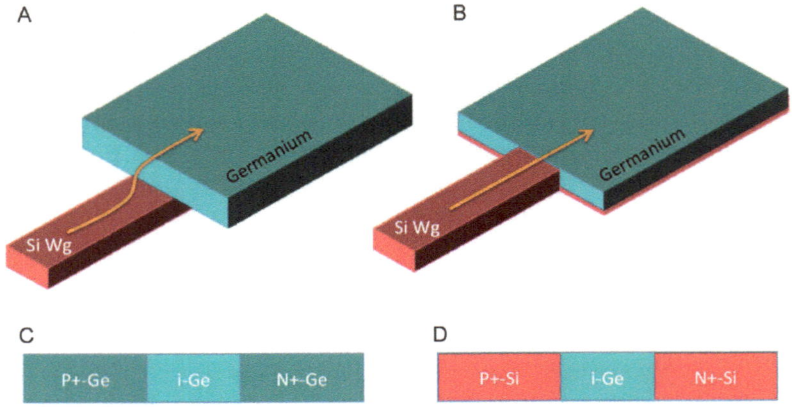

Fig. 7 Schematic view of (A) evanescent coupling and (B) butt coupling and the pin diode for photodetection: (C) pure Ge photodiode, (D) double Si/Ge/Si heterojunction photodiode.

Byrd et al., 2017; Chen et al., 2015a, 2016b; DeRose et al., 2011; Going et al., 2015; Li et al., 2012; Liao et al., 2011; Lischke et al., 2015; Michel et al., 2010; Nam et al., 2015; Okyay et al., 2006; Virot et al., 2017; Vivien et al., 2009, 2012; Zhang et al., 2014).

Both Ge integration schemes have demonstrated similar photodetection performances in terms of speed and responsivity. However, even though there are no fundamental differences, there are some slight variations, which could be detrimental for a few applications. Indeed, the coupling efficiency is more robust for the butt-coupling configuration in terms of wavelength sensitivity and Ge thickness variation. Furthermore, such integration leads to an almost flat surface promoting hybrid integration. However, the detection linearity with input optical power in the evanescent coupling configuration is better than in the butt-coupling configuration as the light is continuously distributed along the germanium layer. This avoids any kind of absorption screening due the strong Ge absorption. According to the target applications, these characteristics have to be taken into the account.

Two kinds of electrical configuration for the photodiodes have emerged, schematically shown in Fig. 7C and D: (i) a pure (homojunction) Ge photodiode with P + and N + Ge-doped contacts and (ii) a double Si/Ge/Si heterojunction photodiode where doped regions are formed in silicon (Virot et al., 2017). Both approaches exhibit comparable detection characteristics. The advantages of a double Si/Ge/Si heterojunction photodiode are: (i) a reduction of the number of technological steps (e.g., a reduction of the fabrication cost and complexity), (ii) a decrease of the contact access resistance thanks to the use of a NiSi silicide, and (iii) an improvement of the optoelectronic device integration because identical silicide-based contacts are used for all optoelectronic Si devices including optical modulators and photodetectors. However, a challenging selective Ge epitaxial growth in a narrow Si trench has to be performed.

Numerous successful demonstrations have been obtained on Ge photodetectors integrated into Si waveguides. Most of these photodetectors reached very high performances in terms of speed (>40 GHz) and responsivity (>1 A/W) and dark current (below 100 nA). Even after a reduction of four orders of magnitude of the dark current from the first generation of Ge detectors, the dark current is still 1–2 orders of magnitude higher than the state-of-the-art III–V detectors. The key to solving this problem is the quality of the epitaxial growth of germanium on silicon in order to reduce the dislocation concentration and the traps at the interface between the Ge and Si.

As stated above, receivers in integrated circuits drive the minimum optical power needed to ensure error-free detection. Therefore, the more sensitivity of the detector, the lower required optical power from the source. The use of avalanche photodiodes (APD) have appeared as a promising way to push the responsivity beyond the limit imposed by external quantum efficiency. Avalanche operation in Germanium based photodiodes has been firstly demonstrated in Separate Absorption Charge Multiplication (SACM) (Kang et al., 2008) and Metal Semiconductor Metal (MSM) structures (Assefa et al., 2010b). SACM avalanche photodiodes provide an optimized marriage between the strong germanium absorption and the low multiplication noise in silicon. However, a high bias is needed (>−10 V) to reach high multiplication gain, and the fabrication is complex due to the need to perfectly control the electric field into Ge and Si in order to reduce the electrical noise. Conversely, internal gain in MSM structures can be achieved at low bias (<−3 V); however, the dark current in MSM structures is well-known to be very high (typically of 50 μA for unity gain), thereby substantially degrading the sensitivity of the avalanche photo-receiver. Recently, a new approach has been developed based on a compact p-i-n Ge photodiode biased in avalanche mode with a thin multiplication Ge layer for dead space enhanced noise reduction (Assefa et al., 2010b; Kang et al., 2008). This approach allows for simplified integration in silicon photonics technology. Moderate avalanche gain, about 10–20, and relatively low bias voltage of about 6 V have been demonstrated.

These preliminary results on Ge receivers are promising; however, they call for more research efforts in order to obtain efficient, low-voltage, low-noise and reliable avalanche photo-detectors directly integrated into Si waveguides.

2.3 Light source

An integrated electrically pumped Si-based optical source has long been considered as a Holy Grail in silicon photonics as long as silicon is an indirect band gap semiconductor that exhibits light emission with poor efficiency. Numerous works have been performed on bulk Si, Si nano-crystals, and Rare-Earth Doped Silicon, to make this material a good candidate for on-chip lasing functionality. Although a lot of scientific breakthroughs have been accumulated over the years, no Si laser was demonstrated so far. However, from band engineering of highly doped Ge, promising demonstrations of germanium-on-silicon lasers have been achieved. Despite numerous efforts on the development of Ge lasers, the threshold is still too high to be considered in today's applications and the emitting wavelength is also limited to wavelengths above 1.55 μm. Therefore, integrating another material on Silicon seems to be unavoidable. A lot of materials have been considered to generate light into Si. This encompasses III–V compounds, carbon nanotubes, quantum dots, rare earth elements, etc. From all results published to date, the most efficient lasers on Si have been developed from III–V semiconductors. Indeed, InP-based materials remain one of the best choices to fabricate efficient laser sources in the range of 1.3–1.55 μm. However, in order to keep a low-cost, these devices must be integrated onto a Si substrate. Two different approaches are then possible: molecular layer bonding or direct epitaxy of III–V on Si. While the latter has been an exciting research topic for decades, the layer bonding is a pragmatic solution, which has allowed several demonstrations of III–V/Si hybrid devices, including light sources.

Furthermore, InP-based materials can also have several advantages for modulator integration. Indeed, Electroabsorption modulators can be achieved through absorption using the InGaAsP quantum well system via the quantum confined Stark effect. Such devices are classically used in electroabsorption modulated laser (EML) systems, and their integration has been demonstrated on Silicon (Tang et al., 2012). But InP and InGaAsP also have interesting properties of refractive index variation with carrier concentration modulation, which is found to be much more efficient than that in Silicon (Han et al., 2016; Weber, 1994). This can be interesting to fabricate carrier-concentration-based modulators, with better efficiency than Si-only modulators.

3. III–V on Si photonics

III–V semiconductors are the materials of choice for the development of efficient optoelectronic devices for the next generations of photonic-electronic circuits. As a new material in the silicon photonics platform, some special attention has to be considered including material contamination and stability as well as the definition of new integration methodologies. In this section, the hybrid integration challenges are presented first, followed by the possible integration schemes of III–V materials on the silicon photonics platform.

3.1 Hybrid integration strategy

3.1.1 Place of III–V materials on Si photonics circuits

The schematic view shown in Fig. 8 reports on a more exhaustive example of a typical PIC with patterned SOI, Si pn junctions for modulation, integrated Ge photodiodes, patterned SiN waveguides for alternative passive devices, a 4-level Cu BEOL (back end of line) for electrical routing, and Cu pillars for 3D bonding (Baudot et al., 2017).

The rather complex PIC configuration shown in Fig. 8A leaves little room for III–V integration without making some compromises. First, it is incompatible with a co-integration of SiN devices. Besides excluding the second photonic level, the challenge is to succeed in dealing with both the topology of the above-mentioned gain medium and the Si waveguide thickening. Moreover, in order to achieve large-scale integration of the III–V material on silicon, one must avoid downsizing the wafer. In this case, the process steps following the die bonding must be done using CMOS foundry compatible process and material. Such integration has been demonstrated (Szelag et al., 2017) on 200 mm wafer using localized waveguide thickening and CMOS compatible contact (Jain et al., 2012) and process, as demonstrated.

The other process to be adapted is the contact level, which must open on four different levels and three materials (Fig. 8B).

An alternative integration consists of flipping the wafer, bonding it to a new handle, etching the backside handle and bonding the III–V dies on the backside (Durel et al., 2017). In the proposed integration, the PIC is

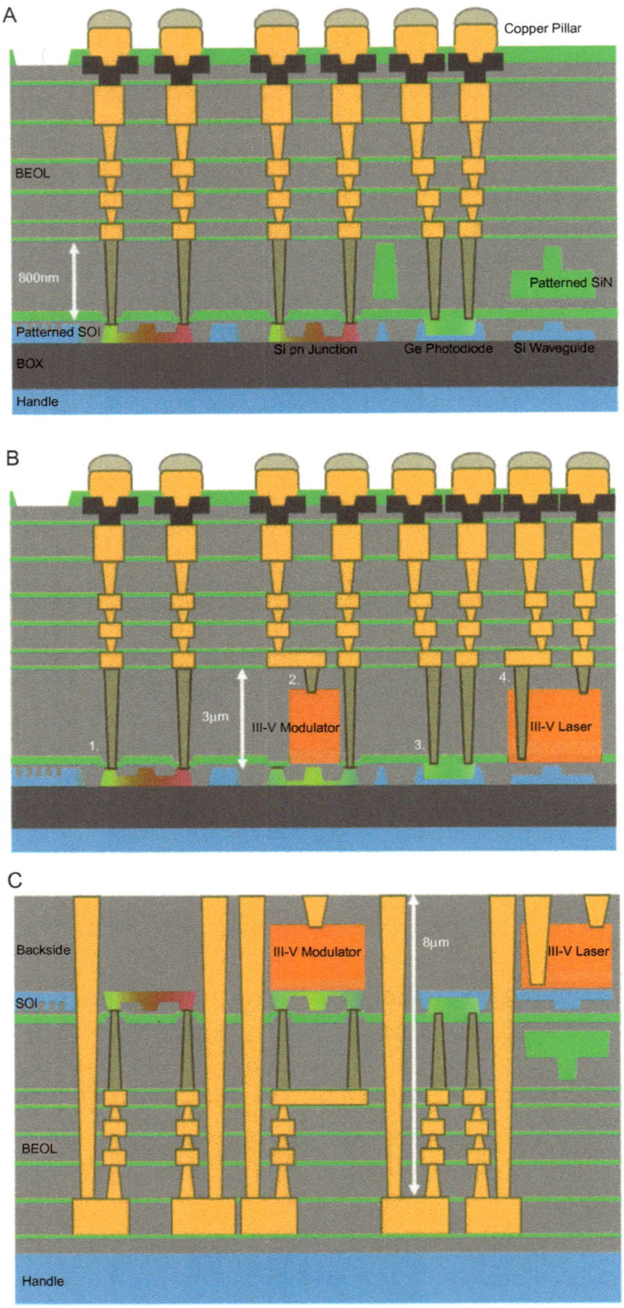

Fig. 8 Schematic view of (A) a PIC showing the patterned SOI photonic devices and a second level of SiN photonic devices within the 4-level BEOL, (B) the co-integration challenge of having a 3 μm high InP die to co-exist within the BEOL of the PIC. The contact level must open on silicon devices (1), top III–V section (2), germanium devices (3) and bottom III–V devices (4) and (C) Alternative integration strategy explored to overcome co-integration challenges between III–V and BEOL.

processed traditionally without any compromise, and the III–V is bonded subsequently on the other waveguide side (Fig. 8C), followed by an adapted second round of BEOL (Aihara et al., 2018).

Other hybrid source integrations were proposed in the literature. For example, to overcome the challenges faced with thick gain regions, developments of thin III–V gain regions with lateral metallization were reported (Crosnier et al., 2017a). Another alternative is to suppress the Si waveguide cavity and patterning photonic crystals within the III–V to generate a resonant cavity (Crosnier et al., 2017b; Raineri et al., 2018).

3.1.2 III–V material integration on Si

State-of-the-art Si photonic platforms are nowadays fabricated on 200 and 300 mm wafers. However, III–V photonic platforms still use comparatively smaller wafers with sizes of 100 and 150 mm. Moreover, materials such as InP are very brittle compared to Si, which is more resistant mechanically. Pricewise, III–V wafers are more expensive than Si substrates irrespective of their size. Based on such assessment points, it appears that it makes more sense to import small III–V portions on Si rather than the other way around. There are two major ways to proceed, which have been widely reported: transfer bonding (Liang et al., 2008) and direct epitaxy (Holland et al., 2017; Jung et al., 2017). The more matured integration method is transfer bonding and consists of preparing millimeter-size InP dies bonded on localized regions of a patterned and planarized SOI wafers (Fig. 9).

A faster and more efficient method than pick-and-place is to do a collective transfer of III–V dies using a die holder (Corbett, 2018; Sanchez et al., 2012). The technique consists of positioning the dies on an adapted surface and proceeding with the wafer bonding in a single step (Fig. 10).

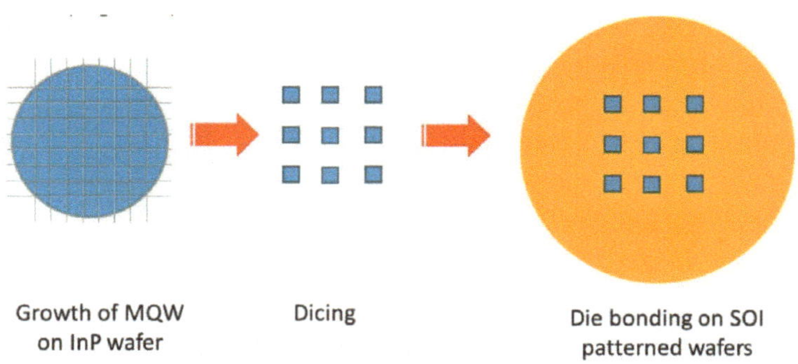

Fig. 9 Hybrid integration strategy by die-to-wafer bonding of InP on SOI.

Fig. 10 Experimental demonstration of collective die-to-wafer bonding on 200 mm patterned SOI. Left picture is showing a silicon-based holder. Right picture presents III–V die bonder on patterned SOI wafer (in this case the Si-holder was not fully filled).

Transferring localized portions of III–V on a Si wafer corresponds to a major advancement. This crucial step enables the co-integration of Si devices, III–V devices, and hybrid devices within the same platform. In this case, alignment between the patterned Si and III–V device is managed with available lithography capabilities, which are definitely better than those of packaging level III–V device integration. Consequently, information transport (electrical or optical) is enhanced due to device proximity. Moreover, devices working within the same environmental conditions have almost the same operation shifts. Finally, the system footprint is reduced due to the redundant individual device packages. However, even if hybrid bonding opens the door to new possibilities, there are still other challenges to be met and satisfied. Typically, the difference in device aspect ratio between III–V and Si can result in both electrical and optical interconnection incompatibilities, depending on the integration scheme.

Hybrid laser integration is extensively explored and specific device architectures lead to tangible arguments about new challenges. A hybrid laser device usually consists of an InP gain medium pumped electrically using a Multiple Quantum Well (MQW) typically made up of stacks of InGaAsP (Fig. 11). The MQW is designed such that the electrical band gap corresponds to the targeted optical bandwidth emission. It is addressed electrically within a p-doped and n-doped InP sandwich. Metal interconnections then connect the doped InP to the power supply. The other section of the laser consists of a Si waveguide cavity, which uses two cavity mirrors at both ends. The mode transition then occurs by favoring an interface between the gain medium and the cavity where adiabatic coupling can take place.

In a proposed integration (Menezo et al., 2014) the metal interconnection is done directly on top of the stacked p-doped InP (Fig. 11).

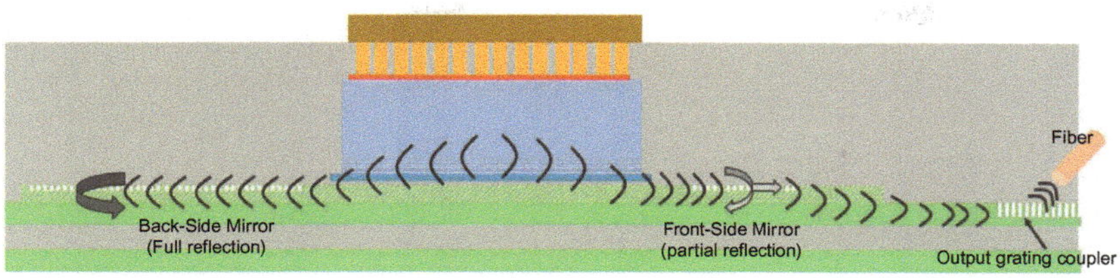

Fig. 11 Schematics of integrated laser using InP dies as gain medium and SOI as resonant cavity.

The p-doped region is thus limited to a minimum thickness of 2 μm to avoid absorption of the optical modes by the metal plate. Consequently, a device topology of about 3 μm and an optically active region of about 2 μm must be considered.

The thick optical region in the gain section has a direct impact on the adiabatic coupling with the underlying Si cavity. Lossless and low reflection coupling can only be achieved if a good effective index transition is performed around a matching point. Consequently, this implies that either very fine transition tips must be etched in InP (<50 nm) or the Si waveguide has to be thicker. In the current integration, the SOI thickness is 500 nm, which is not compatible with conventional silicon photonics platforms mostly based on 220-nm- or 310-nm-thick SOI substrates.

Most of the demonstrations of integrated III–V laser on silicon photonics platform have been obtained by combining a silicon fabrication line and a conventional III–V device foundry to process the wafer with bonded III–V die. This solution implies to downsize the SOI wafer to a diameter supported by the III–V line, typically 4 in. A preliminary version of a co-integrated Si modulator, together with a hybrid laser source and its associated BEOL, was demonstrated (Ben Bakir et al., 2011) using this approach.

As the trend in silicon photonics is to go toward a complete integration of the photonic integrated circuit (PIC) and electronic IC (EIC), strategies have to be developed. One possible way, avoiding strong modifications of both photonics and electronics circuits, is to realize independent developments of the PIC with respect to the EIC. The two sub-systems are then combined using 3D integration by bonding one to another (Fig. 12).

3.2 Hybrid optical modulators

As described in Section 2.2, plasma-dispersion based phase-modulators are the most common way in silicon photonics to fabricate transmitters. The carrier concentration modulation is achieved using either a PN junction

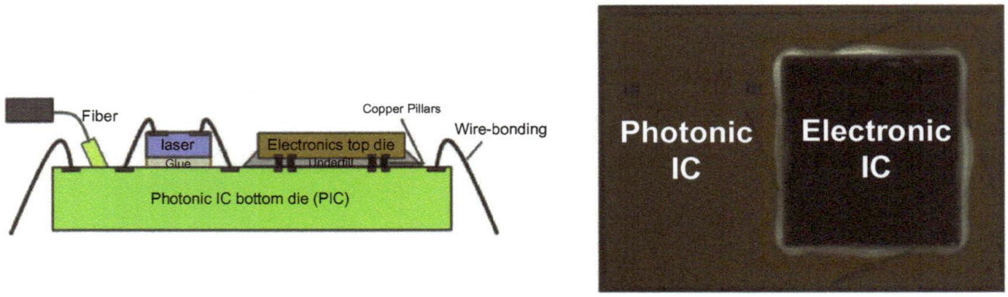

Fig. 12 System level 3D assembly packaging strategy showing an EIC bonded on a PIC.

or a SIS (semiconductor-insulator-semiconductor) capacitor, and most of today's products are based on PN junction phase modulator working up to 56 Gbit/s (Ferrotti et al., 2017). As also pointed out, some challenges remain including efficiency and power consumption, which are too low to achieve with the current approaches. One possible way to overcome the limitation of optical modulators is to take the benefits of III–V materials in terms of *carrier*-induced property changes (Eqs. 4 and 5). Especially, the effective mass of electrons and holes which are smaller in III–V materials compared to Si. For instance, a closer look to the refractive index variation with carrier concentration in InP reveals that the real part of the refractive index is improved with electrons in III–V materials, while the absorption and the hole mobility are degraded relative to silicon (for more information see Temporiti et al., 2016).

The comparison and the evaluation of the performance of a modulator can be roughly governed by the β ratio of modulator's phase shift $\Delta\varphi$ and absorption α. Maximizing β, while keeping a good electrical bandwidth (e.g., low access resistance and low capacitance) is therefore a good criterion. Due to the low hole mobility and the higher free-hole absorption in III–Vs, fabricating a PN homojunction or a SIS in III–V materials would lead to non-optimal optical modulation amplitude (OMA) and may degrade RF performance. Therefore, using either Si or SiGe to fabricate the p-side while using III–V on the n-side has been proposed to integrate modulators. The comparison of β ratio for Si, SiGe and InGaAsP depletion based PN modulator is shown in Fig. 13.

Despite its theoretical interest, a hybrid III–V/Si PN junction phase shifter is difficult to integrate in practice due to the strong lattice mismatch between Si or SiGe and InP or InGaAsP. Nonetheless, SIS structure can be made using molecular bonding (Temporiti et al., 2016). A typical example is shown in Fig. 14.

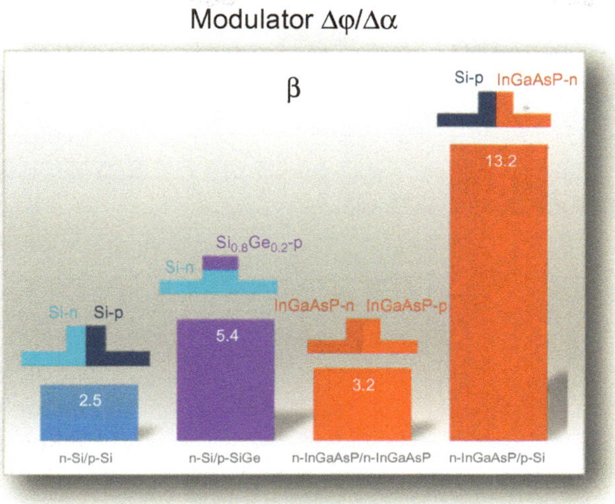

Fig. 13 β ratio calculated for PN junction modulator configurations at the optimal OMA.

The efficiency and loss of such hybrid SIS modulator have been evaluated (Weber, 1994) using a fixed structure (each of the layers are 120-nm-thick, waveguide width is $W = 0.45\,\mu m$) for various configurations of materials, as a function of the oxide thickness T_{ox} (Fig. 14). The hybrid InGaAsP/Si and InP/Si SIS modulators are showing the highest efficiency and the lowest losses, even for oxide thicknesses between 10 and 20 nm. Compared to a typical PN junction modulator, the $V\pi L\pi$ can be reduced by a factor of 10–50 at the optimal OMA point.

This first example is a perfect demonstration of the strong benefit to combine both III–V and Si properties.

3.3 Hybrid III–V/Si transmitter

As reported in the introduction, monolithic integration of a transmitter including a hybrid laser together with a MZM based Si-modulator is a priority for the development of optical communication systems. For this purpose, a hybrid III–V/silicon laser was bonded on a silicon rib waveguide, with a 100-nm-thick SiO_2 gap between the III–V and Si layers. The III–V active region was 700-μm-long and 5-μm-wide. An intrinsic InGaAsP multiple quantum well (MQW), designed to exhibit maximum gain around 1.3 μm, surrounded by n- and p-doped InP layers is used as the gain medium. Light coupling in the silicon is realized by 100 μm long adiabatic

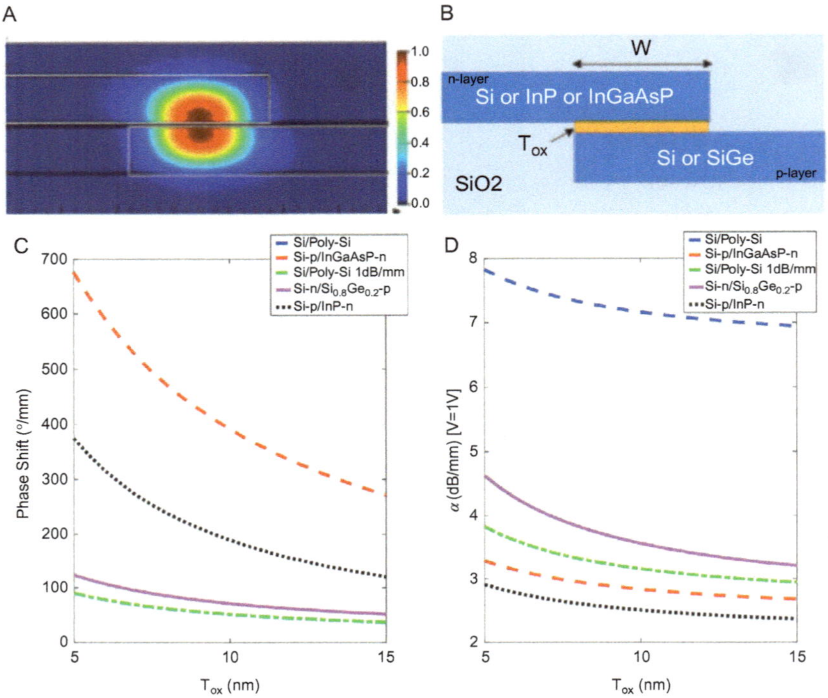

Fig. 14 Example of SIS phase shifter structure: (A) optical mode profile and (B) schematic view. Simulated phase shift (C) and optical loss (D) as a function of the oxide gate thickness T_{ox} for several structure configurations.

tapers patterned in the silicon rib waveguide, located at each end of the III–V waveguide. Their simulated coupling efficiency is above 90%. Two shallowly etched Bragg gratings (10-nm etch depth, 3-μm-wide and 50% duty-cycle) are used to form the optical cavity. The Bragg gratings were 150 and 700-μm long with a grating strength $\kappa = 78\,\text{cm}^{-1}$, giving reflectivity of 70% and 99%, respectively. The period of the gratings was 195 nm, to allow an emission centered at 1302.6 nm. Optical modulation was performed by a silicon MZM modulator based on a *pn* junction in the depletion regime. Additionally, thermal shifters were used to control the static phase difference between the MZM arms and to shift the peak wavelength of the reflectors to tune the lasing wavelength. At a laser wavelength of 1303.5 nm, a 25 Gb/s transceiver was then demonstrated. The measured extinction ratio was 4.7 dB, and no signal distortion was observed after 10 km propagation. All these results are reported and detailed in Boeuf et al. (2017).

4. Conclusion

Thanks to its transparency range and compatibility with CMOS processes, after 30 years of research silicon-based materials have emerged as an enabler of low-cost photonics systems. As optical transmission standard is becoming more and more complex, heading toward fully integrated chip-to-chip communications, and applications of integrated photonics are going beyond the Datacom realm, the development of efficient and versatile optical devices is needed. That includes low power modulators, efficient light sources, and athermal devices.

As it was mentioned, the optical properties of silicon itself do not address all requirements for all applications, especially for light sources. Consequently, the monolithic or hybrid integration of other materials into the silicon photonics structures is a promising approach that allows a judicious combination of efficient properties in a common Si platform. Among the numerous possibilities, the integration of III–V semiconductors on Si is very powerful and allows achieving efficient optoelectronic devices. Even so, there are still some open questions and challenges to address including the Si compatibility and the integration approach: bonding vs epitaxial growth and back-side vs back-end integration, the few demonstrations presented here of efficient laser and modulators on Si, already prove the benefit of such an approach.

Acknowledgments

Authors thank the support from European commission: the European Research Council (ERC) under the European Union's Horizon 2020 research and innovation program (ERC POPSTAR—grant agreement No. 647342) and the H2020-ICT-27-2017 (COSMICC) under Grant 688516 and the ANR (National Research Agency) under the MIR-SPEC and FOIST project. D.M.-M. acknowledges support by the Institut Universitaire de France.

Authors thank P. Chaisakul, M.-S. Rouifed and W. Zhang for fruitful discussion on silicon photonics.

References

Abraham, A., Olivier, S., Marris-Morini, D., Vivien, L., 2014. Evaluation of the performances of a silicon optical modulator based on a silicon-oxide-silicon capacitor. In: 2014 IEEE 11th International Conference on Group IV Photonics (GFP). IEEE, pp. 3–4.

Aihara, T., Hiraki, T., Takeda, K., Hasebe, K., Fujii, T., Tsuchizawa, T., Kakitsuka, T., Matsuo, S., 2018. Lateral current injection membrane buried heterostructure lasers integrated on 200-nm-thick Si waveguide. In: OFC.

Almeida, V.R., Panepucci, R.R., Lipson, M., 2003. Nanotaper for compact mode conversion. Opt. Lett. 28 (15), 1302–1304.
Almeida, V.R., Xu, Q., Barrios, C.A., Lipson, M., 2004. Guiding and confining light in void nanostructure. Opt. Lett. 29 (11), 1209–1211.
Alonso-Ramos, C., Cheben, P., Ortega-Moñux, A., Schmid, J.H., Xu, D.-X., Molina-Fernández, I., 2014. Fiber-chip grating coupler based on interleaved trenches with directionality exceeding 95%. Opt. Lett. 39 (18), 5351–5354.
Assefa, S., Xia, F., Bedell, S.W., Zhang, Y., Topuria, T., Rice, P.M., Vlasov, Y.A., 2010a. CMOS-integarted high-speed MSM germanium waveguide photodetector. Opt. Express 18 (5), 4986–4999.
Assefa, S., Xia, F., Vlasova, Y.A., 2010b. Reinventing germanium avalanche photodetector for nanophotonic on-chip optical interconnects. Nature 464, 80–84.
Atabaki, A.H., Moazeni, S., Pavanello, F., Gevorgyan, H., Notaros, J., Alloatti, L., Wade, M.T., Sun, C., Kruger, S.A., Meng, H., Al Qubaisi, K., Wang, I., Zhang, B., Khilo, A., Baiocco, C.V., Popović, M.A., Stojanović, V.M., Ram, R.J., 2018. Integrating photonics with silicon nanoelectronics for the next generation of systems on a chip. Nature 556, 349–354.
Baudot, C., Dutartre, D., Souhaité, A., Vulliet, N., Jones, A., Ries, M., Mekis, A., Verslegers, L., Sun, P., Chi, Y., Cremer, S., Gourhant, O., Benoit, D., Courgoulet, G., Perrot, C., Broussous, L., Pinguet, T., Siniviant, J., Boeuf, F., 2014. Low cost 300 mm double-SOI substrate for low insertion loss 1D & 2D grating couplers. In: Proceedings of IEEE 11th International Conference on Group IV Photonics, pp. 137–138.
Baudot, C., Douix, M., Guerber, S., Crémer, S., Vulliet, N., Planchot, J., Blanc, R., Babaud, L., Alonso-Ramos, C., et al., 2017. Developments in 300mm silicon photonics using traditional CMOS fabrication methods and materials. In: IEDM.
Ben Bakir, B., Descos, A., Olivier, N., Bordel, D., Grosse, P., Augendre, E., Fulbert, L., Fedeli, J.-M., 2011. Electrically driven hybrid Si/III-V lasers based on adiabatic mode transformers. Opt. Express 19 (11), 10317–10325.
Benedikovic, D., Cheben, P., Schmid, J.H., Xu, D.-X., Lamontagne, B., Wang, S., Lapointe, J., Halir, R., Ortega-Moñux, A., Janz, S., Dado, M., 2015a. Subwavelength index engineered surface grating coupler with sub-decibel efficiency for 220-nm silicon-on-insulator waveguides. Opt. Express 23 (17), 22628–22635.
Benedikovic, D., Alonso-Ramos, C., Cheben, P., Schmid, J.H., Wang, S., Xu, D.-X., Lapointe, J., Janz, S., Halir, R., Ortega-Moñux, A., Wangüemert-Pérez, J.G., Molina-Fernández, I., Fédéli, J.-M., Vivien, L., Dado, M., 2015b. High-directionality fiber-chip grating coupler with interleaved trenches and subwavelength index-matching structure. Opt. Lett. 40 (18), 4190–4193.
Benedikovic, D., Alonso-Ramos, C., Perez-Galacho, D., Guerber, S., Vakarin, V., Marcaud, G., Le Roux, X., Cassan, E., Marris-Morini, D., Cheben, P., Boeuf, F., Baudot, C., Vivien, L., 2017. L-shaped fiber-chip grating couplers with high directionality and low reflectivity fabricated with deep-UV lithography. Opt. Lett. 42 (17), 3439–3442.
Berciano, M., Marcaud, G., Damas, P., Le Roux, X., Crozat, P., Ramos, C.A., Galacho, D.P., Benedikovic, D., Marris-Morini, D., Cassan, E., Vivien, L., 2018. Fast linear electro-optic effect in a centrosymmetric semiconductor. Commun. Phys. 1, 64.
Bock, P.J., Cheben, P., Schmid, J.H., Lapointe, J., Delâge, A., Xu, D.-X., Janz, S., Densmore, A., Hall, T.J., 2010. Subwavelength grating crossings for silicon wire waveguides. Opt. Lett. 18 (15), 16146–16155.
Boeuf, F., Cremer, S., Temporiti, E., Fere, M., Shaw, M., Baudot, C., Vulliet, N., Pinguet, T., Mekis, A., Masini, G., Petiton, H., Le Maitre, P., Traldi, M., Maggi, L., 2016. Silicon photonics R&D and manufacturing on 300-mm wafer platform. J. Lightwave Technol. 34 (2), 286–295.

Boeuf, F., Han, J.-H., Takagi, S., Takenaka, M., 2017. Benchmarking Si, SiGe, and III-V/Si hybrid SIS optical modulators for datacenter applications. IEEE, J. Light. Technol. 35, 4047–4055.

Bogaerts, W., De Heyn, P., Van Vaerenbergh, T., De Vos, K., Kumar Selvaraja, S., Claes, T., Dumon, P., Bienstman, P., Van Thourhout, D., Baets, R., 2012. Silicon microring resonators. Laser Photonics Rev. 6 (1), 47–73.

Byrd, M.J., Timurdogan, E., Su, Z., Poulton, C.V., Fahrenkopf, N.M., Leake, G., Coolbaugh, D.D., Watts, M.R., 2017. Mode-evolution-based coupler for saturation power Ge-on-Si photodetectors. Opt. Lett. 42 (4), 851–854.

Camacho-Aguilera, R.E., Cai, Y., Patel, N., Bessette, J.T., Romagnoli, M., Kimerling, L.C., Michel, J., 2012. An electrically pumped germanium laser. Opt. Express 20, 11316–11320.

Chaisakul, P., Marris-Morini, D., Frigerio, J., Chrastina, D., Rouifed, M.-S., Cecchi, S., Crozat, P., Isella, G., Vivien, L., 2014. Integrated germanium optical interconnects on silicon substrates. Nat. Photonics 8, 482–488. juin.

Cheben, P., Schmid, J.H., Delâge, A., Densmore, A., Janz, S., Lamontagne, B., Lapointe, J., Post, E., Waldron, P., Xu, D.-X., 2007. A high-resolution silicon-on-insulator arrayed waveguide grating microspectrometer with sub-micrometer aperture waveguides. Opt. Lett. 15 (5), 2299–2306.

Cheben, P., Schmid, J.H., Wang, S., Xu, D.-X., Vachon, M., Janz, S., Lapointe, J., Painchaud, Y., Picard, M.-J., 2015. Broadband polarization independent nanophotonic coupler for silicon waveguides with ultra-high efficiency. Opt. Express 23 (17), 22553–22563.

Chen, H., Verheyen, P., De Heyn, P., Lepage, G., De Coster, J., Absil, P., Roelkens, G., Van Campenhout, J., 2015a. High responsivity low-voltage 28Gb/s Ge p-i-n photodetector with silicon contacts. J. Lightwave Technol. 33 (4), 820–824.

Chen, Y., Halir, R., Molina-Fernández, I., Cheben, P., He, J.-J., 2016a. High-efficiency apodized-imaging chip-fiber grating coupler for silicon nitride waveguides. Opt. Lett. 41 (21), 5059–5062.

Chen, H., Verheyen, P., De heyn, P., Lepage, G., De Coster, J., Balakrishnan, S., Absil, P., Yao, W., Shen, L., Roelkens, G., Van Campenhout, J., 2016b. -1 V bias 67 GHz bandwidth Si-contacted germanium waveguide p-i-n photodetector for optical links at 56 Gbps and beyond. Opt. Express 24 (5), 4622–4631.

Chen, X., Thomson, D.J., Crudginton, L., Khokhar, A.Z., Reed, T.G., 2017a. Dual-etch apodised grating couplers for efficient fibre-chip coupling near 1310 nm wavelength. Opt. Express 25 (15), 17864–17871.

Chen, Y., Dominguez-Bucio, T., Khokhar, A.Z., Banakar, M., Grabska, K., Gardes, F.Y., Halir, R., Molina-Fernández, I., Cheben, P., He, J.-J., 2017b. Experimental demonstration of an apodized-imaging chip-fiber grating coupler for Si3N4 waveguides. Opt. Lett. 42 (18), 3566–3569.

Cocorullo, G., Della Corte, F.G., Rendina, I., 1999. Temperature dependence of the thermo-optic coefficient in crystalline silicon between room temperature and 550 K at the wavelength of 1523 nm. Appl. Phys. Lett. 74 (22), 3338–3340.

Corbett, B., 2018. Micro-transfer printing for advanced scalable hybrid photonic integration. In: ECIO.

Crosnier, G., Sanchez, D., Monnier, P., Sagnes, I., Bouchoule, S., Beaudoin, G., Raj, R., Raineri, F., 2017a. InP-on-SOI electrically injected nanolaser diodes. In: SPIE Photonics West.

Crosnier, G., Sanchez, D., Bouchoule, S., Monnier, P., Beaudoin, G., Sagnes, I., Raj, R., Raineri, F., 2017b. Hybrid indium phosphide-on-silicon nanolaser diode. Nat. Photonics 11, 297–300.

Dakss, M.L., Kuhn, L., Heidrich, P.F., Scott, B.A., 1970. Grating coupler for efficient excitation of optical guides waves in thin films. Appl. Phys. Lett. 16 (12), 523–525.

Damas, P., Marris-Morini, D., Çassan, E., Vivien, L., 2016. Bond orbital description of the strain-induced second-order optical susceptibility in silicon. Phys. Rev. B 93, 16.

Damas, P., Berciano, M., Marcaud, G., Alonso-Ramos, C., Marris-Morini, D., Cassan, E., Vivien, L., 2017. Comprehensive description of the electro-optic effects in strained silicon waveguides. J. Appl. Phys. 122, 153105.

Dangel, R., Hofrichter, J., Horst, F., Jubin, D., La Porta, A., Meier, N., Soganci, I.M., Weiss, J., Offrein, B.J., 2015. Polymer waveguides for electro-optical integration in data centers and high-performance computers. Opt. Express 23 (4), 4736.

David, A., Miller, B., 1997. Physical reasons for optical interconnection. J. Optoelectron. 11, 155–168.

de Valicourt, G., Leliepre, A., Vacondio, F., Simonneau, C., Jany, C., Accard, A., Lelarge, F., Lamponi, M., Make, D., Poingt, F., Duan, G.H., Fedeli, J.-M., Messaoudene, S., Bordel, D., Lorcy, L., Antona, J.-C., Bigo, S., 2012. Directly modulated and fully tunable hybrid silicon lasers for fu- ture generation of coherent colorless ONU. Opt. Express 20, B552–B557.

Densmore, A., Vachon, M., Xu, D.X., Janz, S., Ma, R., Li, Y.-H., Lopinski, G., Delâge, A., Lapointe, J., Luebbert, C., Y Liu, Q., Cheben, P., H Schmid, J., 2009. Silicon photonic wire biosensor array for multiplexed real-time and label-free molecular detection. Opt. Lett. 34, 3598.

DeRose, C.T., Trotter, D.C., Zortman, W.A., Starbuck, A.L., Fisher, M., Watts, M.R., Davids, P.S., 2011. Ultra compact 45 GHz CMOS compatible Germanium photodiode with low dark current. Opt. Express 19 (25), 24897–24904.

Ding, J., Ji, R., Zhang, L., Lin, Y., 2013. Electro-optical response analysis of a 40 Gb/s silicon Mach-Zehnder optical modulator. J. Lightwave Technol. 31, 2434–2440.

Ding, Y., Peucheret, C., Ou, H., Yvind, K., 2015. Fully etched apodized grating coupler on the SOI platform with -0.58 dB coupling efficiency. Opt. Lett. 39 (8), 5348–5350.

Dong, P., 2016. Silicon photonic integrated circuits for wavelength-division multiplexing applications wavelength division multiplexing systems. IEEE J. Sel. Top. Quantum Electron. 22, 6100609.

Douix, M., Baudot, C., Marris-Morini, D., Valéry, A., Fowler, D., Acosta-Alba, P., Kerdilès, S., Euvrard, C., Blanc, R., Beneyton, R., Souhaité, A., Crémer, S., Vulliet, N., Vivien, L., Boeuf, F., 2018. Low loss poly-silicon for high performance capacitive silicon modulators. Opt. Express 26 (5), 5983–5990.

Durán-Valdeiglesias, E., Zhang, W., Noury, A., Alonso-Ramos, C., Hoang, T.H.C., Serna, S., Le Roux, X., Cassan, E., Izard, N., Sarti, F., Torrini, U., Biccari, F., Vinattieri, A., Balestrieri, M., Keita, A.-S., Yang, H., Bezugly, V., Cuniberti, G., Filoramo, A., Gurioli, M., Vivien, L., 2016. Integration of carbon nanotubes in silicon strip and slot waveguide micro-ring resonators. IEEE Trans. Nanotechnol. 15 (4), 583–589.

Durel, J., Ben, B.B., Jany, C., et al., 2017. Back-side integration of hybrid III–V on silicon DBR lasers. In: International Symposium on VLSI Technology.

El Kurdi, M., Prost, M., Ghrib, A., Elbaz, A., Sauvage, S., Checoury, X., Beaudoin, G., Sagnes, I., Picardi, G., Ossikovski, R., Boeuf, F., Boucaud, P., 2016. Tensile-strained germanium microdisks with circular Bragg reflectors. Appl. Phys. Lett. 108, 091103.

Fang, A.W., Park, H., Cohen, O., Jones, R., Paniccia, M.J., Bowers, J.E., 2006. Electrically pumped hybrid AlGaInAs-silicon evanescent laser. Opt. Express 14 (20), 9203–9210.

Feng, D., Qian, W., Liang, H., Kung, C.-C., Zhou, Z., Li, Z., Levy, J.S., Shafiiha, R., Fong, J., Luff, B.J., Asghari, M., 2013. High-speed GeSi electro-absorption modulator on the SOI waveguide platform. IEEE J. Sel. Top. Quantum Electron. 19 (6), 3401710.

Ferrotti, T., Duprez, H., Jany, C., Chantre, A., Seassal, C., Ben Bakir, B., 2016a. O-Band III–V-on-amorphous-silicon lasers integrated with a surface grating coupler. IEEE Photon. Technol. Lett. 28, 1944–1947.

Ferrotti, T., Blampey, B., Jany, C., Duprez, H., Chantre, A., Boeuf, F., Seassal, C., Ben Bakir, B., 2016b. Co-integrated 13μm hybrid III-V/silicon tunable laser and silicon Mach- Zehnder modulator operating at 25gb/s. Opt. Express 24, 30379.

Ferrotti, T., Blampey, B., Jany, C., Duprez, H., Chantre, A., Boeuf, F., Seassal, C., Ben Bakir, B., 2017. *Proceedings Volume 10106, Integrated Optics: Devices, Materials, and Technologies XXI*, 101060J, https://doi.org/10.1117/12.2252760.

Friedman, E.G., Haurylau, M., Associate Member, Chen, G., Chen, H., Zhang, J., Friedman, E.G., Fauchet, P.M., 2005. On-chip optical interconnect roadmap: challenges and critical directions. IEEE J. Sel. Top. Quantum Electron. 12 (October), 1699–1705.

Fujikata, J., Takahashi, S., Takahashi, M., Horikawa, T., 2013. High speed and highly efficient Si optical modulator with MOS junction for 1.55μm and 1.3 μm wavelengths. In: 10th International Conference on Group IV Photonics, pp. 65–66, août.

Gao, J., Zhou, H., Jiang, J., Zhou, Y., Sun, J., 2017a. Design of low bias voltage Ge/SiGe multiple quantum wells electro-absorption modulator at 1550 nm. AIP Adv. 7, 035317. mars.

Gao, J., Sun, J., Jiang, J., Zhou, H., Zhou, Y., 2017b. Design and analysis of electro-absorption modulators with uniaxially stressed Ge/SiGe multiple quantum wells. Opt. Express 25, 10874. mai.

Going, R., Seok, T.J., Loo, J., Hsu, K., Wu, M.C., 2015. Germanium wrap-around photodetectors on Silicon photonics. Opt. Express 23 (9), 11975–11984.

Grimmeiss, H.G., Kasper, E., 2009. Today's mainstream microelectronics—a result of technological, market and human enterprise. Mater. Sci. Forum 608, 1–16.

Halir, R., Cheben, P., Janz, S., Xu, D.-X., Molina-Fernández, I., Wangüemert-Pérez, J.G., 2009. Waveguide grating coupler with subwavelength microstructures. Opt. Lett. 34 (9), 1408–1410.

Halir, R., Bock, P., Cheben, P., Ortega-Moñux, A., Alonso-Ramos, C., Schmid, J.H., Lapointe, J., Xu, D.-X., Wangüemert-Pérez, J.G., Molina-Fernández, I., Janz, S., 2014. Waveguide sub-wavelength structures: principles and applications. Laser Photonics Rev. 9 (1), L25.

Han, J.H., Takenaka, M., Takagi, S., 2016. Extremely high modulation efficiency IU-V/Si hybrid MOS optical modulator fabricated by direct wafer bonding. In: 2016 IEEE International Electron Devices Meeting (IEDM), pp. 25.5.1–25.5.4.

Hartmann, J.M., Papon, A.M., Holliger, P., Rolland, G., Billon, T., Rouvière, M., Vivien, L., Laval, S., 2004. Reduced pressure—chemical vapor deposition of Ge thick layers on Si (001) for microelectronics and optoelectronics purposes. MRS Proc. 809, B4.3.

Holland, M., van Dal, M., Duriez, B., Oxland, R., Vellianitis, G., Doornbos, G., Afzalian, A., Chen, T.-K., Hsieh, C.-H., Ramvall, P., Vasen, T., Yeo, Y.-C., Passlack, M., 2017. Atomically flat and uniform relaxed III–V epitaxial films on silicon substrate for heterogeneous and hybrid integration. Sci. Rep. 7, 14632.

Jacobsen, R.S., Andersen, K.N., Borel, P.I., Fage-Pedersen, J., Frandsen, L.H., Hansen, O., Kristensen, M., Lavrinenko, A.V., Moulin, G., Ou, H., Peucheret, C., Zsigri, B., Bjarklev, A., 2006. Strained silicon as a new electro-optic material. Nature 441 (7090), 199–202.

Jain, S.R., Sysak, M.N., Swaidan, M., Bowers, J.E., 2012. Silicon fab-compatible contacts to n-InP and p-InGaAs for photonic applications. Appl. Phys. Lett. 100 (20), 201103.1–201103.4.

Jones, R., Doussiere, P., Driscoll, J.B., Lin, W., Yu, H., Akulova, Y., Komljenovic, T., Bowers, J.E., 2019. Heterogeneously integrated photonics. IEEE Nanotechnology Magazine 17 (April), 17–26, Invited paper.

Jung, D., Zhang, Z., Norman, J., Herrick, R., Kennedy, M.J., Patel, P., Turnlund, K., Jan, C., Wan, Y., Gossard, A., Bowers, J.E., 2017. Highly reliable low threshold InAs quantum dot lasers on on-axis (001) Si with 87% injection efficiency. ACS Photonics 5 (3), 1094–1100.

Kang, Y., Liu, H.-D., Morse, M., Paniccia, M.J., Zadka, M., Litski, S., Sarid, G., Pauchard, A., Kuo, Y.-H., Chen, H.-W., Zaoui, W.S., Bowers, J.E., Beling, A., McIntosh, D.C., Zheng, X., Campbell, J.C., 2008. Monolithic germanium/silicon avalanche photodiodes with 340 GHz gain–bandwidth product. Nat. Photonics 3, 59–63.

Keyvaninia, S., Verstuyft, S., Pathak, S., Lelarge, F., Duan, G.H., Bordel, D., Fedeli, J.M., De Vries, T., Smalbrugge, B., Geluk, E.J., Bolk, J., Smit, M., Roelkens, G., Van Thourhout, D., 2013. III-V-on-silicon multi-frequency lasers. Opt. Express 21 (11), 13675–13683.

Kolahdouz, M., Östling, M., Radamson, H., 2012. High performance infra-red detectors based on Si/SiGe multilayers quantum structure. Mater. Sci. Eng. B 177, 1563–1566.

Koos, C., Vorreau, P., Vallaitis, T., Dumon, P., Bogaerts, W., Baets, R., Esembeson, B., Biaggio, I., Michinobu, T., Diederich, F., Freude, W., Leuthold, J., 2009. All-optical high-speed signal processing with silicon–organic hybrid slot waveguides. Nat. Photonics 3, 216.

Kopp, C., Bernabé, S., Bakir, B.B., Fédéli, J.-M., Orobtchouk, R., Schrank, F., Porte, H., Zimmermann, L., Tekin, T., 2011. Silicon photonic circuits: on-CMOS integration, fiber optical coupling, and packaging. IEEE J. Sel. Top. Quantum Electron. 17 (3), 498–509.

Kuo, Y.-H., Lee, Y.K., Ge, Y., Ren, S., Roth, J.E., Kamins, T.I., Miller, D.A.B., Harris, J.S., 2005. Strong quantum-confined Stark effect in germanium quantum-well structures on silicon. Nature 437, 1334–1336.

Kupijai, S., Rhee, H., Al-Saadi, A., Henniges, M., Bronzi, D., Selicke, D., Theiss, C., Otte, S., Eichler, H.J., Woggon, U., Stolarek, D., Richter, H.H., Zimmermann, L., Tillack, B., Meister, S., 2016. 25 Gb/s silicon photonics interconnect using a transmitter based on a node-matched-diode modulator. J. Lightwave Technol. 34, 2920–2923.

Lamponi, M., Keyvaninia, S., Jany, C., Poingt, F., Lelarge, F., de Vali- court, G., Roelkens, G., Van Thourhout, D., Messaoudene, S., Fedeli, J.-M., Duan, G.H., 2012. Low-threshold heterogeneously integrated InP/SOI laser with a double adiabatic taper coupler. IEEE Photon. Technol. Lett. 24 (1), 76–78.

Lardenois, S., Pascal, D., Vivien, L., Cassan, E., Laval, S., Orobtchouk, R., Heitzmann, M., Bouzaida, N., Mollard, L., 2003. Low-loss submicrometer silicon-on-insulator rib waveguides and corner mirrors. Opt. Lett. 28 (13), 1150–1152.

Le Liepvre, A., Accard, A., Poingt, F., Jany, C., Lamponi, M., Make, D., Lelarge, F., Fedeli, J.-M., Messaoudene, S., Bordel, D., Duan, G.-H., 2013. A wavelength selectable hybrid III-V/Si laser fabricated by wafer bonding. IEEE Photon. Technol. Lett. 25 (16), 1582–1585.

Lever, L., Hu, Y., Myronov, M., Liu, X., Owens, N., Gardes, F.Y., Marko, I.P., Sweeney, S.J., Ikonić, Z., Leadley, D.R., Reed, G.T., Kelsall, R.W., 2011. Modulation of the absorption coefficient at 13 microns in Ge/SiGe multiple quantum well heterostructures on silicon. Opt. Lett. 36, 4158.

Li, G., Luo, Y., Zheng, X., Masini, G., Mekis, A., Sahni, S., Tacker, H., Yao, J., Shubin, I., Raj, K., Cunningham, J.E., Krishnamoorthy, A.V., 2012. Improving CMOS-compatible Germanium photodetectors. Opt. Express 20 (24), 26345–26350.

Li, C., Chee, K.S., Tao, J., Zhang, H., Yu, M., Lo, G.Q., 2014. Silicon photonics packaging with lateral fiber coupling to apodized grating coupler embedded circuit. Opt. Express 22 (20), 24235–24240.

Li, R., Patel, D., Samani, A., El-Fiky, E., Xing, Z., Sowailem, M., Zhong, Q., Plant, D.V., 2017. An 80 Gb/s silicon photonic modulator based on the principle of overlapped resonances. IEEE Photonics J. 9, 1–11.

Liang, D., Fang, A.W., Park, H., Reynolds, T.E., Warner, K., Oakley, D.C., Bowers, J.E., 2008. Low-temperature, strong SiO_2-SiO_2 covalent wafer bonding for III–V compound semiconductors-to-silicon photonic integrated circuits. J. Electron. Mater. 37, 1552–1559.

Liao, S., Feng, N.-N., Feng, D., Dong, P., Shafiiha, R., Kung, C.-C., Liang, H., Qian, W., Liu, Y., Fong, J., Cunningham, J.E., Luo, Y., Asghari, M., 2011. 36 GHz submicron silicon waveguide germanium photodetector. Opt. Express 19 (11), 10967–10972.

Lischke, S., Knoll, D., Mai, C., Zimmermann, L., Peczek, A., Kroh, M., Trush, A., Krune, E., Voigt, K., Mai, A., 2015. High bandwidth, high responsivity waveguide-coupled germanium p-i-n photodiode. Opt. Express 23 (21), 27213–27220.

Liu, J., Beals, M., Pomerene, A., Bernardis, S., Sun, R., Cheng, J., Kimerling, L.C., Michel, J., 2008. Waveguide-integrated, ultralow-energy GeSi electro-absorption modulators. Nat. Photonics 2 (7), 433–437.

Liu, Q., Ramirez, J.M., Vakarin, V., Le Roux, X., Ballabio, A., Frigerio, J., Chrastina, D., Isella, G., Bouville, D., Vivien, L., Ramos, C.A., Marris-Morini, D., 2018. Mid-infrared sensing between 5.2 and 6.6 μm wavelengths using Ge-rich SiGe waveguides. Opt. Mater. Express 8, 1305–1312.

Maegami, Y., Okano, M., Cong, G., Ohno, M., Yamada, K., 2016. Completely CMOS compatible SiN-waveguide- based fiber coupling structure for Si wire waveguides. Opt. Express 24, 264270.

Marris-Morini, D., Vakarin, V., Chaisakul, P., Frigerio, J., Rahman, M., Ramírez, J.-M., Rouifed, M.-S., Chrastina, D., Le Roux, X., Isella, G., et al., 2016. Silicon photonics based on Ge/SiGe quantum well structures. In: 2016 18th International Conference on Transparent Optical Networks (ICTON). IEEE, pp. 1–3.

Martin, A., Dodane, D., Leviandier, L., Dolfi, D., Naughton, A., O'Brien, P., Spuessens, T., Baets, R., Lepage, G., Verheyen, P., De Heyn, P., Absil, P., Feneyrou, P., Bourderionnet, J., 2018. Photonic integrated circuit based FMCW coherent LiDAR. J. Lightwave Technol. 36, 4640–4645. https://doi.org/10.1109/JLT.2018.2840223.

Mekis, A., Gloeckner, S., Masini, G., Narasimha, A., Pinguet, T., Sahni, S., De Dobbelaere, P., 2011. A gratingcoupler-enabled CMOS photonics platform. IEEE J. Sel. Top. Quantum Electron. 17 (3), 597–608.

Menezo, S., Duprez, H., Descos, A., Bordel, D., Sanchez, L., Brianceau, P., Fulbert, L., Carron, V., Bakir, B.B., 2014. Hybrid III-V on silicon lasers: heterogeneous 200mm-wafer-level Integration for WDM-dense-optical-interconnects. In: 2014 IEEE Compound Semiconductor Integrated Circuit Symposium (CSICS).

Michel, J., Liu, J., Kimerling, L.C., 2010. High-performance Ge-on-Si photodetectors. Nat. Photonics 4, 527–534.

Miller, D., 2009. Device requirements for optical interconnects to silicon chips. Proc. IEEE 97, 1166–1185.

Moore, G.E., 1965. Cramming more components onto integrated circuits. Electronics 38 (8), 114.

Nam, J.H., Afshinmanesh, F., Nam, D., Jung, W.S., Kamins, T.I., Brongersma, M.L., Saraswat, K.C., 2015. Monolithic integration of germanium-on-insulator p-i-n photodetector on silicon. Opt. Express 23 (12), 15816–15823.

Okyay, A.K., Nayfeh, A.M., Saraswat, K.C., 2006. Stain enhanced high efficiency germanium photodetectors in the near infrared for integration with Si. In: Proceedings of the 19th Annual Meetings of the IEEE LEOS. IEEE, pp. 460–461.
Passaro, V.M.N., La Notte, M., 2012. Optimizing SOI slot waveguide fabrication tolerances and strip-slot coupling for very efficient optical sensing. Sensors 12 (3), 2436–2455.
Pérez, D., Gasulla, I., Crudgington, L., Thomson, D.J., Khokhar, A.Z., Ke, L., Cao, W., Mashanovich, G.Z., Capmany, J., 2017. Multipurpose silicon photonics signal processor core. Nat. Commun. 8, 636.
Pérez-Galacho, D., Marris-Morini, D., Cassan, E., Baudot, C., Fédéli, J.-M., Olivier, S., Boeuf, F., Vivien, L., 2015. Comparison among silicon modulators based on free-carrier plasma dispersion effect. In: 2015 17th International Conference on Transparent Optical Networks (ICTON). IEEE, pp. 1–4.
Perez-Galacho, D., Baudot, C., Hirtzlin, T., Messaoudène, S., Vulliet, N., Crozat, P., Boeuf, F., Vivien, L., Marris-Morini, D., 2017. Low voltage 25gbps silicon Mach-Zehnder modulator in the O-band. Opt. Express 25 (10), 11217–11222.
Raineri, F., Crosnier, G., Fitsios, D., Manegatti, F., Raj, R., Vu, N., Sanchez, D., 2018. InP-on SOI nanophotonic optoelectronic devices. In: SPIE Photonics West.
Roelkens, G., Van Thourhout, D., Baets, R., 2006. High efficiency silicon-on-insulator grating coupler based on a poly-silicon overlayer. Opt. Express 14 (24), 11622–11630.
Romero-García, S., Moscoso-Mártir, A., Azadeh, S.S., Müller, J., Shen, B., Merget, F., Witzens, J., 2017. High-speed resonantly enhanced silicon photonics modulator with a large operating temperature range. Opt. Lett. 42, 81.
Ron, H.O., Mai, K.W., Fellow, A., 2001. The future of wires. Proc. IEEE 89 (4), 490–504.
Rönn, J., Zhang, W., Autere, A., Leroux, X., Pakarinen, L., Alonso-Ramos, C., Säynätjoki, A., Lipsanen, H., Vivien, L., Cassan, E., Sun, Z., 2019. Ultra-high on-chip optical gain in erbium-based hybrid slot waveguides. Nat. Commun. 10, 432.
Ross, I.M., 1998. The invention of the transistor. Proc. IEEE 86 (1), 7–28.
Sanchez, L., Bally, L., Montmayeul, B., Fournel, F., Dafonseca, J., Augendre, E., Di Cioccio, L., Carron, V., Signamarcheix, T., Taibi, R., 2012. Chip to wafer direct bonding technologies for high density 3D integration. In: Electronic Components and Technology Conference (ECTC).
Schumann, P.A., Phillips, R.P., 1967. Comparison of classical approximations to free carrier absorption in semiconductors. Solid-State Electron 10 (9), 943–948.
Shao, S., Ding, J., Zheng, L., Zou, K., Zhang, L., Zhang, F., Yang, L., 2017. Optical PAM-4 signal generation using a silicon Mach-Zehnder optical modulator. Opt. Express 25, 23003.
Sharif Azadeh, S., Müller, J., Merget, F., Romero-García, S., Shen, B., Witzens, J., 2014. Advances in silicon photonics segmented electrode Mach-Zehnder modulators and peaking enhanced resonant devices. Proc. SPIE 9288, 928817.
Shoji, T., Tsuchizawa, T., Watanabe, T., Yamada, K., Morita, H., 2002. Low loss mode size converter from 0.3 μm square Si wire waveguides to single mode fibres. Electron. Lett. 38 (25), 1669–1670.
Sibson, P., Kennard, J.E., Stanisic, S., Erven, C., O'Brien, J.L., Thompson, M.G., 2017. Integrated silicon photonics for high-speed quantum key distribution. Optica 4, 172–177.
Singh, A., 2010. Free charge carrier induced refractive index modulation of crystalline silicon. In: 7th IEEE International Conference on Group IV Photonics.
Snyder, B., O'Brien, P., 2013. Packaging process for grating-coupled silicon photonic waveguides using angle-polished fibers. IEEE Trans. Compon. Packag. Manuf. Technol. 3 (6), 954–959.

Sodagar, M., Hosseinnia, A.H., Isautier, P., Moradinejad, H., Ralph, S., Eftekhar, A.A., Adibi, A., 2015. Compact, 15 Gb/s electro-optic modulator through carrier accumulation in a hybrid Si/SiO_2/Si microdisk. Opt. Express 23, 28306.

Soref, R.A., Bennett, B.R., 1987. Electrooptical effects in silicon. IEEE J. Quantum Electron. 23 (1), 123–129.

Soref, R., Schmidtchen, J., Petermann, K., 1991. Large single-mode rib waveguides in GeSi-Si and Si-on-SiO_2. IEEE J. Quantum Electron. 27 (8), 1971–1974.

Srinivasan, S., Moreira, R., Blumenthal, D., Bowers, J.E., 2014. Design of integrated hybrid silicon waveguide optical gyroscope. Opt. Express 22, 24988.

Streshinsky, M., Ding, R., Liu, Y., Novack, A., Yang, Y., Ma, Y., Tu, X., Chee, E.K.S., Lim, A.E.-J., Lo, P.G.-Q., Baehr-Jones, T., Hochberg, M., 2013. Low power 50 Gb/s silicon traveling wave Mach-Zehnder modulator near 1300 nm. Opt. Express 21, 30350.

Sun, C., Wade, M.T., Lee, Y., Orcutt, J.S., Alloatti, L., Georgas, M.S., Waterman, A.S., Shainline, J.M., Avizienis, R.R., Lin, S., Moss, B.R., Kumar, R., Pavanello, F., Atabaki, A.H., Cook, H.M., Ou, A.J., Leu, J.C., Chen, Y.-H., Asanović, K., Ram, R.J., Popović, M.A., Stojanović, V.M., 2015. Single-chip microprocessor that communicates directly using light. Nature 528, 534–538.

Szelag, B., Hassan, K., Adelimini, L., Ghegin, E., et al., 2017. Hybrid III-V/Si DFB laser integration on a 200 mm fully CMOS-compatible silicon photonics platform. In: IEDM.

Taillaert, D., Bogaerts, W., Bienstman, P., Krauss, T.F., Van Daele, P., Moerman, I., Verstuyft, S., De Mesel, K., Baets, R., 2002. An out-of-plane grating coupler for efficient butt-coupling between compact planar waveguides and single mode fibers. IEEE J. Quantum Electron. 38 (7), 949–955.

Taillaert, D., Bienstman, P., Baets, R., 2004. Compact efficient broadband grating coupler for silicon-on-insulatorwaveguides. Opt. Lett. 29 (23), 2749–2751.

Tamir, T., Peng, S.T., 1977. Analysis and design of grating couplers. Appl. Phys. 14 (3), 235–254.

Tang, Y., Peters, J.D., Bowers, J.E., 2012. Over 67 GHz bandwidth hybrid silicon electroabsorption modulator with asymmetric segmented electrode for 1.3 μm transmission. Opt. Express 20 (10), 11529–11535.

Temporiti, E., Ghilioni, A., Minoia, G., Orlandi, P., Repossi, M., Baldi, D., Svelto, F., 2016. Insights into silicon photonics Mach -Zehnder-based optical transmitter architectures. IEEE J. Solid-State Circuits 51 (12), 3178–3191.

Texas Instruments The chip that jack built. http://www.ti.com/corp/docs/kilbyctr/jackbuilt.shtml. Accessed: 24-04-2016.

Thomson, D.J., Gardes, F.Y., Liu, S., Porte, H., Zimmermann, L., Fedeli, J.-M., Hu, Y., Nedeljkovic, M., Yang, X., Petropoulos, P., Mashanovich, G.Z., 2013. High performance Mach-Zehnder-based silicon optical modulators. IEEE J. Sel. Top. Quantum Electron. 19 (6), 85–94.

Van Campenhout, J., Pantouvaki, M., Verheyen, P., Selvaraja, S., Lepage, G., Yu, H., Lee, W., Wouters, J., Goossens, D., Moelants, M., et al., 2012. Low-voltage, low-loss, multi-Gb/s silicon micro-ring modulator based on a MOS capacitor. In: 2012 Optical Fiber Communication Conference and Exposition (OFC/NFOEC) and the National Fiber Optic Engineers Conference. IEEE, pp. 1–3.

Vermeulen, D., Selvaraja, S., Verheyen, P., Lepage, G., Boagerts, W., Absil, P., Van Thourhout, D., Roelkens, G., 2010. High-efficiency fiber-to-chip grating couplers realized using an advanced CMOS-compatible silicon-on-insulator platform. Opt. Express 18 (17), 18278–18283.

Vermeulen, D., Aroca, R., Chen, L., Pellach, L., McBrien, G., Doerr, C., 2016. Demonstration of silicon photonics push–pull modulators designed for manufacturability. IEEE Photon. Technol. Lett. 28 (10), 1127–1129.

Virot, L., Benedikovic, D., Szelag, B., Alonso-Ramos, C., Karakus, B., Hartmann, J.-M., Le Roux, X., Crozat, P., Cassan, E., Marris-Morini, D., Baudot, C., Boeuf, F., Fédéli, J.-M., Kopp, C., Vivien, L., 2017. Integrated waveguide PIN photodiodes exploiting lateral Si/Ge/Si heterojunction. Opt. Express 25 (16), 19487–19496.

Vivien, L., Osmond, J., Fédéli, J.-M., Marris-Morini, D., Crozat, P., Damlencourt, J.F., Cassan, E., Lecunff, Y., Laval, S., 2009. 42 GHz p.i.n Germanium photodetector integrated in a silicon-on-insulator waveguide. Opt. Express 17 (8), 6252–6257.

Vivien, L., Polzer, A., Marris-Morini, D., Osmond, J., Hartmann, J.M., Crozat, P., Cassan, E., Kopp, C., Zimmermann, H., Fédéli, J.-M., 2012. Zero-bias 40Gbit/s germanium waveguide photodetector on silicon. Opt. Express 20 (2), 1096–1101.

Vivien, L., Marris-Morini, D., Cassan, E., Alonso-Ramos, C., Baudot, C., Bœuf, F., Szelag, B., 2018. Circuits intégrés photonique silicium. Photoniques 93. https://doi.org/10.1051/photon/20189318.

Wada, H., Kamijoh, T., 1996. Room-temperature CW operation of InGaAsP lasers on Si fabricated by wafer bonding. IEEE Photon. Technol. Lett. 8 (2), 173–175.

Wang, Z., Tsang, Y., Wosinki, L., He, S., 2010. Experimental demonstration of high-efficiency polarization splitter based on a one-dimensional grating with a Bragg reflector underneath. IEEE Photon. Technol. Lett. 22 (21), 1568–1570.

Wang, J., Qiu, C., Li, H., Ling, W., Li, L., Pang, A., Sheng, Z., Wu, A., Wang, X., Zou, S., Gan, F., 2013. Optimization and demonstration of a large-bandwidth carrier-depletion silicon optical modulator. J. Lightwave Technol. 31, 4119–4125.

Wang, Z., Xu, X., Fan, D., Wang, Y., Chen, R.T., 2016. High quality factor subwavelength grating waveguide micro-ring resonator based on trapezoidal silicon pillars. Opt. Lett. 41 (14), 3375–3378.

Wang, M., Zhou, L., Zhu, H., Zhou, Y., Zhong, Y., Chen, J., 2017a. Low-loss high-extinction-ratio single-drive push-pull silicon Michelson interferometric modulator. Chin. Opt. Lett. 15, 042501–042505.

Wang, Z., Van Gasse, K., Moskalenko, V., Latkowski, S., Bente, E., Kuyken, B., Roelkens, G., 2017b. A III-V-on-Si ultra-dense comb laser. Light Sci. Appl. 6, e16260.

Watanabe, T., Ayata, M., Koch, U., Fedoryshyn, Y., Leuthold, J., 2017. Perpendicular grating coupler based on a blazed antiback-reflection structure. J. Lightwave Technol. 35 (21), 4663–4669.

Weber, J.P., 1994. Optimization of the carrier-induced effective index change in InGaAsP waveguides-application to tunable Bragg filters. IEEE J. Quantum Electron. 30 (8), 1801–1816.

Webster, M., Gothoskar, P., Patel, V., Piede, D., Anderson, S., Tummidi, R., Adams, D., Appel, C., Metz, P., Sunder, S., et al., 2014. An efficient MOS-capacitor based silicon modulator and CMOS drivers for optical transmitters. In: 2014 IEEE 11th International Conference on Group IV Photonics (GFP). IEEE, pp. 1–2.

Xiao, X., Li, M., Wang, L., Chen, D., Yang, Q., Yu, S., 2017. High speed silicon photonic modulators. J. Lightwave Technol. 106, 2158–2182.

Xiong, C., Gill, D., Proesel, J., Orcutt, J., Haensch, W., Green, W.M., 2015. A monolithic 56 Gb/s CMOS integrated nanophotonic PAM-4 transmitter. In: Optical Interconnects Conference (OI), pp. 16–17.

Xu, H., Li, X.-Y., Xiao, X., Li, Z.-Y., Yu, Y.-D., Yu, J.-Z., 2013. High-speed and broad optical bandwidth silicon modulator. Chin. Phys. B 22, 114212.

Xu, D.-X., Schmid, J.H., Reed, G.T., Mashanovich, G.Z., Thomson, D.J., Nedeljkovic, M., Chen, X., Van Thourhout, D., Keyvaninia, S., Selvaraja, S.K., 2014. Silicon photonic integration platform—have we found the sweet spot? IEEE J. Sel. Top. Quantum Electron. 20 (4), 189–205.

Yang, S., Zhang, Y., Baehr-Jones, T., Hochberg, M., 2014. High efficiency germanium-assisted grating coupler. Opt. Express 22 (25), 30607–30612.

Yong, Z., Sacher, W.D., Huang, Y., Mikkelsen, J.C., Yang, Y., Luo, X., Dumais, P., Goodwill, D., Bahrami, H., Lo, P.G.-Q., Bernier, E., Poon, J.K.S., 2017. U-shaped PN junctions for efficient silicon Mach-Zehnder and microring modulators in the O-band. Opt. Express 25, 8425.
Zang, K., Lu, C.-Y., Chen, X., Fei, E., Xue, M., Claussen, S., Morea, M., Chen, Y., Dutt, R., Huo, Y., Kamins, T.I., Harris, J.S., 2017. Germanium quantum well QCSE waveguide modulator with tapered coupling in distributed modulator-detector system. J. Lightwave Technol. 35, 4629–4633.
Zaoui, W.S., Kunze, A., Vogel, W., Berroth, M., Butschke, J., Letzkus, F., Burghartz, J., 2014. Bridging the gap between optical fibers and silicon photonic integrated circuits. Opt. Express 22 (2), 1277–1286.
Zhang, Y., Yang, S., Yang, Y., Grould, M., Ophir, N., Lim, A.E.-J., Lo, G.-Q., Magill, P., Bergman, K., Baehr-Jones, T., Hochberg, M., 2014. A high-responsivity photodetector absent metal-germanium direct contact. Opt. Express 22 (9), 11367–11375.
Zheng, L., Ding, J., Shao, S., Zhang, L., Yang, L., 2017. Silicon PAM-4 optical modulator driven by two binary electrical signals with different peak-to-peak voltages. Opt. Lett. 42, 2213.

Further reading

Chen, H., Verbist, J., Verheyen, P., De Heyn, P., Lepage, G., De Coster, J., Absil, P., Yin, X., Bauwelinck, J., Van Campenhout, J., Roelkens, G., 2015b. High sensitivity 10 Gb/s Si photonic receiver based on a low-voltage waveguide-coupled Ge avalanche photodetector. Opt. Express 23, 815–822.
Virot, L., Crozat, P., Fédéli, J.-M., Hartmann, J.-M., Marris-Morini, D., Cassan, E., Boeuf, F., Vivien, L., 2014. Germanium avalanche receiver for low power interconnects. Nat. Commun. 5, 4957.

CHAPTER TWO

Heterogeneously integrated III–V photonic devices on Si

Shinji Matsuo*
NTT Device Technology Laboratories, NTT Corporation, Atsugi, Kanagawa, Japan
*Corresponding author: e-mail address: shinji.matsuo.vm@hco.ntt.co.jp

Contents

1. Introduction

With increasing data traffic over long to very short distances (chip-to-chip, on-chip), further increases in transmission capacity and reductions in transmission distance are required where optical technologies are employed. Fig. 1 shows the relationship between transmission distance and capacity. The transition from electrical to optical links has been observed when the

Semiconductors and Semimetals, Volume 101
ISSN 0080-8784
https://doi.org/10.1016/bs.semsem.2019.07.001

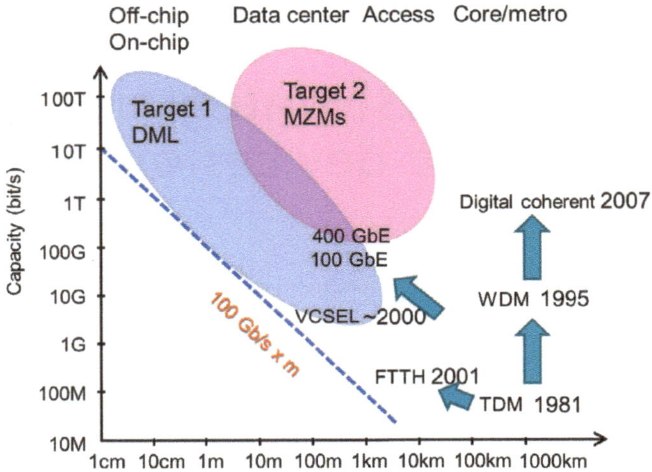

Fig. 1 Relationship between transmission distance and capacity.

product of capacity and distance exceeds 100 Gbit/s × m (Krishnamoorthy et al., 2011). Ever since the first installation of 100-km optical links with a modulation speed of 100 Mbit/s, optical technologies have been increasingly employed for shorter distance and higher capacity, as shown in Fig. 1.

When optical links are used for shorter distances, the number of transmitters dramatically increases; therefore, power consumption and cost are critical issues (Miller, 2009). Currently, directly modulated lasers (DMLs) using multimode vertical-cavity surface-emitting lasers (VCSELs) are widely used inside datacenters because they reduce power consumption and cost by employing multimode fibers. VCSELs also increase capacity by increasing the number of parallel channels and employing coarse wavelength-division-multiplexing (WDM) technologies. For further increases in transmission capacity, it is important to reduce the channel spacing of WDM systems. It is also important to develop transmitters for the 1300- and 1550-nm wavelength bands because they can extend the transmission reach up to 40 km. Therefore, directly modulated in-plane lasers are quite important because their lasing wavelength can be easily controlled by controlling the grating pitch. However, the operating energy of directly modulated distributed feedback (DFB) lasers is typically 1 pJ/bit for 25-Gbit/s NRZ signals (Kobayashi et al., 2013; Matsuda et al., 2012; Nakahara, 2015; Tadokoro et al., 2009), which is 10 times larger than that of VCSELs (Heroux et al., 2015; Moser et al., 2012). To reduce the operating energy, it is important to increase the optical confinement factor in the active region in addition to reduce the cavity length, as described in Section 2. In this context, membrane buried heterostructure (BH) lasers, whose structure is 10 times thinner but has optical confinement factor 3 times larger than those of conventional in-plane lasers, will play a key role inside datacenter networks.

The cost of both fabrication and assembly are other important issues. To reduce them, large-scale photonic integrated circuits (PICs) including lasers, photodetectors, and optical filters have attracted much attention because they can reduce the assembly cost compared with transmitters constructed by discrete devices. PICs based on Si photonics are quite suitable for constructing compact high-performance optical components, such as wavelength multiplexers (MUXs), demultiplexers (DEMUXs), low-loss waveguides connecting each element, and spot-size convertors (SSCs). In addition, Si photonics devices are fabricated by using mature CMOS fabrication technologies, which enables us to reduce the fabrication cost while achieving high-yield and high-uniformity. Since III–V compound semiconductor photonic devices are indispensable for constructing PICs including lasers, many research groups have pursued heterogeneous integration of lasers and Si photonics devices, for which direct bonding and adhesive bonding are key technologies. In previous demonstrations (Fang et al., 2006; Roelkens et al., 2006), the devices have long cavity length around 1 mm with relatively low optical confinement factor, but exhibited good characteristics such as single-mode lasing with high output power. Therefore, these lasers are intended for use as bias lights for Si Mach-Zehnder modulators (MZMs). However, they are not suitable for use in a direct modulation scheme inside datacenters. To heterogeneously integrate membrane BH lasers on SiO_2/Si substrate, we have developed a fabrication process where buried regrowth of an InP layer is carried out by using a directly bonded InP layer on SiO_2/Si substrate, which is described in Section 3. In Section 4, we will describe the device characteristics of the membrane BH lasers.

To further reduce the energy cost when optical links are employed in computercom chip-to-chip and on-chip interconnects, we must reduce the cavity length to a wavelength scale, i.e., a few micrometers. For this purpose, we have proposed and demonstrated photonic crystal (PhC) lasers, which are described in Section 5.

External modulators with advanced modulation formats are also required to further increase transmission capacity and/or extend transmission reach, as shown in Fig. 1. Currently, Si-based MZMs are used for 100-Gbit/s and 400-Gbit/s Ethernet with non-return-to-zero (NRZ) signal and pulsed amplitude modulation 4 (PAM4). However, when considering the need for further increases in capacity, Si is not suitable for constructing efficient phase modulators because it can only use the carrier plasma effect, which results in a tradeoff relationship between efficiency and absorption loss. On the other hand, III–V compound semiconductors, such as InP and InGaAsP, can use not only the carrier plasma effect but also the band-filling effect, Franz-Keldysh (FK) effect, quantum confined Stark effect (QCSE), and so on. Therefore, we can expect high modulation efficiency with low absorption loss. In Section 6, we will describe MZMs incorporating InP-based compound semiconductors.

In Section 7, we will describe photodetectors incorporating InP-based materials, which enable us to achieve high-response speed. Since we use the laser active region as an absorption region, it is easy to integrate the photodetectors with lasers.

Finally, Section 8 will briefly summarize this chapter.

2. Membrane photonic devices

2.1 Structure of membrane photonic devices

Energy cost EC, the energy for transferring a single bit, is the key issue for directly modulated lasers (DMLs) and is given by

$$EC = \frac{V_{bias} \cdot I_{bias}}{BR} = \frac{V_{bias}}{BR} i_{bias} \cdot V \tag{1}$$

where V_{bias} is bias voltage, I_{bias} is bias current, BR is the bit rate of the signal, i_{bias} is bias current density, and V is the active volume. The VCSELs now widely used in datacenters consume approximately 100 fJ/bit. Therefore, this is the target energy cost for in-plane lasers used in WDM networks inside datacenters. Further reduction of the energy cost is required for computercom networks, where predicted required energy costs for chip-to-chip and on-chip interconnections are 34 and 7 fJ/bit, respectively (Miller, 2009).

To reduce the energy cost, reducing bias current I_{bias} is important, as shown in (1). On the other hand, the bit rate is proportional to relaxation oscillation frequency f_r described by

$$f_r = \frac{1}{2\pi}\sqrt{\frac{\Gamma v_g}{q}\frac{dg}{dn}\eta_i\frac{(I_{bias} - I_{th})}{V}} = \frac{1}{2\pi}\sqrt{\frac{\Gamma v_g}{q}\frac{dg}{dn}\eta_i(i_{bias} - i_{th})} \tag{2}$$

where Γ is the optical confinement factor in the active region, v_g is group velocity, q is electric charge, dg/dn is differential gain, I_{th} is threshold current, and i_{th} is threshold current density. As shown in (2), to maintain modulation speed when decreasing bias current, both the bias current densities and the threshold current densities have to maintained by reducing the cavity volume. As shown in (1), when the relationship between bit rate and bias current density can be maintained, the energy cost is proportional to the active volume. Thus, reducing the cavity length is critical in achieving low operating energy. Eq. (2) also shows the importance of a large optical confinement factor to improve modulation efficiency.

To achieve lasing with reduced cavity length, we have to maintain the threshold gain. The relationship between threshold gain g_{th} and the photon lifetime in the cavity, τ_p, is given by

$$\Gamma v_g g_{th} = \frac{1}{\tau_p} = \frac{\omega}{Q} \tag{3}$$

where ω is the angular frequency and Q is the cavity Q-factor. To maintain the threshold gain, it is important to maintain the photon lifetime, i.e., the cavity Q-factor, when reducing the cavity length. For example, in the case of a Fabry-Perot cavity, the photon lifetime is given by

$$\tau_p = N_{ref} L / c \tag{4}$$

where N_{ref} is the number of reflections, and L is cavity length. The calculated Q-factor as a function of reflectivity for various cavity lengths is shown in Fig. 2. The cavity Q-factor increases as the cavity length and reflectivity increase. As described in Matsuo et al. (2014b), the required Q-factor for achieving both high modulation speed and high output power ranges from 10^3 to 10^4. When we assume a cavity length of 1 μm, the reflectivity values needed to obtain Q-factors of 1×10^3 and 1×10^4 are 99.53% and 99.95%, respectively. This makes it difficult to fabricate an extremely short-cavity DFB laser with a buried grating because the buried grating consists of different materials with a small difference in their refractive index, i.e., InGaAsP and InP.

To overcome these problems, we proposed a membrane BH laser on SiO_2/Si, in which the active region is buried within a thin InP layer, typically about 0.25-μm thick, as shown in Fig. 3. A membrane laser enables us to obtain a high optical confinement factor and a high coupling coefficient when a surface grating is employed. In this structure, we can expect an

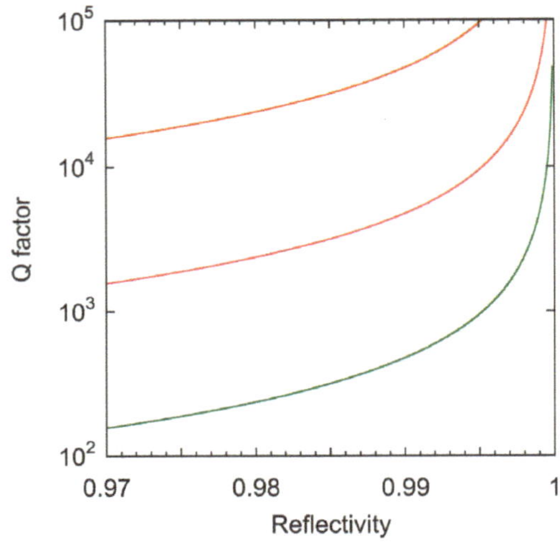

Fig. 2 Calculated Q-factor of a Fabry-Perot cavity as a function of reflectivity.

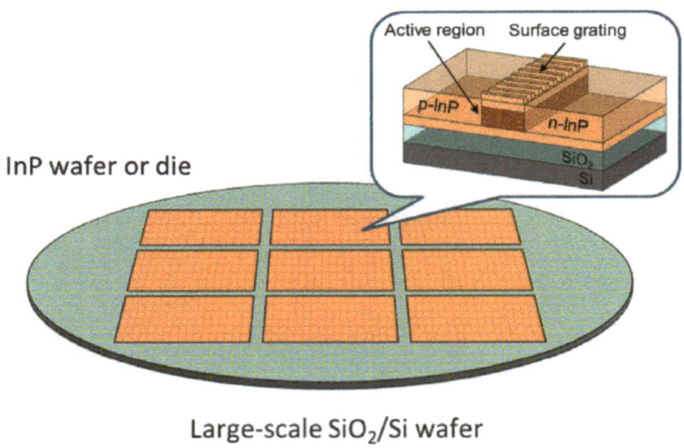

Fig. 3 Schematic diagram of lateral current injection DFB laser, which is integrated on large-scale SiO_2/Si wafer.

optical confinement factor that is three times as large as that with the conventional vertical current injection structure, in which the active region is entirely buried with InP. The surface grating enables us to achieve a high coupling coefficient because it can use the large difference in the refractive index between InP and low-refractive-index materials such as air or SiO_2. These features are advantageous for constructing short-cavity DFB lasers, i. e., low-operating energy DMLs.

Furthermore, it is important to employ a BH to achieve high-performance photonic devices because it provides both good carrier and photon confinements. However, in terms of fabrication, precise alignment between it and a Si waveguide is required. We have proposed fabricating the BH after bonding III–V layers on Si to overcome this problem (details will be described in Section 3).

2.2 Spot-size convertor and optical coupling with Si waveguide

Reducing assembly cost is important because it is the main factor in transmitter cost. This is because there is optical mode field size mismatch between a laser and a single-mode fiber. Lenses are required in the form of a spot-size convertor (SSC), and precise alignments are required. However, in the case of the membrane laser on a SiO_2 layer, it is easy to construct an SSC consisting of an inverse-taper InP waveguide and a SiO_x waveguide, as shown in Fig. 4 (Nishi et al., 2015; Shoji et al., 2002).

Fig. 5 shows calculated mode profile changes in the SSC. In this calculation, we assumed that the output InP waveguide with 250-nm height is laterally tapered and its width is reduced from 1.5 to 0.1 μm over a length of 300 μm. The InP taper waveguide is embedded in the SiO_x core with a 3-μm-square cross section. The SiO_x core is covered with an SiO_2 overcladding layer so that the relative refractive index difference becomes about 3%. Fig. 5A and B shows calculated mode profiles for 1.5-μm-wide InP and 3-μm-square SiO_x waveguides, respectively. Fig. 5C and D shows side-view and top-view images of mode conversion from the laser to SiO_x waveguide. The SSC loss is calculated to be 1.7 dB by the eigenmode-expansion method

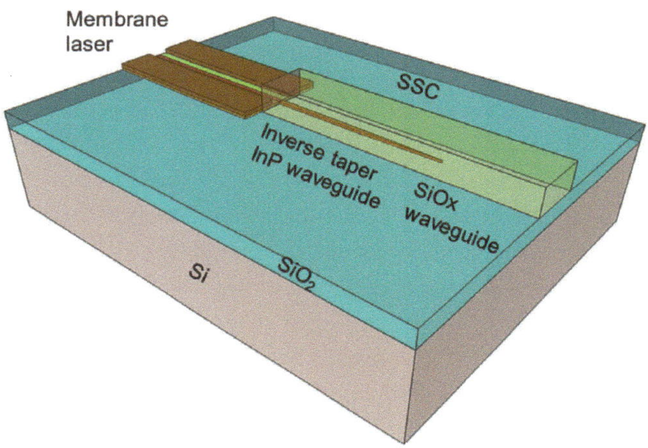

Fig. 4 Schematic of SSC, where an inverse taper InP waveguide and SiO_x waveguide are used.

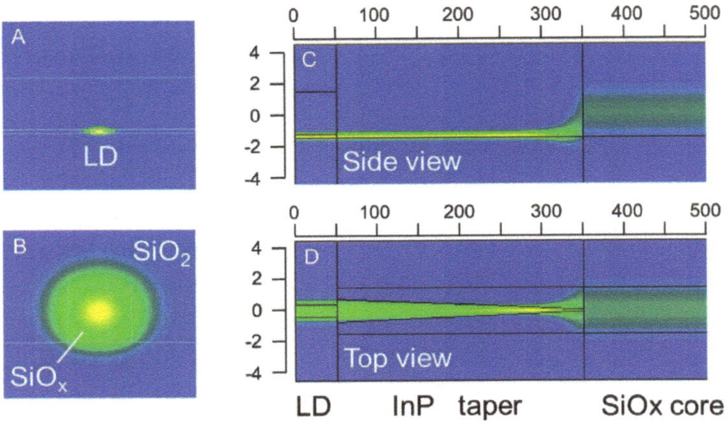

Fig. 5 Calculated mode profile change in SSC. (A and B) Mode fields at the DFB region in the membrane laser and at the SiO_x waveguide. (C and D) Side- and top-view images of mode conversion from the membrane laser to SiO_x waveguide via the SSC.

(Nishi et al., 2010). It should be noted that the SSC loss can be reduced to 0.1 dB by optimizing the taper structure (Nishi et al., 2015). Another important point is that a laser with the SSC can be butt-coupled with a fiber because there are no facets of III–V compound semiconductors, which results in low assembly cost.

The membrane structure is also easy to be heterogeneously integrated with Si photonics devices that employ 200-nm-thick SOI wafers, which are widely used in Si photonics foundries. Fig. 6 shows the effective refractive index of a Si waveguide as a function of Si thickness, where the ridge height and width are 100 and 600 nm, respectively. Dotted lines show the effective refractive indices of InP waveguides with a thickness of 200 nm and 2 μm. At the thickness near the cross point, it is easy to control the optical confinement factor of each waveguide. Thus, a Si waveguide around 430-nm thick is widely used for heterogeneous integration with conventional vertical injection lasers. For integrating 200-nm-thick SOI wafers using conventional vertical injection lasers, complicated taper structure or fabrication process is required (Dong et al., 2014; Wang et al., 2012).

Fig. 7A shows schematics of top- and cross-sectional views of a heterogeneously integrated III–V device on a Si waveguide. The InP taper waveguide is used to transfer the optical mode field between the Si waveguide and a hybrid waveguide consisting of InP and Si waveguides. For the calculation, we assumed that the InP taper width was decreased from 2 μm to 200 nm and the Si rib waveguide was 600-nm wide and 200-nm thick. Fig. 7B shows coupling loss for InP thicknesses of 200 nm and 2 μm as a function of

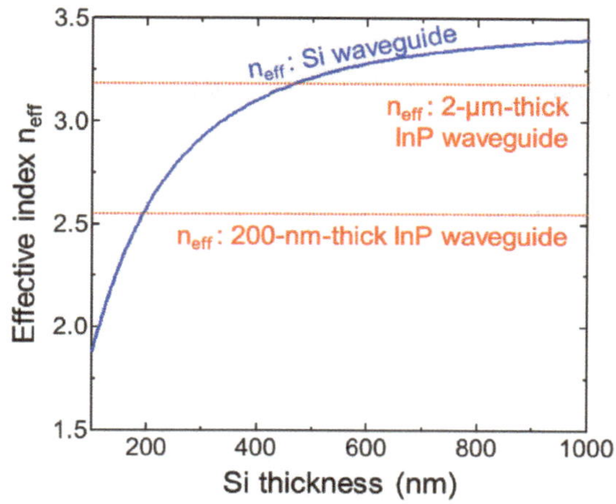

Fig. 6 Effective refractive index of Si waveguide as a function of Si thickness. Dotted lines show the effective refractive indices of InP waveguides with 200-nm and 2-μm thickness.

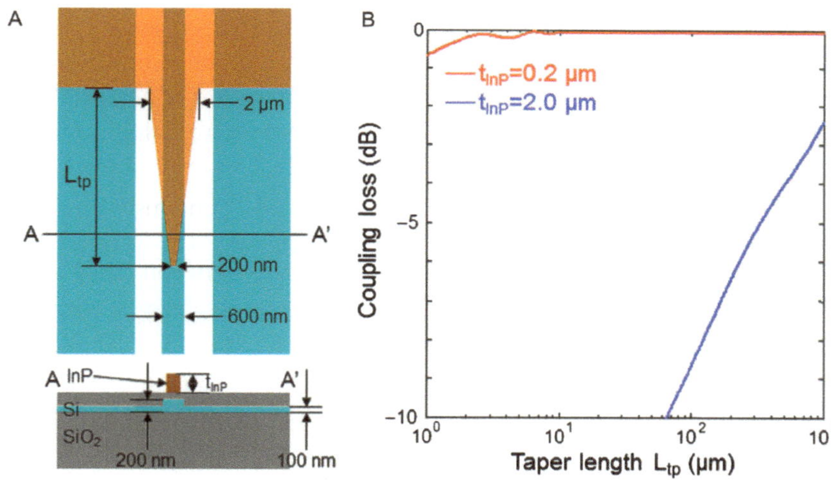

Fig. 7 (A) Schematic cross-sectional view of heterogeneously integrated III-V device on Si waveguide. (B) Calculated taper length dependence of coupling loss for InP thicknesses of 200 nm and 2 μm.

calculated taper length. As shown in this, when we use a membrane laser with a thickness of approximately 200 nm, the output light is easy to be coupled to the Si waveguide by using a 10-μm-long taper. On the other hand, taper with lengths over 1 mm are required when using the conventional laser structure. These results clearly show that the membrane structure is quite suitable to be integrated with Si photonics devices that use 200-nm-thick SOI wafers.

3. Fabrication of heterogeneously integrated III–V devices on SOI wafer

3.1 Wafer-, die-, and chip-scale integration

The fabrication procedure used for heterogeneous integration depends on whether the scale of the integration of III–V devices is wafer, die, or chip. Wafer and die bonding require the fabrication of III–V devices after integration on a Si wafer. The advantages are fabrication using large Si wafers, which results in low-cost fabrication, high uniformity, and the ability to use the state-of-the-art Si CMOS fabrication. Precise alignment between III–V devices and Si photonics devices is easy to achieve because the position of III–V devices is defined by process markers on the Si wafer. On the other hand, we cannot use the optimum fabrication process for III–V compound semiconductor in some cases due to the limitation imposed by the Si fabrication process, such as process temperature and dry and wet etching. Therefore, we have to carefully choose the fabrication process.

In chip-scale integration, both the III–V devices are bonded to SOI wafer after the fabrication is finished. Thus, the fabrication of both the III–V devices and Si photonics devices can be optimized separately. Another advantage is that we can select the devices before bonding. On the other hand, chip-scale integration requires fine alignment between III–V and Si photonics devices.

In addition, we must consider the footprint ratio between III–V and Si photonics devices. When it is small, chip-scale integration is preferable because if we employ wafer-scale fabrication, most parts of the III–V layers are removed. Thus, die and wafer-scale integration must be used when this ratio becomes high.

In the next section, we will describe our proposed procedures for fabricating membrane BH photonic devices. We propose employing wafer-scale fabrication because the BH must be finely aligned with the Si waveguide. In addition, the footprint ratio will most likely be high because Si photonics technology can provide compact optical devices due to its ability to provide high optical confinement in Si waveguides.

3.2 Buried regrowth on thin-InP template

Since the membrane structure requires a BH for effective confinements of both carriers and photons, the main issue is how to fabricate a BH that can utilize the advantages of Si photonics technologies. Fig. 8 shows the fabrication procedure for membrane BH devices, in which the BH is

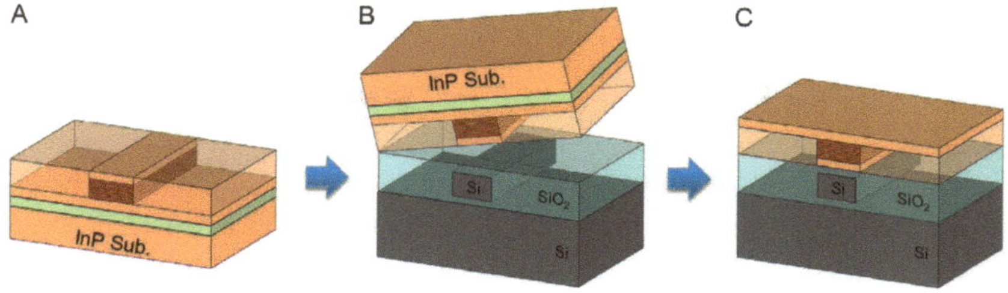

Fig. 8 Fabrication procedure for membrane BH devices, in which the BH is fabricated on InP substrate. (A) Fabriccation of BH on InP substrate. (B) Direct bonding. (C) Removal of InP substrate.

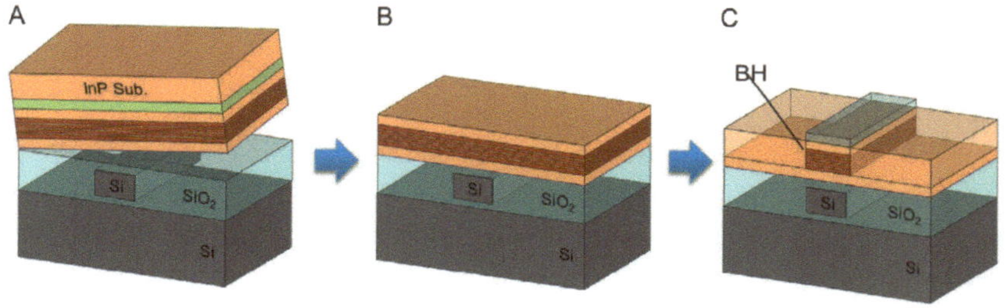

Fig. 9 Fabrication procedure for membrane BH devices, in which the BH is fabricated on Si substrate. (A) Direct bonding. (B) Removal of InP substrate. (C) Fabrication of BH on Si substrate.

fabricated on an InP substrate (Matsuo et al., 2014a). This is the easiest way to fabricate the BH because growth under lattice constant mismatch and thermal expansion coefficient mismatch conditions is not required. However, when the InP wafer is bonded to the Si photonics wafer, precise alignments are necessary. In addition, fabrication throughput and the equipment available are limited by the InP wafer size.

To overcome these problems, we have proposed the fabrication procedures shown in Fig. 9. First, we fabricate a Si waveguide using SOI substrate. The active layer is sandwiched between the thin InP layers, and a sacrificial layer is grown on an InP substrate (Fig. 9A). Then, the two substrates are directly bonded to each other, and the InP substrate and sacrificial layer are selectively removed using selective wet chemical etching (Fig. 9B). Next, an SiO_2 mask to fabricate the mesa stripe is defined by using process markers on the Si substrate. Thus, the position of the active region and Si waveguide are precisely aligned with the accuracy of photolithography. After removal of the active layer, except for the mesa stripe, an InP layer is grown on a thin InP layer (Fig. 9C). Therefore, this fabrication does not require epitaxial growth in a lattice mismatch condition.

However, the problem of thermal expansion coefficient mismatch remains. The growth of III–V compound semiconductors on an InP layer bonded to Si has been demonstrated (Fontcuberta i Morral et al., 2003; Matsumoto et al., 2013; Zhang et al., 2013), but the thermal expansion coefficient mismatch degraded the quality of the grown layers. To overcome this problem, the thickness of the III–V layers on Si should be reduced. In the proposed fabrication procedures, thermally induced strain is zero at the bonding temperature because the two substrates are free-standing before bonding. Thermal strain is induced at the growth temperature during the growth of the III–V layer, and it is also induced at room temperature after the growth. Thermally induced strain ε is calculated by

$$\varepsilon = (a_{\mathrm{Si}} - a_{\mathrm{InP}})\Delta T, \tag{5}$$

where a_{Si} and a_{InP} are the thermal expansion coefficients of Si and InP, for which we assume 2.62 and 4.56 ppm/K, respectively. For simplicity, we ignore the SiO_2 layer because the thermal expansion coefficient of a SiO_2/Si substrate is mainly determined by the Si when considering their thicknesses. With 610 and 200 °C assumed for the growth and bonding temperatures, thermally induced strain is calculated to be 795.4 ppm. Fig. 10 shows the calculated critical thickness as a function of induced strain (Fujii et al., 2015).

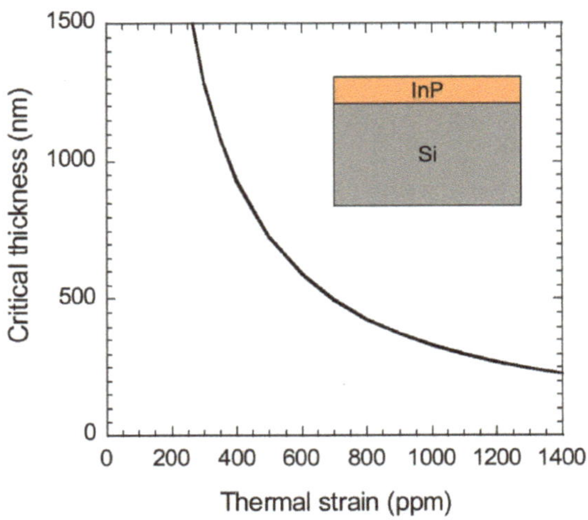

Fig. 10 Calculated critical thickness as a function of induced strain, with the mechanical equilibrium model and strain-energy-density equilibrium model.

The model based on strain-energy-density equilibrium expresses the critical thickness h_c as follows according to Van De Leur et al. (1988):

$$h_c = \frac{\sqrt{6}b(2-\nu)}{8\sqrt{3/2}\pi f(1+\nu)}\left\{\ln\left(\frac{8\sqrt{3/2}ah_c}{b}\right)-2\right\} \tag{6}$$

where b is Burger's vector of dislocation, ν is Poisson's ratio for strained material, f is the strain, and a is a parameter related to the dislocation energy which is typically about 4. The critical thickness of InP on a Si substrate is calculated to be 430 nm for a thermal strain of 795.4 ppm. Therefore, dislocation should not occur during MOVPE growth unless the InP membrane thickness exceeds 430 nm.

We have experimentally determined the critical thickness as follows (Fujii et al., 2015; Matsuo et al., 2014a). We grew MQWs sandwiched by InP layers and a sacrificial layer on a 2-in. InP(100) substrate. Then, the InP substrate including MQW layer was bonded to a 2-in. Si substrate with 1-μm thermal oxide using O_2 plasma at 200 °C. After the InP substrate and sacrificial layer had been removed, three samples, which included the same number of MQWs but had different total thicknesses (d = 250, 500, and 850 nm), were annealed in the MOVPE reactor at 610 °C in a PH_3 atmosphere for 30 min. Fig. 11 shows the PL spectra for the MQWs before (solid lines) and after (dashed lines) annealing. The peak PL intensity for the samples with total thickness of 850 and 500 nm were reduced to 32% and 69%, respectively, after annealing. However, when the total thickness was 250 nm, the degradation was dramatically reduced.

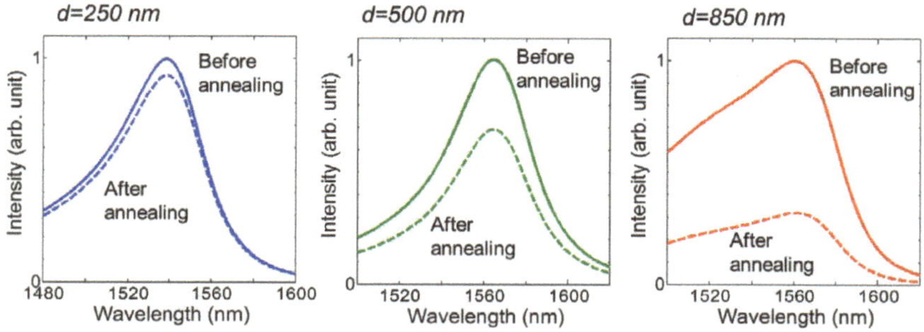

Fig. 11 PL spectra for the MQWs after bonding (solid lines) and annealing at 610 °C in PH_3 atmosphere for 30 min (dashed lines). Thicknesses of III–V layers, including the MQW layer, are 250, 500, and 850 nm.

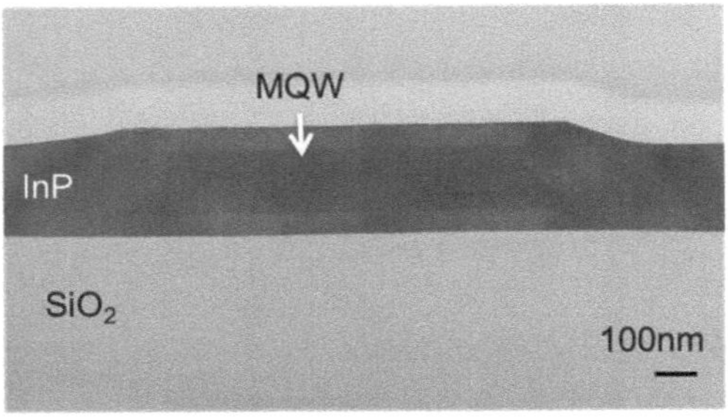

Fig. 12 Cross-sectional SEM image after fabricating BH.

We then fabricated a BH on SiO_2/Si substrate using 250-nm-thick III–V layers including the MQW layer. After defining the mesa structure, we grew an InP layer by MOVPE at 610°C to bury the active layer. Fig. 12 shows a cross-sectional SEM image after fabricating BH. As shown in this, we have successfully grown InP layer without dislocations and cracks.

3.3 Fabrication of lateral current injection structure

The lateral current injection structure shown in Fig. 13 is suitable for constructing a *p-i-n* junction for a thin-membrane structure. Unlike a vertical *p-i-n* junction structure, it does not require a thick layer (several micrometers) between the active region and electrode to avoid absorption. Zn thermal diffusion and Si ion implantation are used for the *p*- and *n*-type doping to an undoped InP layer, respectively. Silicon ions are implanted in open areas patterned in the photoresist for *n*-type doping with implantation energy of 240 keV and a dose of $1.2 \times 10^{14}\,\mathrm{cm}^{-2}$. After removing the photoresist, we deposit a SiO_2 layer and then anneal the wafer at 650°C for 3 min to activate the dopants, which results in the doping concentration of $\sim 1 \times 10^{18}\,\mathrm{cm}^{-3}$. Then, a *p*-type InP region is formed by Zn thermal diffusion at 550°C for 1 min. A patterned SiO_2 layer is used to define the *p*-type region. The doping concentration of Zn is typically $\sim 1 \times 10^{18}\,\mathrm{cm}^{-3}$ for the *p*-InP layer. Note that, for *p*-type dopant ion implantation, the activation ratio is poor, so Zn thermal diffusion is suitable for InP. Then, the grating is formed at the top surface, which enables us to obtain a high coupling coefficient. Finally, *p*- and *n*-electrodes are deposited, as shown in Fig. 13B.

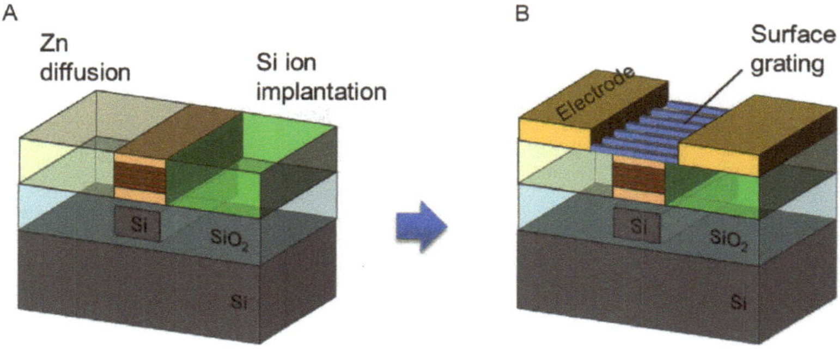

Fig. 13 Fabrication of (A) lateral current injection structure and (B) surface grating and electrodes.

4. Membrane BH lasers

4.1 Membrane DR lasers

Distributed reflector (DR) lasers have been proposed and demonstrated using vertical *p-i-n* junction structure (Komori et al., 1989; Matsuda et al., 2015). We have fabricated membrane distributed reflector (DR) laser in which cavity length ranges from 20 to a few hundred micrometers. Fig. 14A shows the typical structure of the DR laser we fabricated, which consists of a DFB laser section, a rear DBR section, and a front waveguide (Nishi et al., 2016). Since the rear DBR has a large reflectivity, the output light is mainly achieved from the front InP waveguide. To suppress spatial hole-burning effect, we use a DFB laser section with a uniform grating. As shown in Fig. 14B, both sides of the stop band have large transmittance, denoted as DFB spectrum. The center wavelength of the DBR is shifted to a shorter or longer wavelength compared with that of the DFB grating. In the case of Fig. 14B, the longer stop band in the DFB spectrum is selected to lase. Compared with quarter-wavelength-shift DFB laser, the DR laser with uniform-grating DFB laser section is expected to suppress spatial-hole burning at high injection current density. An InP output waveguide is used for constructing an SSC combined with an SiO_x waveguide, as described in Section 2.2. The front output InP waveguide with 250-nm-height is laterally tapered, and its width is reduced from 1.5 to 0.1 μm over a length of 300 μm. This InP taper is embedded in the SiO_x core with a 3-μm-square cross section. We deposited the SiO_x core layer using an electron-cyclotron-resonance (ECR) PECVD method at less than 200 °C, which prevents thermal damage to the DR laser during deposition. After that, we partly

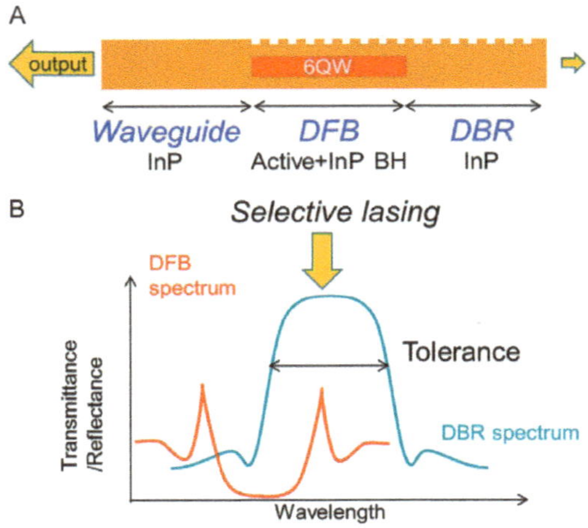

Fig. 14 (A) Schematic diagram of DR laser, consisting of DFB laser section, DBR section, and output waveguide. (B) Operating principle of DR laser.

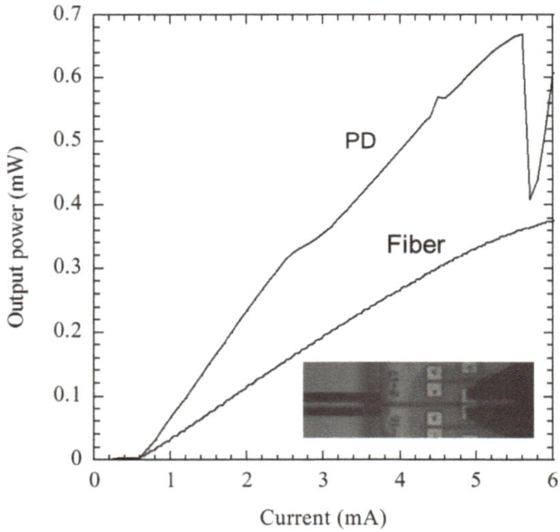

Fig. 15 I-L characteristics of DR laser with 50-μm-long active region.

removed the SiO_x by reactive ion etching and defined the SiO_x core, and then deposited a 4-μm-thick SiO_2 overcladding layer. Finally, we opened contact holes for probing by dry etching of the SiO_2 and SiO_x.

Fig. 15 shows measured current-light output (I-L) curves in room-temperature continuous-wave (RT-CW) operation for a DR laser with a 50-μm-long DFB laser section. We used a large-area photodetector (denoted as PD) for direct detection and butt-coupled single-mode fiber (denoted as Fiber). As shown in the photograph, a single-mode high numerical-aperture fiber (HNAF) with a mode-field diameter of 4.1 μm was physically contacted to the SiO_x waveguide facet. The curve measured by the photodetector placed in front of the SiO_x waveguide facet shows some kinks. This is due to the reflection at the SiO_x waveguide facet, because there is refractive index mismatch between the air and the SiO_x waveguide. On the other hand, the curve measured by the HNAF exhibits no kinks thanks to the suppression of reflection at the boundary between the SiO_x waveguide facet and HNAF. The threshold current and the maximum fiber output power were 0.8 mA and 0.4 mW, respectively. The fiber coupling loss was 2.2 dB, which, by using a lensed fiber, is reduced by about 6 dB from the coupling loss in a previous device without an SSC. The lasing spectrum with the current of 3 mA is shown in Fig. 16. We achieved stable single-mode lasing at the longer stop-band edge with side-mode suppression ratio (SMSR) of 42 dB. We estimated the κ of the DFB laser section to be about 1500 cm^{-1} from the DFB stop-band width of about 30 nm.

Dynamic responses were also measured. Fig. 17 shows relaxation oscillation frequency f_r as a function of the square root of bias current above threshold. The f_r was measured from the small-signal frequency response at 25 °C. We also measured membrane DFB lasers with cavity lengths of 73 and 120 μm. As shown in this figure, the modulation efficiency increases with decreasing cavity length and is 9.4 $GHz/mA^{0.5}$ for a DR laser with a 50-μm length.

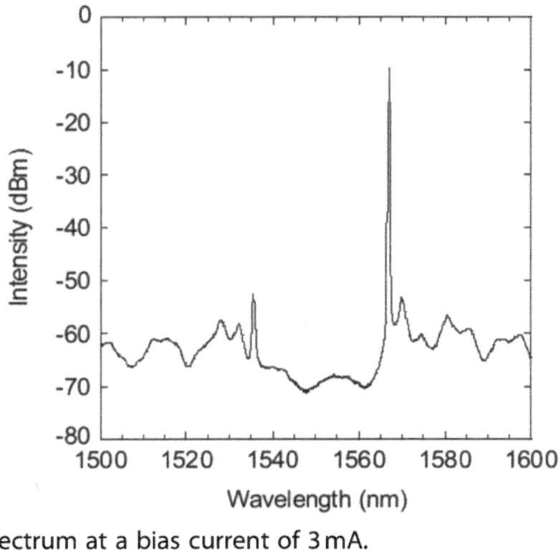

Fig. 16 Lasing spectrum at a bias current of 3 mA.

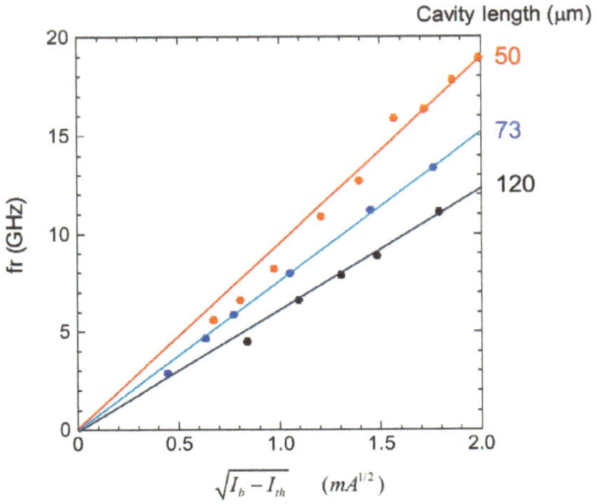

Fig. 17 Dynamic characteristics of the DR laser with 50-mm DFB laser section and membrane DFB lasers with cavity length of 73 and 120 μm; f_r estimated from small-signal frequency response as a function of square root of bias current above threshold.

Fig. 18 shows eye diagrams for 25.8-Gbit/s non-return-to-zero (NRZ) modulation signals with a pseudo-random bit sequence (PRBS) of $2^{31}-1$ at bias currents of 3.5 and 5.0 mA. The energy cost was calculated to be 212 and 358 fJ/bit, respectively. However, the eye-opening is not very clear due to noise, which is probably caused by back reflection from the abrupt end of the rear DBR section.

In further reducing cavity length, it is important to maintain the cavity Q-factor, i.e., increasing the coupling coefficient of grating, as described in Section 2.1. A straightforward way to increase the coupling coefficient is to increase the etching depth. However, this seems to increase unwanted scattering loss. Therefore, we have introduced a twin-mirror membrane DR laser, in which the DFB laser section is sandwiched by DBRs. Fig. 19 shows the calculated threshold gain, Γg_{th}-α_{loss}, as a function of the coupling coefficient for the DFB laser, single-mirror DR laser, and twin-mirror DR laser structures (Kanno et al., 2018). In these calculations, the length of the DFB laser section was set to be 20 μm with a phase shift at its center. For the DR laser, the length of the rear DBRs was set to be 50 μm. For the twin-mirror DR laser, the length of the rear DBR was set to be 50 μm, and front DBR length was changed from 10 to 30 μm. In these calculations, the loss of all waveguides was assumed to be 0. As shown in Fig. 19, the twin-mirror DR laser dramatically decreases the required coupling coefficient compared with the single-mirror DR laser and DFB laser. When we assume the

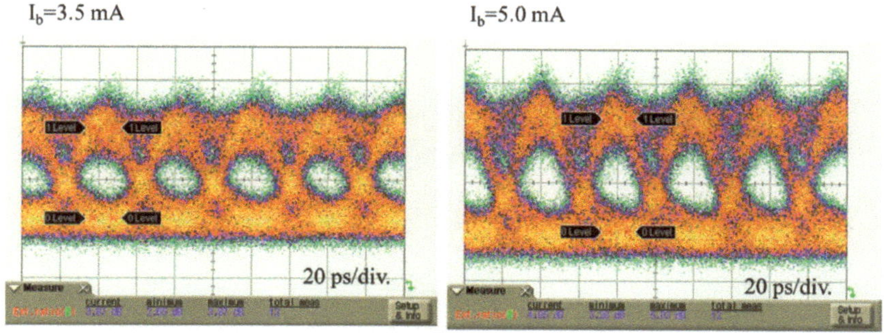

Fig. 18 Eye diagrams for 25.8-Gbit/s NRZ direct modulation with a bias current of 2.5 and 5.0 mA.

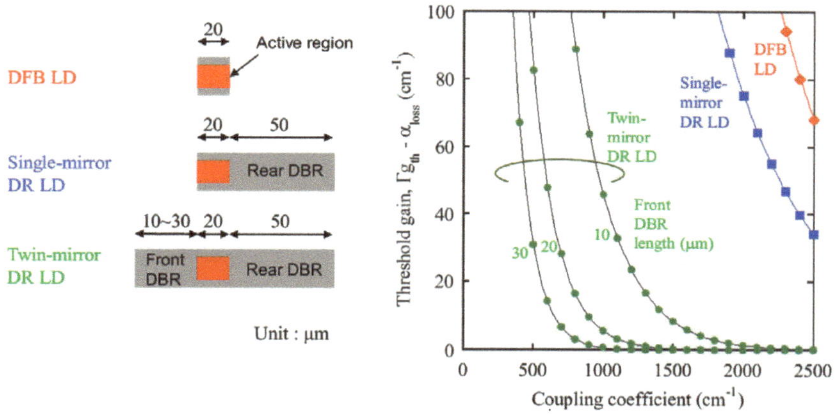

Fig. 19 Calculated threshold gain Γg_{th}-α_{loss} as a function of grating coupling coefficient for DFB laser, single-mirror DR laser, and twin-mirror DR laser structures.

threshold gain of 50 cm^{-1}, the required coupling coefficients are 2700 and 2200 cm^{-1} for DFB lasers and single-mirror DR lasers, respectively. On the other hand, in twin-mirror DR lasers, the required coupling efficiency is less than 1000 cm^{-1}, and the required coupling coefficient decreases with increasing front DBR length. These values are comparable to those of the fabricated membrane DR lasers described above.

I-L-V characteristics in RT-CW operation of fabricated twin-mirror DR laser are shown in Fig. 20. We used nine QWs for the active layer. The DFB laser section was 20-μm long, the rear DBR was 50-μm long, and front DBR lengths were 5, 10, and 20 μm. The I-V characteristics of three lasers are almost the same, and differential resistance at bias current of 1 mA is 830 Ω. Threshold current decreased with increasing front

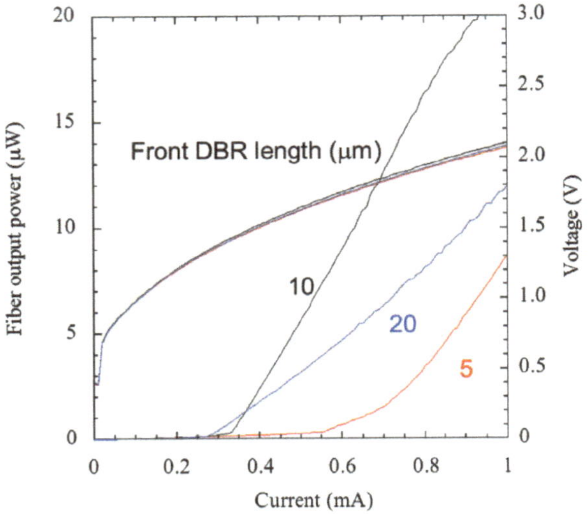

Fig. 20 I-L-V characteristics in RT-CW operation of fabricated twin-mirror DR laser. We used 9QWs for the active layer. The DFB laser section was 20-μm long, the rear DBR was 50-μm long, and front DBR lengths were 5, 10, and 20 μm.

DBR length, and were 0.67, 0.33, and 0.275 mA for front DBR lengths of 5, 10, and 20 μm. On the other hand, in terms of slope efficiency, the front DBR length of 10 μm was optimum. From the stop-band width, the grating coupling coefficient in the DFB laser section was estimated to be 1031 cm^{-1}.

We measured the dynamic characteristics of the twin-mirror DR lasers. Fig. 21A shows the relative intensity noise (RIN) spectrum of the twin-mirror DR lasers, which had a DFB and a front DBR length of 20 and 10 μm, respectively. The relaxation oscillation frequency was determined by fitting RIN spectra with a numerical model, and the result is shown in Fig. 21B. The modulation efficiency of the relaxation oscillation frequency was 11.4 $GHz/mA^{0.5}$.

As shown in these experiment results, the DR laser provides low energy cost thanks to the large optical confinement factor. Integration of the SSC was successfully demonstrated by using butt-coupling, which is promising for reducing assembly cost. The results also indicate that twin-mirror DR laser enables us to reduce the cavity length to a few tens of micrometers with a coupling coefficient of around 1000 cm^{-1}.

4.2 Integration with SiN waveguide

The refractive index of SiN of about two is moderate compared with low-refractive-index materials, such as SiO_2, and high-refractive-index materials,

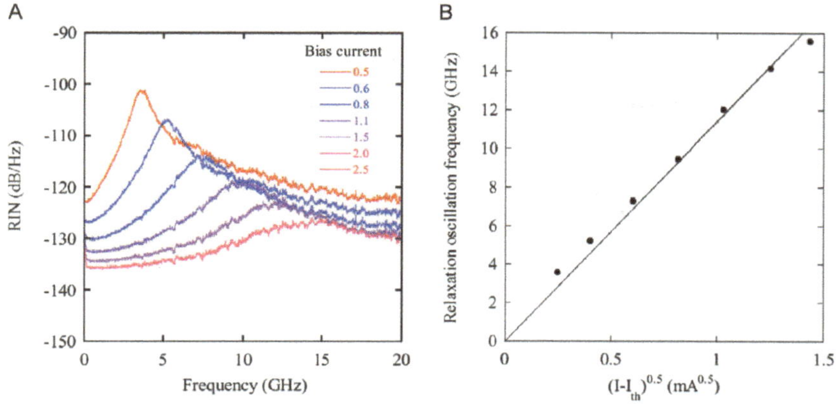

Fig. 21 Dynamic characteristics of the twin-mirror DR laser. (A) Relative intensity noise (RIN) spectra with bias current of 0.5–2.5 mA. (B) Relaxation oscillation frequency f_r determined by RIN spectra as a function of the square root of the injected current minus the threshold current.

such as Si and InP. Therefore, SiN is suitable for constructing low-propagation-loss waveguides and compact optical filters. In addition, the deposition temperature of SiN is relatively low, which makes it suitable for use in the backend process. However, when SiN is used in the C-band, post annealing at up to 1100 °C is required to remove N—H bonds, which have large absorption in the C-band. To avoid this issue, we have proposed using SiD_4 gas instead of SiH_4 gas for the plasma-enhanced chemical vapor deposition (PE-CVD) method. Fig. 22 shows the transmission spectrum of 1.8-cm-long SiN waveguides fabricated by using the SiD_4 gas and SiH_4 gas (Hiraki et al., 2017a). The refractive index of SiN film was 1.87. As shown in this, SiN film deposited by using SiD_4 gas has successfully demonstrated to suppress the absorption in the O/C/L telecom bands. Here, the residual O—H bond peak at around 1350 nm would be caused by the SiO_2 overcladding. We believe it can be also eliminated by using SiD_4 gas for depositing the overcladding film.

Fig. 23 shows a photograph of a fabricated device consisting of an eight-channel membrane DR laser array, an SSC to connect InP and SiN waveguides, and a SiN arrayed waveguide grating (AWG) filter. The overall footprint of the integrated device is $1.8 \times 2.0\,mm^2$. Each membrane DR laser is separated with a 250-μm pitch. The membrane DR laser consists of an 80-μm-long DFB laser section and a 40-μm-long rear DBR. The active region of the DFB laser consists of six InGaAsP-based QWs, which are embedded in a 280-nm-thick InP layer. The coupling coefficient of the

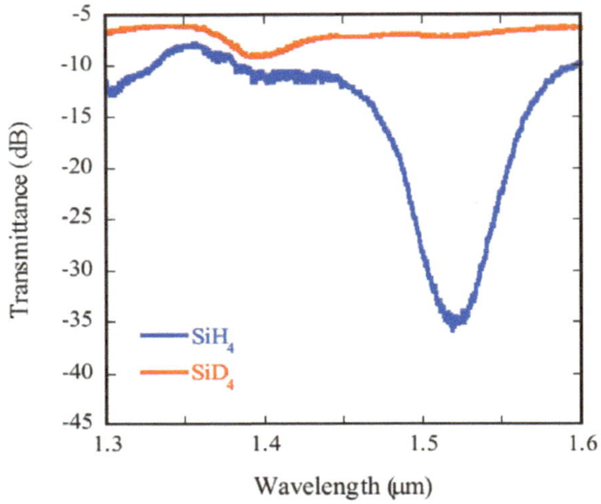

Fig. 22 Transmission spectrum of 1.8-cm-long SiN:D and SiN:H waveguides using a core with the refractive indices of 1.87.

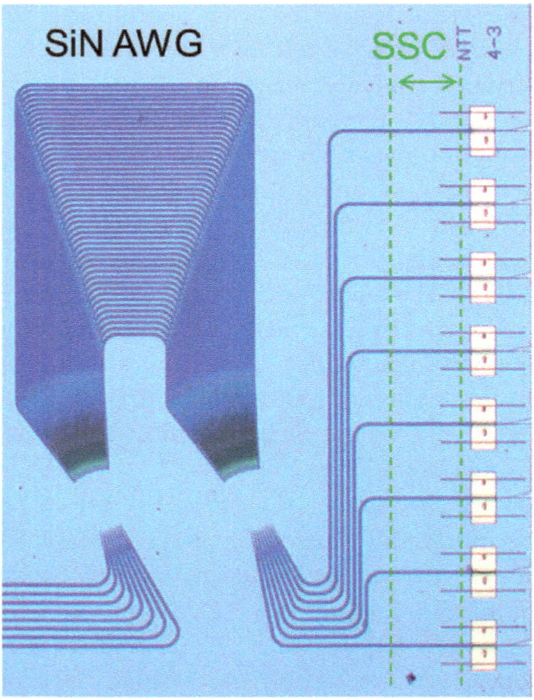

Fig. 23 Photograph of eight-channel DR laser array integrated with SIN-based AWG filter.

grating was designed to be 600 cm^{-1}. The output InP waveguide has an inverse taper shape for adiabatic mode conversion from the InP to the SiN waveguide, and the straight SiN core covers the 200-μm-long InP tapers. The SiN core is 0.9-μm wide and 0.45-μm thick. The AWG filter with 40 waveguides was designed to have a 380-GHz channel spacing and to ensure a free spectrum range (FSR) with a cyclic configuration. In the arrayed waveguides, each curved waveguide has the same total length, and the straight waveguides have different lengths for 380-GHz channel spacing. In the straight waveguide, SiN core width is increased to 2.0 μm using taper waveguides to reduce phase errors (Bogaerts et al., 2006; Okayama et al., 2016). At the chip facet, the SiN core width is also enlarged to 3.0 μm to reduce the coupling loss with lensed fiber.

A SiN film was deposited at less than 200 °C by using electron-cyclotron-resonance (ECR) PE-CVD technology with a source of a SiD_4/N_2 gas mixture, which can avoid thermal damage to the fabricated lasers (Hiraki et al., 2017a). The SiN waveguide and the AWG filter were formed by i-line lithography and reactive-ion etching. The SiO_2 overcladding layer was then deposited by using Tetraethyl orthosilicate (TEOS) PECVD technology.

Fig. 24 shows a spectrum measured from a single AWG output with the eight channels driven simultaneously by using multiple DC probes. At the

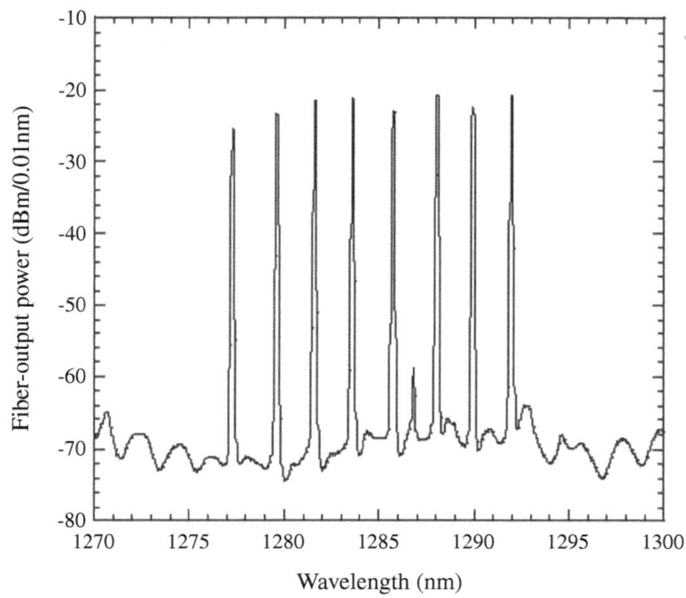

Fig. 24 Spectrum measured from a single AWG output with the eight-channels operated simultaneously.

output facet, a lensed fiber was aligned to a single AWG output channel. We confirmed stable single-mode lasing of all eight channels with 380-GHz spacing with bias current of ~8 mA. The threshold currents for all channels were 1.4 ± 0.1 mA. We estimated the loss of the passive components, including the InP-SiN SSC, the SiN AWG filter, and fiber coupling to be about 20 dB. Among them, the fiber coupling loss was 2.0 dB. We also fabricated the SiN AWG filter on a SiO_2/Si substrate without the laser integration process as a reference. We estimated the on-chip losses of the SiN AWG to be 2.3 and 4.4 dB at the center and edge channels, respectively. Thus, it seems that the integration increased the losses of the passive components. One considerable origin of the additional loss was the InP-SiN SSC. Instead of the ideal case of the designed 0.2-dB loss, the InP-SiN SSC seems to have a significant loss due to lithography imperfections in the InP taper tip, as we used double (side-by-side) lithography through an i-line stepper. We expect that the InP-SiN SSC loss can be decreased by using an improved lithography process such as electron-beam (EB) lithography, which has the potential to form better InP tapers. In addition, wafer warpage due to the heterogeneous integration will be considered in the future. Reducing it will lead to reduced loss in passive-device fabrication.

We performed 56-Gbit/s PAM-4 direct modulation in all eight channels. We set the DC bias current from 5.1 to 18.3 mA, depending on the channel, to compensate for the mismatch between the lasing wavelength and the AWG filter's passband. During the measurements, the AWG output channel was fixed. At the receiver side, the optical signal was fed to a praseodymium-doped fiber amplifier, filtered by a tunable optical band-pass filter, and changed to an electrical signal by a pin photodiode with an RF bandwidth of over 30 GHz. The electrical data were directly corrected by a digital storage oscilloscope with an 80-GSa/s sampling rate and processed offline after equalization. On this integrated device, the received signal was distorted by the mixed linear and nonlinear characteristics of the DR-laser response and pass-band shape on the integrated SiN AWG filter. To compensate for that, we applied a T/2-spaced, reduced-complexity second-order Volterra nonlinear equalizer with memory lengths of 40 samples for the linear part and 50 samples for the nonlinear part. The eye diagrams obtained after the offline processing are shown in Fig. 25, which also indicates the swing voltage and the DC bias current on each channel. We confirmed clear eye openings at all eight channels using directly modulated optical signals after multiplexing by the integrated SiN AWG filter. The estimated back-to-back bit-error rate (BER) for each channel is shown in

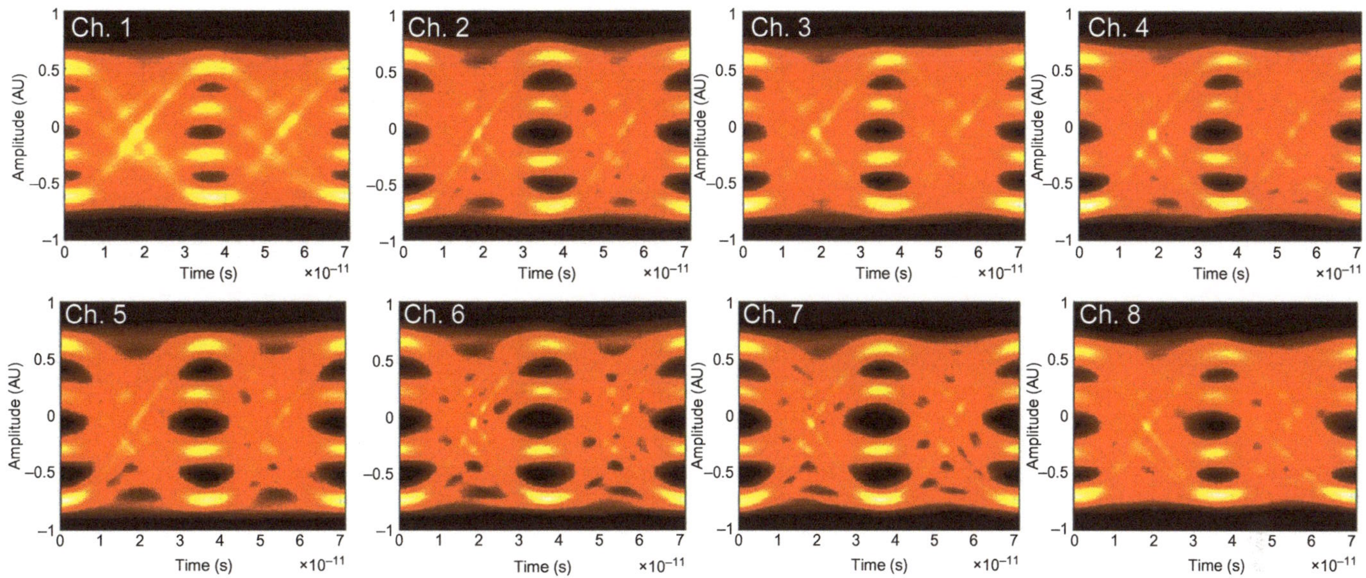

Fig. 25 Eye diagrams for 56 Gbit/s PAM-4 signals after offline processing.

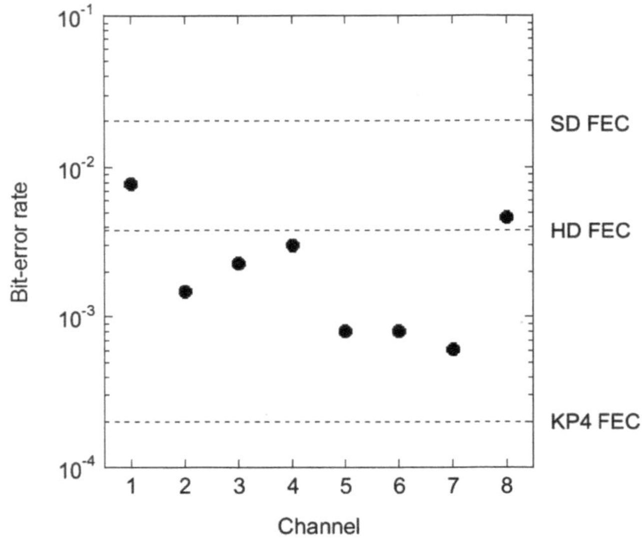

Fig. 26 Estimated back-to-back bit-error rate (BER) for each channel. Right *y*-axis shows allowed bit-error rates for soft-decision (SD), hard-decision (HD), and KP4 FEC, respectively.

Fig. 26. Considering a combined 400 G forward error correction (FEC) architecture, the total BER of our transmitter was 2.7×10^{-3}, which is below the 7% overhead hard-decision FEC limit of 3.8×10^{-3}. We expect that improving the loss characteristics in PIC and increasing the fiber output power will improve the BER characteristics.

4.3 Heterogeneous integration with Si waveguide

Si photonics provides high-performance optical filters and compact MZMs with compact size and is suitable for fabricating large-scale PICs. Thus, heterogeneous integration of III–V photonic devices with Si-based PICs is desirable. As described in Section 2.2, membrane devices are quite advantageous for heterogeneous integration. Fig. 27 shows the fabrication procedure for heterogeneous integration. Si waveguides are fabricated using an SOI wafer. Then, the SiO_2 layer is covered with a Si waveguide, and chemical mechanical polishing (CMP) is performed to achieve a flat top surface (Fig. 27A). This is important for high-yield direct bonding between III–V and Si substrates. Controlling the SiO_2 thickness on Si waveguide is required to achieve the designed optical confinement factors for III–V and Si waveguides (Fig. 27C). After removing the substrate, we fabricate the BH by employing epitaxial growth on thin InP layer. Then, we fabricate *p*- and *n*-type doped InP regions and deposit SiN film for fabricating the grating

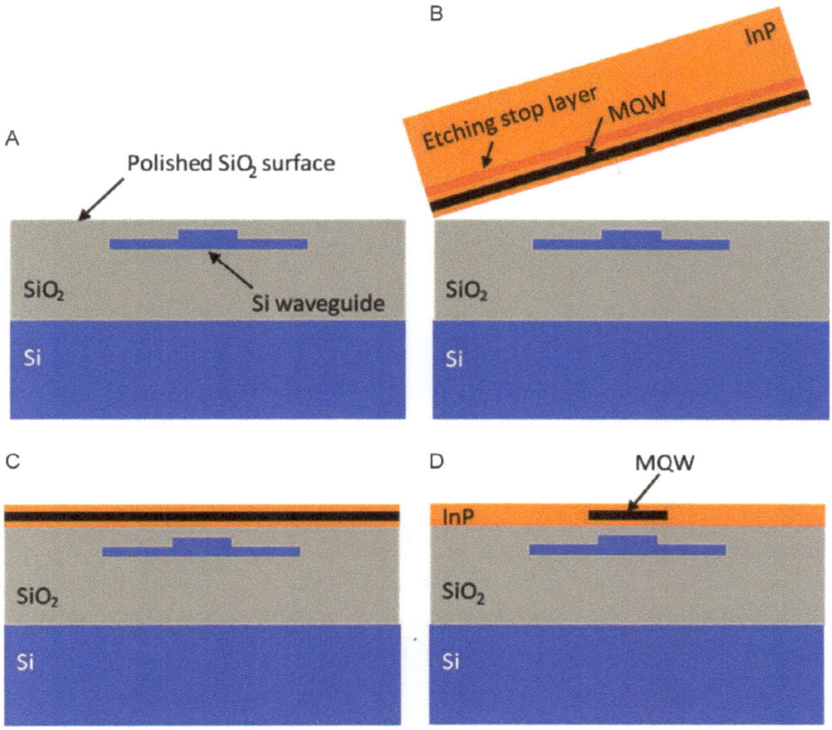

Fig. 27 Fabrication procedure for heterogeneous integration of Si waveguide and membrane BH laser. (A) CMP to achieve a flat top surface. (B) Direct bonding. (C) Removal of InP substrate. (D) Fabrication of BH.

and the electrode. In this experiment, we designed DFB laser length to be 500 μm. Therefore, we fabricated the surface grating using SiN film, which results in smaller coupling coefficient compared with previous results.

We fabricated a heterogeneously integrated DFB laser with active region length of 500 μm and width of 0.68 μm. This laser is intended for integration with Si-based MZMs, and it will be used as a bias light for MZMs, not for DMLs. The active region consists of six layers of InGaAsP MQWs with a PL peak of 1543 nm. The grating was fabricated using a 50-nm-thick SiN waveguide to achieve a coupling coefficient of 90 cm^{-1}. Fig. 28A shows the fiber-coupled output power from the Si waveguide versus injection current for stage temperatures ranging from 25 to 130 °C. The threshold current was 6 mA, and the maximum fiber-coupled output power was 4.6 mW at 25 °C. The coupling loss between the Si waveguide and lensed fiber was estimated to be about 2 dB; therefore, the optical power within the Si output

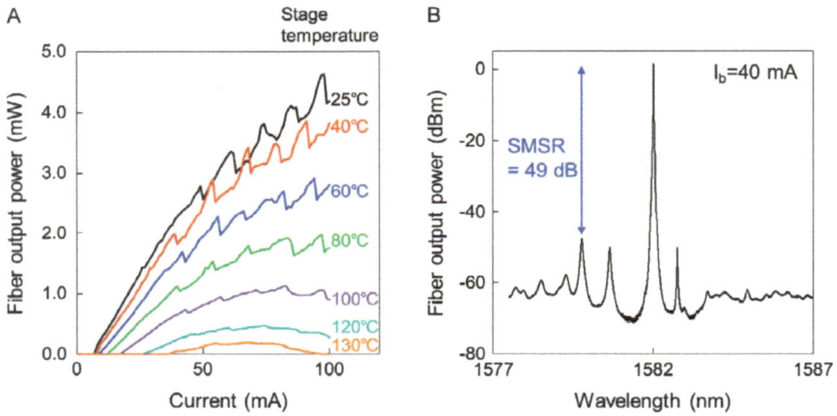

Fig. 28 (A) I-L characteristics of membrane BH laser integrated with Si waveguide for various stage temperatures. (B) Lasing spectrum with bias current of 40 mA at stage temperature of 25 °C.

waveguide was 7.4 mW. Lasing was achieved at up to 130 °C. The lasing spectrum with injection current of 40 mA at the stage temperature of 25 ° C is shown in Fig. 28B. The lasing wavelength and single-mode suppression ratio were 1582 nm and 49 dB, respectively.

5. λ-Scale embedded active region photonic crystal (LEAP) lasers

The demand for constructing computercom optical networks, as shown in Fig. 29, is increasing because chip-to-chip and on-chip data transmission is now hampering the further development of CPUs in terms of transmission capacity and power consumption. As described in Section 1, optical links represent the potential solution to these problems; however, a critical issue to develop an DML is the requirement of extremely small operating energy. The predicted energy costs for chip-to-chip and on-chip interconnects are approximately 36 and 7 fJ/bit, respectively (Miller, 2009). When the data speed of 10 Gbit/s is assumed, the power consumption is 0.36 and 0.07 mW for chip-to-chip and on-chip interconnects, respectively. For simplicity, if we assume a bias voltage of 1 V, bias currents are 360 and 70 μA, respectively. These extremely small bias currents are difficult to achieve even for state-of-the-art VCSEL technologies. Since the energy cost of a DML is proportional to the active volume, we need to develop a wavelength-scale cavity laser.

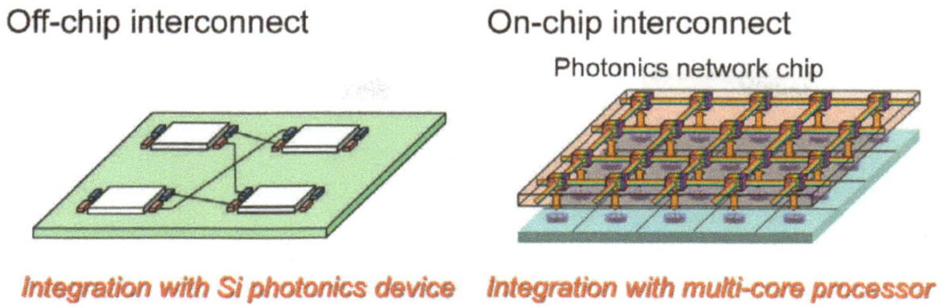

Fig. 29 Schematic diagrams of chip-to-chip and on-chip interconnects.

5.1 LEAP lasers on InP substrate

To achieve a high-performance DML with a wavelength-scale cavity, we employ a two-dimensional PhC cavity because it provides a high-Q factor in a wavelength-scale cavity on a SOI wafer (Takahashi et al., 2007; Tanabe et al., 2007). Since two-dimensional PhC cavities typically use a 200-nm-thick Si slab, a lateral current injection structure is suitable for fabricating an electrically driven laser without degrading Q-factor. The first PhC laser was demonstrated by optical pumping in 1999, with the sample temperature set at 143 K and the laser pumped with 10-ns pulses (4% duty cycle) (Painter et al., 1999). Since then, many research groups have fabricated optically pumped PhC lasers. However, RT-CW operation has been difficult to achieve due to the increase in the active region temperature.

Fig. 30 shows a schematic diagram of a two-dimensional PhC laser consisting of a gain medium, where the PhC cavity is created by removing some air-holes. It also shows the calculated temperature distribution in the active region, where we assumed a 100-μW heat source on a 3.0-μm-long and 0.5-μm-wide active region. The thermal conductivity of the gain medium was assumed to be 4.2 W/mK (InGaAsP slab), and the temperature of the InP substrate was assumed to be kept constant at 25 °C. As shown in Fig. 30, the active region temperature increases to 77.2 °C. Furthermore, there is no carrier confinement structure, and generated carriers in the cavity escape from the cavity, which also degrades device efficiency, i.e., an increase in the pumping light power is required, which results in a further increase in the active region temperature. These issues make it difficult to achieve lasing even for RT-CW operation.

To overcome these problems, we have proposed a PhC laser whose active region is buried within an InP PhC slab (Matsuo et al., 2010). As shown in Fig. 31A, the small active region is surrounded by InP, whose thermal conductivity (68 W/m/K) is higher than that of InGaAsP

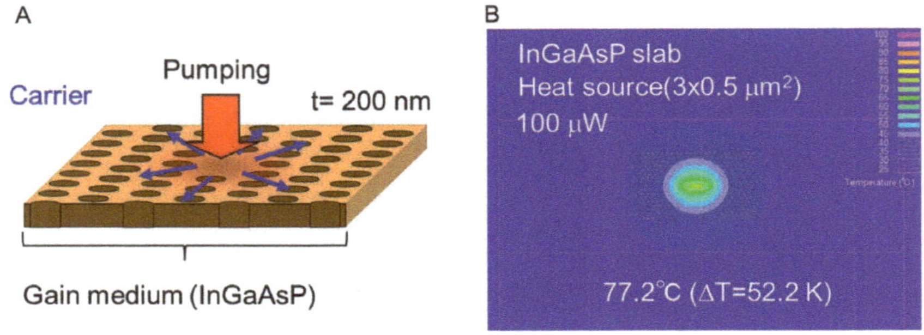

Fig. 30 (A) Schematic of two-dimensional PhC laser using a uniform gain media. (B) Temperature distribution with 100-μW heat source on active region. *Reprinted by permission from Macmillan Publishers Ltd.: Matsuo, S., et al., 2010. High-speed ultracompact buried heterostructure photonic-crystal laser with 13 fJ of energy consumed per bit transmitted. Nat. Photonics 4, 648–654. Copyright 2010.*

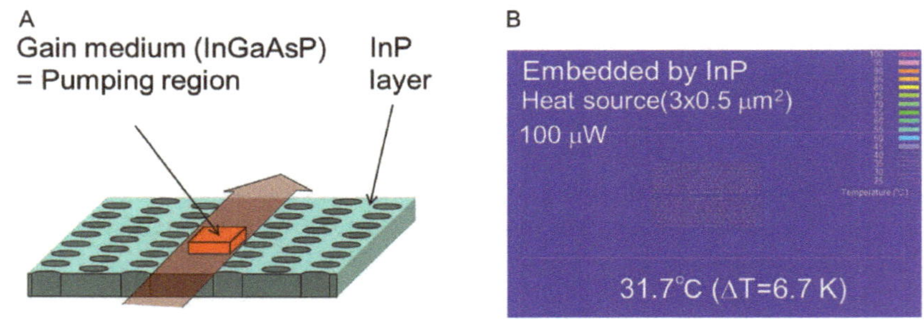

Fig. 31 (A) Schematic of LEAP laser where active region is embedded within line-defect waveguide of InP-based PhC slab. (B) Temperature distribution with 100-μW heat source on active region. *Reprinted by permission from Macmillan Publishers Ltd.: Matsuo, S., et al., 2010. High-speed ultracompact buried heterostructure photonic-crystal laser with 13 fJ of energy consumed per bit transmitted. Nat. Photonics 4, 648–654. Copyright 2010.*

by an order of magnitude. We call this a λ-scale embedded active region PhC laser or LEAP laser. Fig. 31B shows the calculated temperature distribution with a 100-μW heat source. The increase in active region temperature was calculated to be 6.7 K, which is a dramatic decrease from the calculated result for the PhC laser with the uniform gain medium shown in Fig. 30B. This clearly shows the advantage of the LEAP laser.

Since the refractive index of the active region is slightly higher than that of InP, a high Q-factor cavity is easy to create without shifting the air hole position (Matsuo et al., 2011). Fig. 32A shows a schematic of a LEAP laser without an output waveguide. Mode fields calculated by the finite-

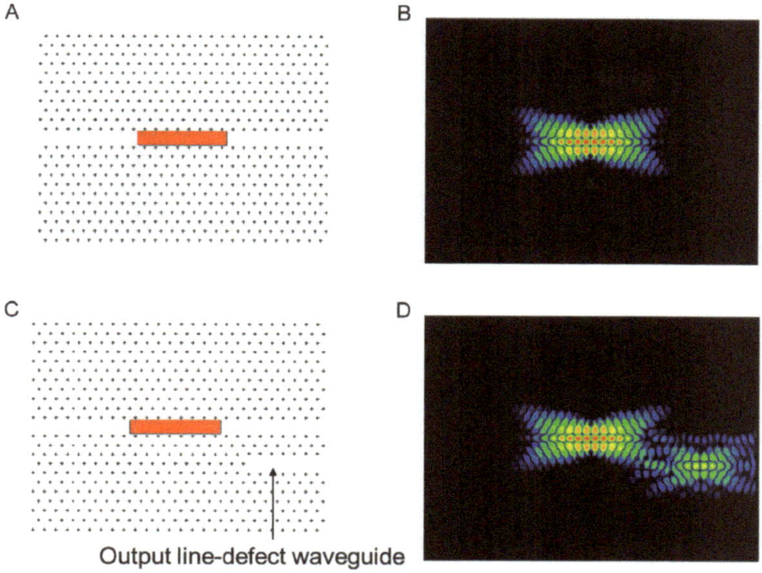

Fig. 32 (A) Schematic diagrams of LEAP laser without output waveguide. (B) Calculated mode profiles of the device. (C) Schematic diagrams of LEAP laser with output waveguide. (D) Calculated mode profiles of the device with output waveguide.

difference time-domain (FDTD) method are shown in Fig. 32B. We assumed that the effective refractive indices of the InGaAsP-based active region and the surrounding InP layer are 2.77 and 2.59, respectively. The LEAP laser without the output waveguide exhibits a high-Q ($\sim 1.1 \times 10^6$) cavity mode with $V_{eff} \sim 0.25\,\mu m^3$. The mode field extends to the Γ-M direction when the cavity has no output waveguide. Fig. 32C shows a schematic of a LEAP laser with an output waveguide. We define the offset as the number of rows between the line-defect waveguide and output waveguide. In the case of Fig. 32C, the offset is 3. Fig. 32D shows the calculated mode profiles of devices with an output waveguide for offset 3. As shown in this, effective coupling can be achieved with the output waveguide when it is placed in the Γ-M direction of the PhC cavity. The coupling between the cavity and output waveguide increases with decreasing offset and cavity Q-factor (Matsuo et al., 2011).

To fabricate a current-driven LEAP laser, we grew an InAlAs sacrificial layer and InGaAlAs-based three QWs sandwiched by thin InP layers on InP substrate. The fabrication procedures for the BH and *p-i-n* diode are the same as described in previous sections. The air holes were formed by using electron-beam lithography and inductive coupled plasma reactive ion

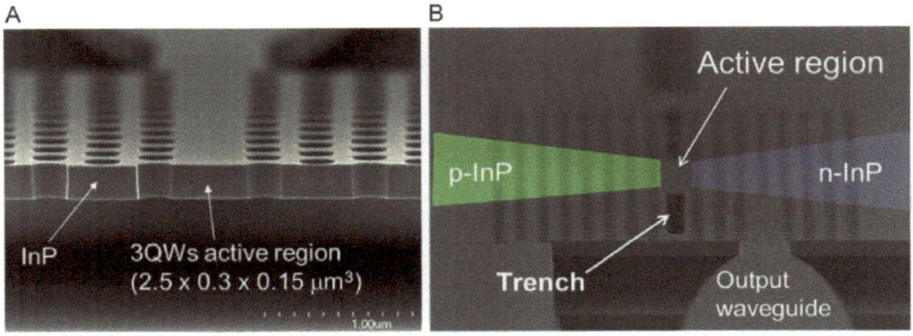

Fig. 33 (A) Cross-sectional SEM image of LEAP laser. (B) Bird's eye view of fabricated LEAP laser with current blocking trenches. *Reprinted by permission from Macmillan Publishers Ltd.: Takeda, K., et al., 2013. A few-fJ/bit data transmissions using directly modulated lambda-scale embedded active region photonic-crystal lasers. Nat. Photonics, 7, 569–575. Copyright 2013.*

etching (ICP-RIE). Finally, contact metals were deposited, and we formed an air-bridge structure by wet-etching the InAlAs sacrificial layer, as shown in Fig. 33. Fig. 33A shows a cross-sectional SEM image of a fabricated LEAP laser, in which a 2.5-μm-long active region was successfully buried in the InP-based PhC slab. We also achieved a flat top surface and smooth dry-etched air holes. These results indicate that we can achieve a high-Q cavity using the buried active region. Fig. 33B shows a bird's eye view of a fabricated LEAP laser, where current blocking trenches were employed in the line-defect waveguide. As shown in Fig. 32C, the optical mode field of the LEAP laser is extended to the Γ-M direction, and it is very low in the line-defect waveguide, except the active region. Therefore, we can fabricate current blocking trenches in a line-defect waveguide, which effectively suppress leakage current. Fig. 34A shows the I-L characteristics of the fabricated device in RT-CW operation. The threshold current was 4.8 μA thanks to the reduction of the active volume and the suppression of leakage current.

The device was modulated with 10-Gbit/s NRZ signal with a bias current of 25 μA and voltage of 1.74 V. The dynamic response is shown in Fig. 34B. Electrical power consumption is 44 μW, which results in the energy cost of 4.4 fJ/bit. This indicates that a laser with a wavelength-scale cavity can be used as a light source for optical links inside CPUs.

To improve the output power and slope efficiency, we increased the number of QWs from three to six and improved the *n*-type doping process by increasing the ion-implantation energy (Sato et al., 2015). Fig. 35A and B

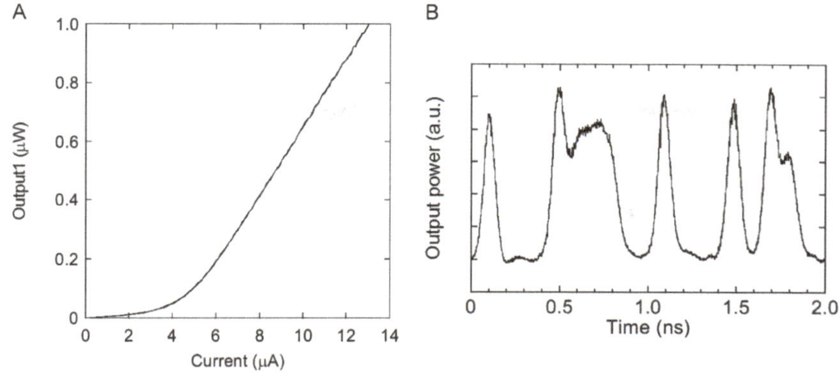

Fig. 34 (A) I-L characteristic of a LEAP laser with current blocking trenches. (B) Dynamic response of directly modulated LEAP laser with bias current of 25 μA. *Reprinted by permission from Macmillan Publishers Ltd.: Takeda, K., et al., 2013. A few-fJ/bit data transmissions using directly modulated lambda-scale embedded active region photonic-crystal lasers. Nat. Photonics, 7, 569–575. Copyright 2013.*

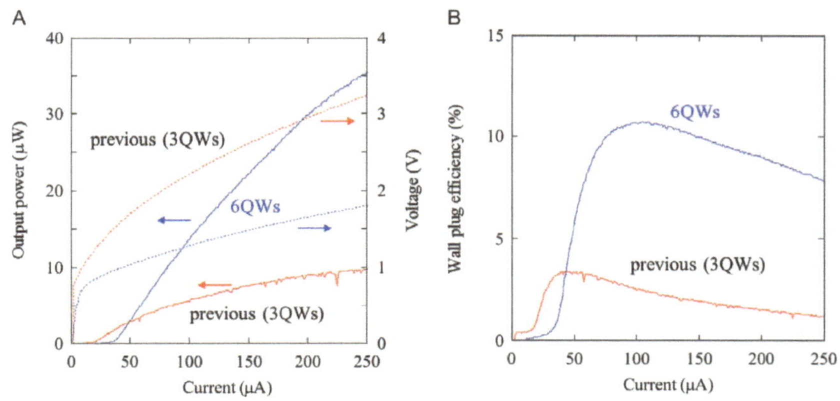

Fig. 35 (A) I-L characteristics of fabricated LEAP laser with 6QWs. For comparison, I-L characteristic of previous device with 3QWs is also shown. (B) Wall-plug efficiency of 6QWs and 3QWs.

shows the I-L characteristics and wall-plug efficiency of LEAP laser with six QWs. Fig. 35 also shows the results for the previously developed three-QW device (Matsuo et al., 2013). From these results, we find that the output power of the six-QW device at 300 μA is increased to 39.4 μW which is two times higher than that of the three-QW device, although the threshold current increases to 36 μA because of the increase in the number of QWs. By increasing the QWs to six, we obtained the maximum wall plug efficiency

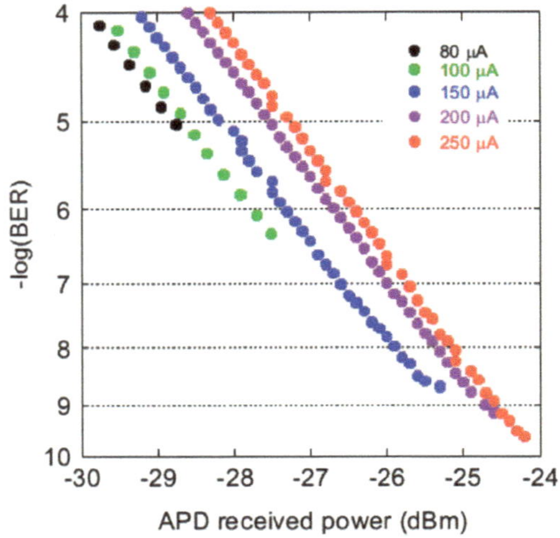

Fig. 36 BER measurement results for various bias currents.

(WPE) of 10.7%, as shown in Fig. 35B. This value is three times larger than that in the previous work. Moreover, WPE greater than 7% is maintained for the high current range.

We measured dynamic responses of the device. In these experiments, an air coplanar probe was used without 50-Ω termination, and a pseudorandom binary sequence (PRBS) of 2^{31}–1 and a 9.953-Gbit/s NRZ signal was used. Fig. 36 shows the results of BER measurements for various bias currents ranging from 80 to 250 μA, corresponding to energy costs from 9.9 to 48.4 fJ/bit. The coupling loss between the PhC waveguide and single-mode fiber was 8 dB, and the optical attenuator had 1.3-dB optical loss. We used an APD, and there was no fiber amplifier in this experiment. The BER characteristics improved as the bias current decreased. This is because the noise at signal "0" was decreased when the bias current was decreased. On the other hand, because of the increase in fiber-coupled output power, the BER of less than 10^{-9} was achieved for bias currents of 200 and 250 μA. Since the coupling loss between the single-mode fiber and output waveguide was estimated to be 8 dB, we can expect to use a *p-i-n* photodiode as a detector when we integrate with an SSC to reduce the coupling loss.

5.2 LEAP laser on SiO_2/Si substrate

Computercom applications require a huge number of compact transmitters. To connect and switch between lasers and PDs, employing Si photonics

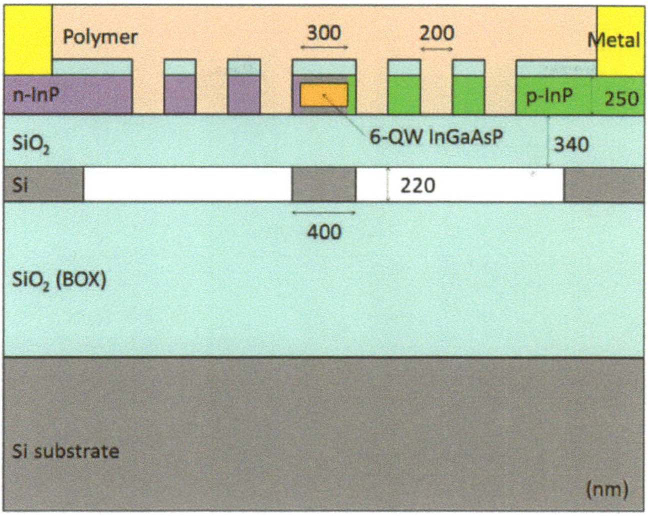

Fig. 37 Cross-sectional view of heterogeneously integrated LEAP laser on Si waveguide.

devices, such as switches and filters, is quite important and therefore, heterogeneous integration plays the main role in this situation. Fig. 37 shows a cross-sectional view of a heterogeneously integrated LEAP laser and a Si waveguide. A LEAP laser and Si waveguide are vertically stacked with a 340-nm-thick SiO_2 layer. For the fabrication, we used a 220-nm thick and 400-nm wide Si waveguide. The thickness of the InP-based slab was 250 nm, including the six QW active region (Takeda et al., 2017).

The device fabrication procedure is shown in Fig. 38. First, Si waveguides were formed using an SOI substrate (Fig. 38A). The active layer (six QWs) and InGaAs sacrificial layers were grown on an InP substrate and, then, SiO_2 layer was deposited on the top surface of the InP substrate (Fig. 38B). Oxygen-plasma-assisted direct bonding was used to directly bond both the InP and Si substrates (Fig. 38C). In this fabrication procedure, we did not cover the Si waveguide; therefore, we can precisely control the distance between it and the active layer. After removing InP substrate (Fig. 38D), we fabricated LEAP lasers on SiO_2 layer by using the same fabrication procedure shown in Figs. 9 and 13. The active region was $3.8 \times 0.3 \times 0.15\,\mu m^3$. Fig. 39 shows (A) a photograph and cross-sectional SEM image and (B) a cross-sectional SEM image of fabricated device. Thanks to the use of photolithography in fabricating the BH on Si substrate, positional alignment between the BH and Si waveguide is well controlled.

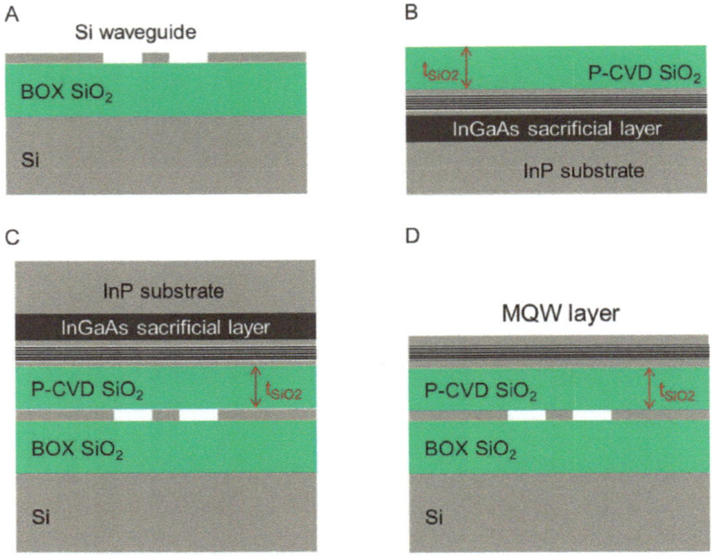

Fig. 38 Fabrication procedure for heterogeneous integration. (A) Fabrication of Si waveguide. (B) Deposition of SiO_2 layer on InP substrate including active layer. (C) Direct bonding. (D) Removal of InP substrate.

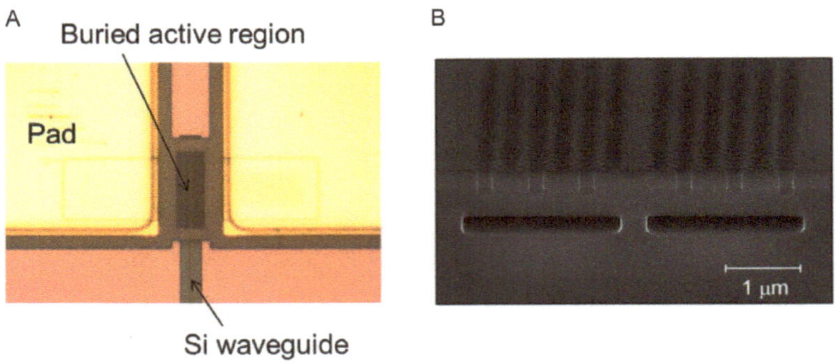

Fig. 39 (A) Photograph and (B) cross-sectional SEM image of fabricated LEAP laser on Si waveguide device.

Static characteristics of a fabricated device in RT-CW operation are shown in Fig. 40. The threshold current was 42 μA, and the fiber coupled output power via Si waveguide was 0.72 μW at an injected current of 0.5 mA. The lasing wavelength was 1541 nm. Although the threshold current is much larger than we expected for the LEAP laser on InP substrate, we have successfully demonstrated the heterogeneous integration. To improve

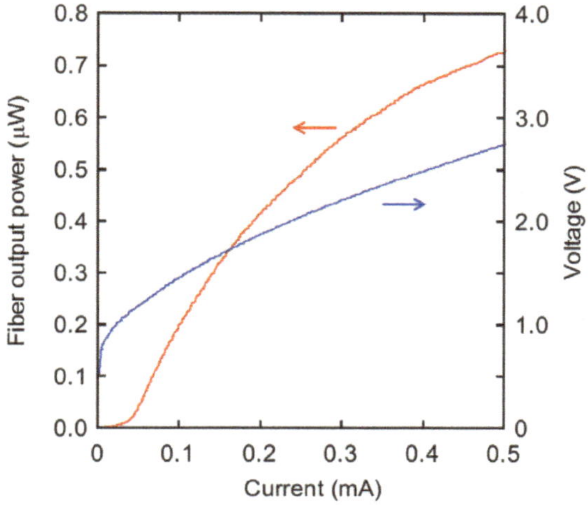

Fig. 40 I-L-V characteristics of fabricated LEAP laser on Si waveguide device in RT-CW operation. Threshold current was 42 μA, and fiber coupled output power was 0.72 μW at an injected current of 0.5 mA.

the device performances, it is important to suppress the leakage currents. It is also important to optimize the device design for high coupling efficiency between the PhC cavity and Si waveguide.

6. MZMs using III–V/Si MOS capacitor phase shifter

The demand for external modulators is increasing due to the increase in Internet traffic, and they have begun to be used in short-distance communications inside datacenters, as described in Section 1. To meet this demand, Si-based MZMs have attracted much attention because Si photonics technologies enable us to integrate a polarization rotator and combiner with MZMs, which are key components for constructing digital coherent transmitters. MZMs on Si platforms have been widely demonstrated by employing free-carrier effects in carrier-injection (Green et al., 2007; Thomson et al., 2008; Xu et al., 2007), carrier-depletion (Feng et al., 2010; Thomson et al., 2012), or carrier-accumulation devices (Liao et al., 2005; Liu et al., 2004; Webster et al., 2015). Carrier injection MZMs provide high modulation efficiency, but the cut-off frequency is limited to ~100 MHz. In contrast, modulators employing carrier depletion and accumulation have gigahertz-class cut-off frequencies due to their fast majority-carrier response.

However, Si MZMs can only use the carrier plasma effect, so their modulation efficiency is limited to the half-wave voltage-interaction length product $V_{\pi}L$ of a few Vcm. To increase the modulation efficiency, carrier density must be increased, but this causes an increase in absorption, which is a tradeoff relationship between efficiency and loss. Since short-distance interconnects require a huge number of transmitters, it is important to make MZMs small by improving the modulation efficiency without increasing absorption loss. In this context, introducing III–V compound semiconductors into Si MZMs is the key to solve this problem because the large electron-induced refractive index change, high electron mobility, and low carrier-plasma absorption of III–V compound semiconductors are beneficial for overcoming the tradeoffs among the $V_{\pi}L$, operation speed, and insertion loss of Si MZMs. We have proposed incorporating an *n*-type III–V material into a MOS-capacitor modulator, which operates by carrier depletion and accumulation (Hiraki et al., 2017b). This provides a high overlap integral between the optical intensity and carrier distribution in both the *n*-type III–V material and *p*-type Si. Fig. 41A shows a schematic diagram of the proposed phase shifter using an InGaAsP/Si MOS capacitor. The MZ interferometer is constructed by Si multimode interferometers (MMIs), Si waveguides, and phase shifters. Fig. 41B shows a cross-sectional image of the phase shifter. A thin insulator is sandwiched by an *n*-type InGaAsP layer and *p*-Si layer. The overlap region of both layers is set to 600 nm. We used an *n*-type $In_{0.72}Ga_{0.28}As_{0.62}P_{0.38}$ layer with photoluminescence (PL) peak wavelength of 1300 nm for the operation wavelength of around 1550 nm.

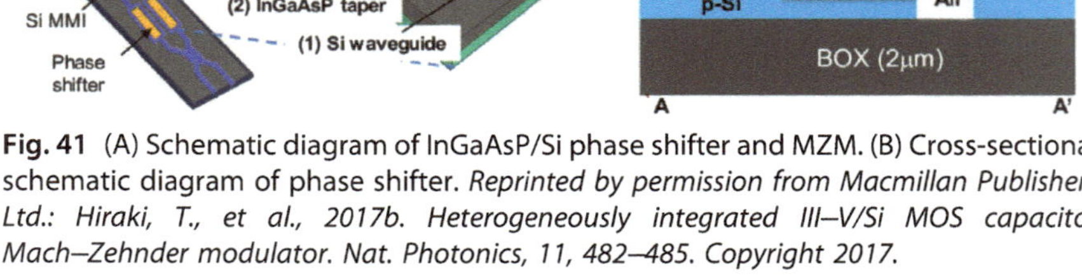

Fig. 41 (A) Schematic diagram of InGaAsP/Si phase shifter and MZM. (B) Cross-sectional schematic diagram of phase shifter. *Reprinted by permission from Macmillan Publishers Ltd.: Hiraki, T., et al., 2017b. Heterogeneously integrated III–V/Si MOS capacitor Mach–Zehnder modulator. Nat. Photonics, 11, 482–485. Copyright 2017.*

Fig. 42A shows the calculated $V_\pi L$ in the accumulation mode at 1550 nm for InGaAsP/Si, InP/Si, and Si/Si MOS-capacitor phase shifters as a function of SiO_2 thickness. The $V_\pi L$ of the InGaAsP/Si MOS capacitor is over three times smaller than that of the Si/Si MOS capacitor at the same insulator thickness and also smaller than that of the InP/Si MOS capacitor. This is because *n*-type InGaAsP can use not only the carrier plasma effect but also the band-filling effect. In a MOS capacitor, a fundamental tradeoff is that smaller SiO_2 thickness provides lower $V_\pi L$ but higher capacitance, which leads to a large RC time delay. Fig. 42B shows the calculated relationship between applied DC voltage and absorption loss for various doping concentrations. The absorption loss decreases with decreasing doping concentration, but there is a tradeoff relationship with efficiency. The results also indicate that the absorption loss in the InGaAsP/Si MOS capacitor phase shifter is smaller than that of the Si/Si MOS capacitor, even if we use higher doping concentrations in the InGaAsP layer. This benefit arises from the lower free-carrier absorption and absence of poly-crystalline material in the InGaAsP/Si MOS capacitor, which reduces insertion loss and the loss imbalance between the two arms. In addition, a higher donor concentration reduces series resistance. When we designed the donor concentration in InGaAsP layer to be $2 \times 10^{18}/cm^3$, the series resistance of the *n*-type InGaAsP layer was over 10 times lower than that of the *p*-Si layer with an acceptor concentration of $5 \times 10^{17}/cm^3$.

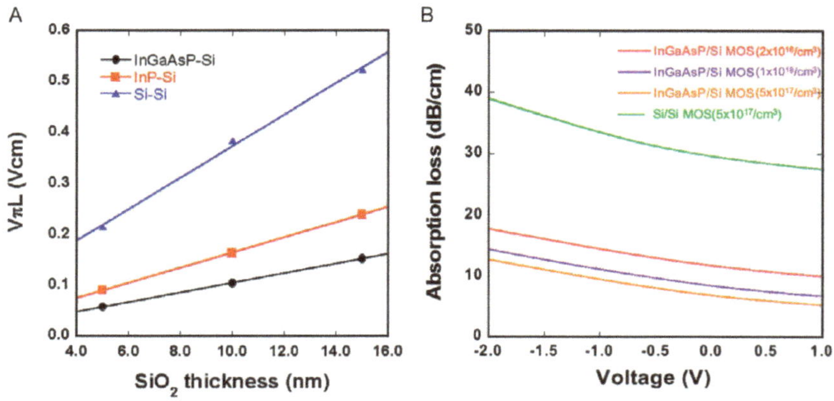

Fig. 42 (A) Calculated $V_\pi L$ as a function of SiO_2 thickness for phase shifter employing InGaAsP/Si, InP/Si, and Si/Si. (B) Calculated absorption loss as a function of applied voltage for various doping concentrations. *Reprinted by permission from Macmillan Publishers Ltd.: Hiraki, T., et al., 2017b. Heterogeneously integrated III–V/Si MOS capacitor Mach–Zehnder modulator. Nat. Photonics, 11, 482–485. Copyright 2017.*

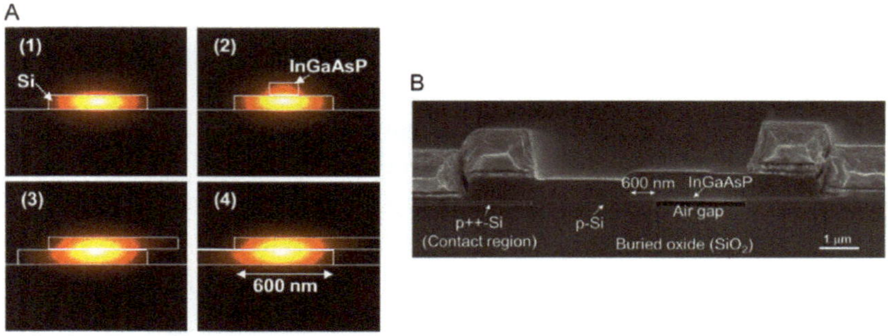

Fig. 43 (A) Calculated cross-sectional mode-field patterns for the waveguide, tapers, and capacitor, numbered (1)–(4) in Fig. 41A. (B) SEM image of a fabricated MOS capacitor. *Reprinted by permission from Macmillan Publishers Ltd.: Hiraki, T., et al., 2017b. Heterogeneously integrated III–V/Si MOS capacitor Mach–Zehnder modulator. Nat. Photonics, 11, 482–485. Copyright 2017.*

Fig. 43A shows the calculated cross-sectional mode-field patterns for the waveguide, tapers, and capacitor, numbered (1)–(4) in Fig. 41A. The 80-μm-long taper design ensures negligible mode-conversion loss and reflection at the tapers with calculated reflection of around −40 dB. We fabricated devices with different phase shifter lengths from 250 to 700 μm. Fig. 43B shows an SEM image of a fabricated MOS capacitor.

The transmission spectra of fabricated MZMs with a 700-μm-long phase shifter as a function of applied voltage is shown in Fig. 44A. The output power was normalized by that of a reference Si waveguide with a propagation loss of 6 dB/cm. When the negative voltage applied to the *n*-type electrode was increased, the spectrum blue-shifted due to a carrier-induced refractive index change. The loss and extinction ratio were degraded in the shorter wavelength range because the optimum operating wavelength of the MMI coupler was shifted to a longer wavelength due to lithography patterning errors. Fig. 44B shows the relationship between V_π and inverse phase-shifter lengths for various phase shifter lengths. From a linear fitting, the $V_\pi L$ in the accumulation mode was estimated to be around 0.09 Vcm, which is consistent with the calculated results shown in Fig. 42A.

Fig. 45 shows the eye diagrams at 25 and 32 Gbit/s, respectively. We used finite impulse response (FIR) filters, and 2- and 3-tap FIR filters were applied for bit rates of 25 and 32 Gbit/s, respectively. We applied RF voltages to one arm of the MZ modulator, and the DC bias of the MOS capacitor was 1 V. The extinction ratios at 25 and 32 Gbit/s were 4.0 and 3.1 dB, respectively. Thanks to the low $V_\pi L$ of the InGaAsP/Si MOS

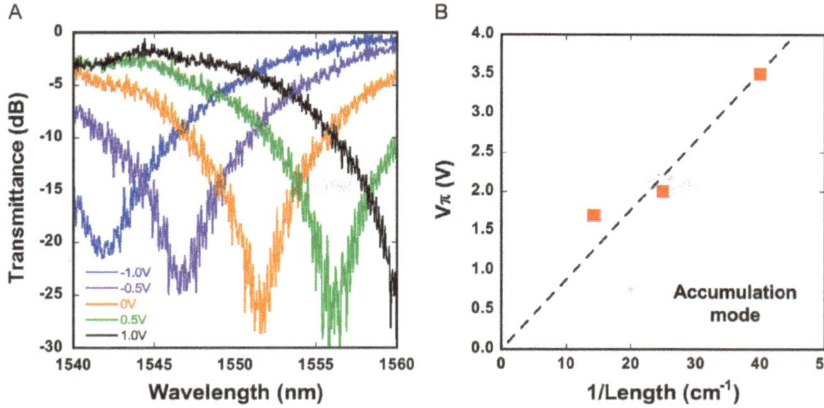

Fig. 44 (A) Transmission spectrum of fabricated MZMs with 700-μm-long phase shifter as a function of applied voltage. (B) Measured V_π for various phase shifter lengths. Estimated $V_\pi L$ in the accumulation mode was around 0.09 Vcm. *Reprinted by permission from Macmillan Publishers Ltd.: Hiraki, T., et al., 2017b. Heterogeneously integrated III–V/Si MOS capacitor Mach–Zehnder modulator. Nat. Photonics, 11, 482–485. Copyright 2017.*

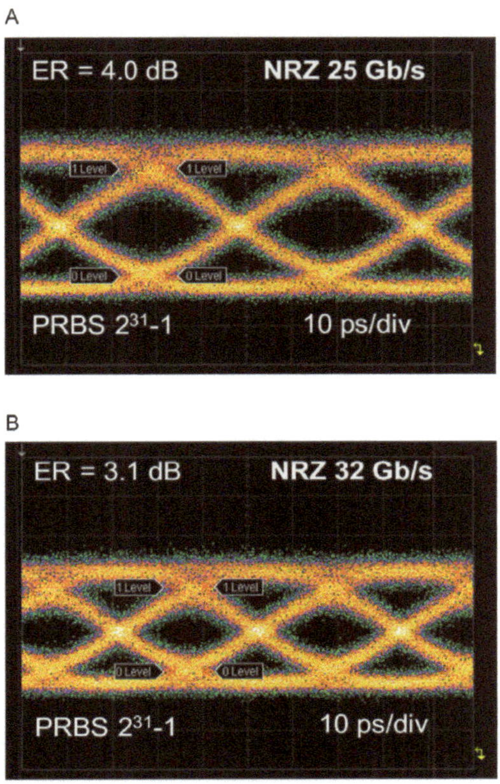

Fig. 45 Eye diagrams at (A) 25 Gbit/s and (B) 32 Gbit/s. Finite impulse response (FIR) filters were used in these experiments. *Reprinted by permission from Macmillan Publishers Ltd.: Hiraki, T., et al., 2017b. Heterogeneously integrated III–V/Si MOS capacitor Mach–Zehnder modulator. Nat. Photonics, 11, 482–485. Copyright 2017.*

capacitor phase shifter, the eyes opened with small voltage swing in one arm. Note that the FIR filters compensated for the RF-power loss in the high-frequency region, because the measured 3-dB cutoff frequency of the fabricated device was 1.0–2.6 GHz at DC voltages. The large RF power loss was mainly caused by series resistances, including large contact resistance at the *p*-Si due to etching damage, and by the long length between the electrode region and MOS capacitor (2.5 μm).

7. Membrane photodetector

Photodetectors are also important for constructing optical links. For extremely short distances, it is important to fabricate high-density PICs including both lasers and detectors because of the limited space to install them. Therefore, we have proposed the use of a laser active region as an absorption region to simplify the fabrication process by reducing the number of regrowth processes.

In addition, the total power consumption of both the transmitters and receivers, including their electrical circuits, must be decreased because it limits device density. In this context, introducing APDs is essential because they dramatically decrease the required output power of the transmitter, which results in a reduction of the power consumption of both the laser and the driver. Note that the large bias voltage of the APD has a small impact on the total power consumption since the power consumption of the receiver is mainly dominated by that of the trans-impedance amplifier (TIA).

We fabricated a membrane p-i-n photodiode with a 15-μm-long and 0.6-μm-wide absorption region on SiO_2/Si substrate (Matsuo et al., 2017). The absorption region consists of nine QWs, and its photoluminescence (PL) peak is 1540 nm, which is used for membrane lasers in the C-band wavelength range. A 1.5-μm-wide InP waveguide was used for the input waveguide, and the facet did not have an anti-reflection coating. The coupling loss between the lensed fiber and InP waveguide was estimated to be 8.5 dB. Fig. 46A shows static characteristics for various input powers at an input-light wavelength of 1530 nm. Dark current increased up to 40 μA when the reverse bias voltage was increased to 15 V. Fig. 46B shows responsivity curves as a function of bias voltage. The responsivity slightly decreases with increasing reverse bias voltage from 0 to 5 V due to the broadening of the exciton peak of the MQWs and the band-edge shift of the Franz-Keldysh effect. Then, it increases with increasing reverse bias voltage due to the avalanche gain. As shown in the figure, the responsivity curves shift to a large reverse bias voltage

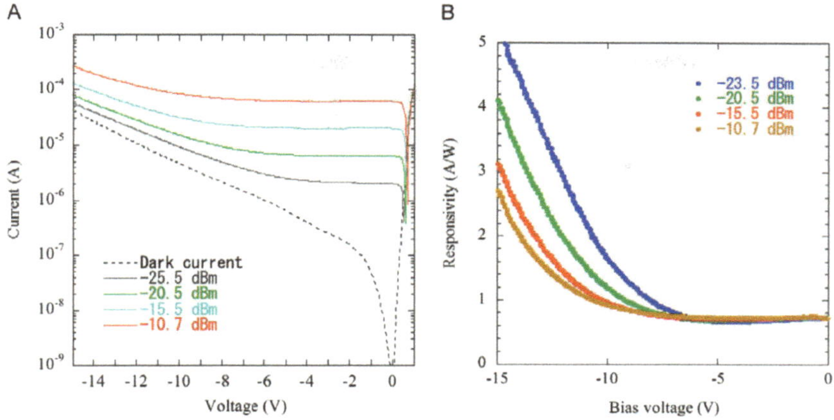

Fig. 46 (A) Static characteristics for various input powers at an input-light wavelength of 1530 nm. (B) Responsivity curves as a function of bias voltage for various input powers.

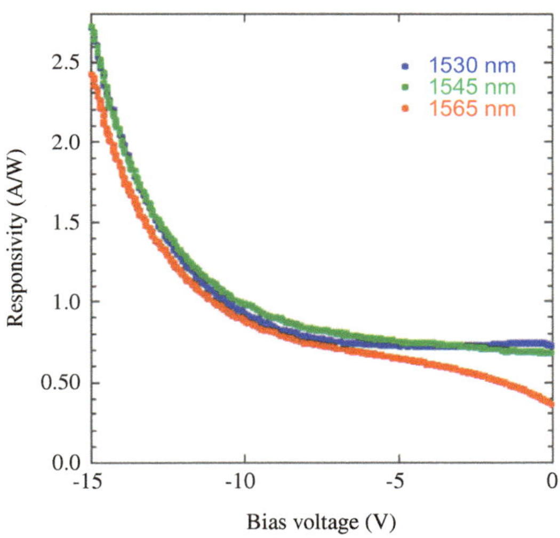

Fig. 47 Wavelength dependence of responsivity as a function of reverse bias voltage. In this experiment, input power was −10.7 dBm.

with increasing input power. This shift causes an increase in the temperature of the active region, which results in a decrease in the band gap. Responsivity of 2.7 A/W, or avalanche gain of 4, was achieved when the input power and bias were −10.2 dBm and −15 V, respectively.

Fig. 47 shows the wavelength dependence of the responsivity as a function of reverse bias voltage. Input power was −10.7 dBm. In the region of lower reverse bias voltage, the responsivity of 1565-nm light

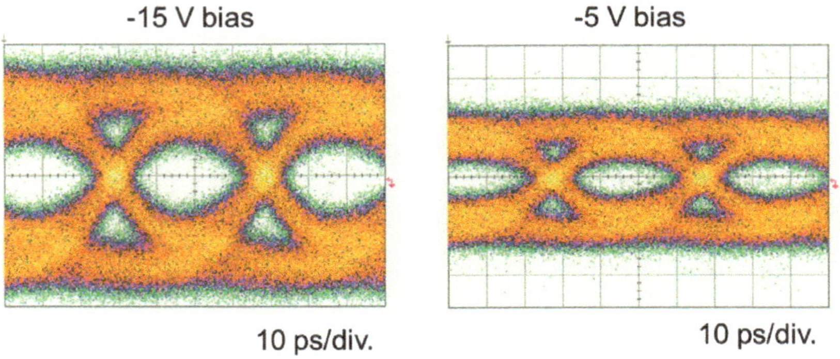

Fig. 48 Eye diagrams of 25-Gbit/s NRZ signal for reverse bias voltages of 15 and 5V, respectively. Average input power was −3.5 dBm, and the wavelength of the input light was 1530 nm.

is smaller than that at other wavelengths because we used QWs in the absorption region with a 1540-nm PL peak. By increasing the reverse bias voltage, the responsivity in the C-band wavelength range becomes almost the same value thanks to the broadening of the exciton peak and the Franz-Keldysh effect.

We also measured the dynamic response of the device. Fig. 48 shows eye diagrams of 25-Gbit/s NRZ signal for reverse bias voltages of 15 and 5 V, respectively. Average input power was −3.5 dBm and wavelength of the input light was 1530 nm. Eye openings for the membrane p-i-n photodiode at both bias voltages and input wavelengths were confirmed and, in the case of bias voltage of 15 V, we achieved avalanche gain.

8. Summary

Heterogeneous integration of III–V compound semiconductors and Si are very important for constructing PICs including lasers and high-efficiency modulators.

Direct modulation of membrane BH lasers will be used in short-range optical interconnects ranging from datacenter to on-chip. This is because the direct modulation scheme for lasers is a promising way to achieve low power consumption. Membrane BH lasers are also suitable for use in WDM networks—they can increase transmission capacity because it is easy to control the lasing wavelength by changing the grating pitch. Heterogeneous integration with Si photonics devices is quite important for reducing the footprint because it provides compact high-performance optical devices, and SSCs and optical filters can reduce assembly cost. Since the power consumption of a DML is proportional to cavity volume, the cavity volume must be decreased down to wavelength scale for extremely short-distance optical interconnects such as chip-to-chip and on-chip. For this purpose, PhC lasers will play an important role.

Introducing III–V compound semiconductors improves the device performance for modulators and photodetectors. We have demonstrated an MZM incorporating an InGaAsP/Si MOS capacitor phase shifter with $V_{\pi}L$ of 0.09 Vcm, which is 10 times smaller than that for the conventional carrier-depletion Si MSM. Membrane BH lasers are also used as photodetectors, which makes it easy to fabricate large-scale PICs. By increasing reverse bias voltage, we can operate a membrane laser as a photodetector with wavelength independent operation and avalanche gain.

To increase wafer size up to 8 or 12 in., multiple bonding technologies are important because of the limitation of InP wafer size. Epitaxial growth of III–V compound semiconductors using large-scale wafers is also important in terms of uniformity. Developments of these technologies are important for heterogeneous integration in Si photonics foundries.

References

Bogaerts, W., et al., 2006. Compact wavelength-selective functions in silicon-on-insulator photonic wires. IEEE J. Sel. Top. Quantum Electron. 12, 1394.

Dong, P., et al., 2014. Novel integration technique for silicon/III-V hybrid laser. Opt. Express 22, 26861–26868.

Fang, A.W., et al., 2006. Electrically pumped hybrid AlGaInAs-silicon evanescent laser. Opt. Express 14, 9203–9210.

Feng, N.-N., et al., 2010. High speed carrier-depletion modulators with 1.4V-cm $V_{\pi}L$ integrated on 0.25 mm silicon-on-insulator waveguides. Opt. Express 18, 7794–7799.

Fontcuberta i Morral, A., et al., 2003. InGaAs/InP double heterostructures on InP/Si templates fabricated by wafer bonding and hydrogen-induced exfoliation. Appl. Phys. Lett. 83, 5413–5415.

Fujii, T., et al., 2015. Epitaxial growth of InP to bury directly bonded thin active layer on SiO_2/Si substrate for fabricating distributed feedback lasers on silicon. IET Optoelectron. 9, 151–157.

Green, W.M.J., Rooks, M.J., Sekaric, L., Vlasov, Y.A., 2007. Ultra-compact, low RF power, 10 Gb/s silicon Mach-Zehnder modulator. Opt. Express 15, 17106–17113.

Heroux, J.B., et al., 2015. Energy-efficient 1060-nm optical link operating up to 28 Gb/s. J. Lightwave Technol. 33, 733–740.

Hiraki, T., et al., 2017a. Deuterated SiN/SiON waveguides on Si platform and their application to C-band WDM filters. IEEE Photonics J. 9, 2500207.

Hiraki, T., et al., 2017b. Heterogeneously integrated III–V/Si MOS capacitor Mach–Zehnder modulator. Nat. Photonics 11, 482–485.

Kanno, E., et al., 2018. Twin-mirror membrane distributed-reflector lasers using 20-μm-long active region on Si substrates. Opt. Express 26, 1268–1277.

Kobayashi, W., et al., 2013. 50-Gb/s direct modulation of a 1.3-μm InGaAlAs-based DFB laser with a ridge waveguide structure. J. Lightwave Technol. 19, 1500908, (8 page).
Komori, K., et al., 1989. Single-mode properties of distributed-reflector lasers. IEEE J. Quantum Electron. 25, 1235–1244.
Krishnamoorthy, A.V., et al., 2011. Progress in low-power switched optical interconnects. IEEE J. Sel. Top. Quantum Electron. 17, 357–376.
Liao, L., et al., 2005. High speed silicon Mach-Zehnder modulator. Opt. Express 13, 3129–3135.
Liu, A., et al., 2004. A high speed silicon optical modulator based on a metal-oxide-semiconductor capacitor. Nature 427, 615–618.
Matsuda, M., et al., 2012. Uncooled, low-driving-current 25.8 Gbit/s direct modulation using 1.3 μm AlGaInAs MQW distributed-reflector lasers. Electron. Lett. 48, 450–452.
Matsuda, M., et al., 2015. 1.3-μm-wavelength AlGaInAs multiple-quantum-well semi-insulating buried-Heterostructure distributed-reflector laser arrays on semi-insulating InP substrate. IEEE J. Sel. Top. Quantum Electron. 21, 1502307.
Matsumoto, K., Makino, T., Kimura, K., Shimomura, K., 2013. Growth of GaInAs/InP MQW using MOVPE on directly-bonded InP/Si substrate. J. Cryst. Growth 370, 133–135.
Matsuo, S., et al., 2010. High-speed ultracompact buried heterostructure photonic-crystal laser with 13 fJ of energy consumed per bit transmitted. Nat. Photonics 4, 648–654.
Matsuo, S., et al., 2011. 20-Gbit/s directly modulated photonic crystal nanocavity laser with ultra-low power consumption. Opt. Express 19, 2242–2250.
Matsuo, S., et al., 2013. Ultra-low operating energy electrically driven photonic crystal lasers. IEEE J. Sel. Top. Quantum Electron. 19, 4900311.
Matsuo, S., et al., 2014a. Directly modulated buried heterostructure DFB laser on SiO_2/Si substrate fabricated by regrowth of InP using bonded active layer. Opt. Express 22, 12139–12147.
Matsuo, S., et al., 2014b. Photonic crystal lasers using wavelength-scale embedded active region. J. Phys. D 47, 023001.
Matsuo, S., et al., 2017. Avalanche Gain in Membrane P-I-N Photodiodes on Si Substrate. In: The 24th Congress of the International Commission for Optics, The International Commission for Optics (ICO), Th1J-5.
Miller, D.A.B., 2009. Device requirements for optical interconnects to silicon chips. Proc. IEEE 97, 1166–1185.
Moser, P., et al., 2012. 56 fJ dissipated energy per bit of oxide-confined 850 nm VCSELs operating at 25 Gbit/s. Electron. Lett. 48, 1292–1294.
Nakahara, K., 2015. Direct modulation at 56 and 50 Gb/s of 1.3-μm InGaAlAs ridge-shaped-BH DFB lasers. IEEE Photon. Technol. Lett. 27, 534–536.
Nishi, H., et al., 2010. Monolithic integration of a silica-based arrayed waveguide grating filter and silicon variable optical attenuators based on p-i-n carrier-injection structures. Appl. Phys. Express 3 (10), 102203.
Nishi, H., et al., 2015. Monolithic integration of InP wire and SiO_x waveguides on Si platform. IEEE Photonics J. 7, 4900308.
Nishi, H., et al., 2016. Membrane distributed-reflector laser integrated with SiOx -based spot-size converter on Si substrate. Opt. Express 24, 18346–18352.
Okayama, H., et al., 2016. Low loss 100 GHz spacing Si arrayed-waveguide grating using minimal terrace at slab-array interface. Electron. Lett. 52, 1545.
Painter, O., et al., 1999. Two-dimensional photonic band-gap defect mode laser. Science 284, 1819–1821.
Roelkens, G., et al., 2006. Laser emission and photodetection in an InP/InGaAsP layer integrated on and coupled to a silicon-on-insulator waveguide circuit. Opt. Express 14, 8154–8159.

Sato, T., et al., 2015. Photonic crystal lasers for chip-to-chip and on-chip optical interconnects. IEEE J. Sel. Top. Quantum Electron. 21, 4900410, (10 pages) (2015).
Shoji, T., et al., 2002. Low loss mode size converter from 0.3 μm square Si wire waveguides to singlemode fibers. Electron. Lett. 38, 1669.
Tadokoro, T., et al., 2009. Operation of a 25-Gb/s direct modulation ridge waveguide MQW-DFB laser up to 85°C. IEEE Photon. Technol. Lett. 21, 1154–1156.
Takahashi, Y., et al., 2007. High-Q nanocavity with a 2-ns photon lifetime. Opt. Express 15, 17206–17213.
Takeda, K., et al., 2017. Si nanowire waveguide coupled current-driven photonic-crystal lasers. In: Proceedings of European Conference on Lasers and Electro-Optics 2017, IEEE, paper CK-9.4.
Tanabe, T., et al., 2007. Trapping and delaying photons for one nanosecond in an ultrasmall high-Q photonic-crystal nanocavity. Nat. Photonics 1, 49–52.
Thomson, D., et al., 2008. Total internal reflection optical switches to restrict carrier diffusion in the guiding layer. J. Lightwave Technol. 26, 1288–1294.
Thomson, D.J., et al., 2012. 50-Gb/s Silicon Optical Modulator. IEEE Photon. Technol. Lett. 24, 234–236.
Van De Leur, R.H.M., Schellingerhout, A.J.G., Tuinstra, F., Mooij, J.E., 1988. Critical thickness for pseudomorphic growth of Si/Ge alloys and superlattices. J. Appl. Phys. 64, 3043–3050.
Wang, Q., et al., 2012. Heterogeneous Si/III-V integration and the optical vertical interconnect access. Opt. Express 20, 16745–16756.
Webster, M., et al., 2015. Low-power MOS-capacitor based silicon photonic modulators and CMOS drivers. In: Proc. in OFC, W4H.3.
Xu, Q., et al., 2007. 12.5 Gbit/s carrier-injection-based silicon microring silicon modulator. Opt. Express 15, 430–436.
Zhang, C., Liang, D., Bowers, J.E., 2013. MOCVD regrowth of InP on hybrid silicon substrate. ECS Solid State Lett. 2, Q82–Q86.

Further reading

Takeda, K., et al., 2013. A few-fJ/bit data transmissions using directly modulated lambda-scale embedded active region photonic-crystal lasers. Nat. Photonics 7, 569–575.

CHAPTER THREE

Quantum dot lasers for silicon photonics

Yasuhiko Arakawa[a,*], Takahiro Nakamura[b], Jinkwan Kwoen[a]
[a]Institute for Nano Quantum Information Electronics, The University of Tokyo, Tokyo, Japan
[b]Photonics Electronics Technology Research Association, The University of Tokyo, Tokyo, Japan
*Corresponding author: e-mail address: arakawa@iis.u-yokyo.ac.jp

Contents

Semiconductors and Semimetals, Volume 101
ISSN 0080-8784
https://doi.org/10.1016/bs.semsem.2019.07.007

1. Introduction

As the interchip transmission bandwidth in the servers increases by a factor of two every 2 years, the optical interconnect technology based on silicon photonics has become an essential technology for solving the bandwidth bottleneck problem. In 2009, the concept of photonics electronics convergence systems to solve the transmission bandwidth bottleneck issue between LSI chips was proposed in one of the national projects at the Funding Program for World-Leading Innovative R&D on Science and Technology (The FIRST program) in Japan (Arakawa et al., 2013). An achievement of the project was the demonstration of a prototype of silicon optical interposers fully integrated with optical components on a silicon substrate and achieved a high bandwidth density of 30 Tbps/cm^2 at room temperature at 1.55 μm (Urino et al., 2013).

However, for practical application, it is necessary to operate the interposer under high temperature conditions without using complicated feedback control to cope with the heat generated by mounted LSI chips. Since LSI bare chips are placed near a semiconductor laser light source, the laser is affected by heat. The system needs a temperature-insensitive semiconductor laser to avoid reduction of optical output power as the temperature rises.

Quantum dot (QD) laser is expected to operate without any degradation under high temperature environment because of its temperature insensitivity of threshold current. To integrate the QD lasers on silicon, three methods have been developed; flip-chip bonding, wafer bonding and direct epitaxial growth.

This chapter describes recent advances in QD lasers for silicon photonics. In Section 2, evolution of InAs/GaAs QD lasers is briefly overviewed including the first proposal in 1982, superior lasing characteristics theoretically forecasted, realization of the QD lasers in 1990s, and demonstration of temperature insensitive threshold current. In Section 3, an assembling method of the QD lasers onto silicon photonics circuit by flip-chip bonding as well as applications of this technique to a 5 mm × 5 mm optical transceiver chip (Optical I/O Core) is discussed. In Section 4, a wafer bonding technique is discussed for forming InAs/GaAs QD lasers on silicon substrates via direct bonding. After the process flow of the wafer bonding for laser fabrication is discussed, the realization of the wafer bonded QD lasers on silicon, including a hybrid silicon III-V QD laser evanescently coupled to a silicon waveguide are described. In the Section 5, direct epitaxial growth for forming InAs/GaAs QD lasers on silicon substrates is discussed. An epitaxial growth technique using the molecular beam epitaxy (MBE) on a CMOS compatible silicon (001) on-axis substrate can form QDs of high quality. After the crystal growth and laser fabrication are described, characteristics of the fabricated laser device are discussed.

2. Progress in quantum dot lasers on GaAs

2.1 Early stage of the quantum dot laser research

After more than a decade since the first development of the super-lattice by Esaki and Tsu (1970), Arakawa and Sakaki proposed the concept of 10 nm-scale QDs that confine electrons three-dimensionally and their application to lasers in 1982 (Arakawa and Sakaki, 1982). Furthermore, the temperature insensitivity of the threshold current was theoretically predicted and the three-dimensional confinement effect was experimentally demonstrated by placing a quantum well laser in a high magnetic field.

The electrons in the bulk semiconductors move freely in three dimensions, and their energy distribution is determined by both the density of states and the Fermi-Dirac distribution function as indicted in Fig. 1. In the semiconductor lasers, with the increase of temperature, the Fermi-Dirac distribution function changes, leading to reduction of the maximum value of the gain spectrum (i.e., the peak gain). In order to keep the lasing condition, the carrier density needs to be increased, resulting in increase of the

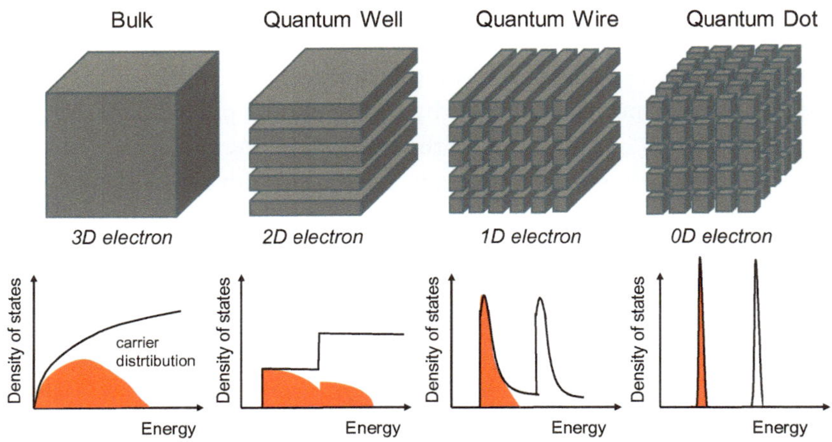

Fig. 1 Evolution of semiconductor nanostructures for lasers.

threshold current. If electrons are confined in three-dimensional nanostructures (i.e., QD), due to spatial separation, all the electrons can be populated at the same energy level despite the nature of the Fermion. As a result, the peak gain does not decrease the temperature rides and the temperature dependence of the threshold current is much more suppressed compared to that of bulk and quantum well lasers. Because of the narrow gain spectrum, the threshold current density itself is also substantially reduced in the QD lasers (Asada et al., 1986).

In the first paper discussing the QDs (Arakawa and Sakaki, 1982), this nanostructure was named "three-dimensional quantum wells" or "quantum boxes." But today, the QD has become the most common name of the nanostructure. In the early 1980s, the QDs were considered to be a dream-like nanostructure that would not be realized by the end of the 20th century. In fact, it was not even easy to grow high quality quantum wells in the 1980s. For this reason, most of our research focused on theoretical analysis to study the basics of QD lasers and high magnetic field experiments described above (Arakawa and Yariv, 1986; Arakawa et al., 1983, 1984).

In addition to the excellent characteristics of the threshold current, we forecasted enhancement of dynamic properties of the semiconductor lasers resulting from to the quantum effects (i.e., the dimensionality of electron motion freedom) in quantum well lasers and QD lasers (Arakawa and Yariv, 1986; Arakawa et al., 1984). Application of semiconductor lasers in optical communication systems requires both broad band modulation and low noise characteristics. Two important parameters that determine such characteristics are the relaxation oscillation frequency f_r and the linewidth enhancement factor a. The f_r sets the direct modulation bandwidth that is proportional to the square root of f_r. The a determines the relationship between AM and FM modulation indices as well as the degree to which spectral purity is reduced by the amplitude phase coupling. Since the representation of f_r and a are of the form involving the density of states, these two parameters are substantially improved with the decrease of the number of dimensions of the degree of freedom of the electron motions. In fact, theoretical calculation showed that modulation performance can be significantly improved (f_r enhanced) with simultaneous reduction of phase noise in the QD lasers as compared to conventional semiconductor lasers. In particular, a becomes zero if population of electrons at higher sub-band effect can be neglected in the QD lasers. Reduced a is also of essential importance to enhance feedback noise tolerance (Arakawa and Yariv, 1986; Arakawa et al., 1984).

The progress in growth technique was much faster than expected. In 1985, Goldstein investigated to grow InAs/GaAs strained quantum wells by MBE, but formation of a three-dimensional structure (i.e., InAs/GaAs QDs), was discovered by the Transmission Electron Microscopy (TEM) (Goldstein et al., 1985). This self-assembling formation of the QDs was soon understood as the Stranski-Krasrtanow (S-K) growth mode. The S-K growth mode for forming QDs widely used today was thus developed by an accidental discovery or so-called serendipity. The advantage of this formation method is that high quality QDs can be formed using only normal thin film crystal growth techniques. Although growth by MBE was the mainstream in the early stage of the QD growth by the S-K mode (Glas et al., 1987; Guha et al., 1990; Leonard et al., 1993; Snyder et al., 1991; Tabuchi et al., 1992). The growth of 10 nm class InGaAs/GaAs QDs was also demonstrated with metal organic chemical vapor deposition (MOCVD) in 1994 (Oshinowo et al., 1994).

2.2 Evolution and practical implementation of quantum dot lasers

An attempt of realization of the QD laser with the S-K growth mode was reported in 1994 by MBE (Kirstaedter et al., 1994). Although it was unclear whether laser oscillation is was actually attributable to the gain of the QDs, the report was a step toward realizing high performance QD lasers with the three dimensional quantum confinement effect. After that, lasing oscillation was reported one after another in various groups in the world. Fujitsu Laboratories reported a 1.3 μm band QD laser by increasing the size via stacking QDs in a columnar shape (Shoji et al., 1995). The record lowest threshold current density of a semiconductor laser is realized by a QD lasers. Furthermore, Fujitsu Laboratories Ltd. and the University of Tokyo achieved high temperature stability of the threshold current as shown in a wide temperature range in Fig. 2. 10 Gb/s direct modulation operation was also demonstrated in the temperature range of 20–70 °C with no current adjustment (Otsubo et al., 2004).

Inspired by the above results, QD Laser Inc. was launched in 2006 as a Fujitsu's curve-out type venture company. One of the major advantages of the InAs/GaAs QD laser technology is that telecom wavelength of 1.3 mm lasers can be realized on a GaAs substrate. As a result, low cost manufacturing process lines developed for red-color lasers used for compact disc (CD) applications can be re-utilized for the production of QD lasers, providing low-cost telecom laser to the commercial market. It should be noted that the concept of p-doped structure in QD lasers has already been discussed in the first paper in 1982 (Arakawa and Sakaki, 1982).

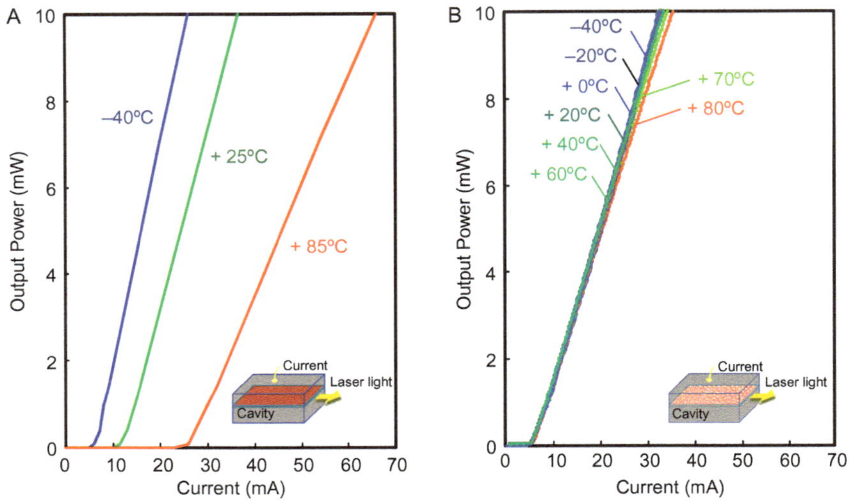

Fig. 2 L-I characteristics of (A) quantum well lasers and (B) quantum dot lasers in a wide temperature range (Otsubo et al., 2004, QD Laser Inc. White Paper).

In addition to the temperature dependence of the threshold current, one of the authors also predicted theoretically the improvement of modulation characteristics and the spectral properties, which have been experimentally proved. At the present time, the QD lasers have the highest performance in most of lasing characteristics except the modulation characteristics, that is due to carrier injection problem. Temperature stability as well as high temperature operation itself are becoming more important for silicon photonics application. As shown in Fig. 3, QD Laser Inc. and the University of Tokyo Co., Ltd. realized high temperature operation of the QD lasers up to 220 °C which is the maximum operating temperature among any kinds of lasers (Kageyama et al., 2011).

2.3 Application to silicon photonics

Silicon photonics is expected to be a major application field of the QD lasers in the near future. Since heat generation from the LSI cannot be avoided in the silicon optical interposer on which the LSI bare chips are mounted, it is necessary to integrate light sources that operate even in a high temperature environment. As already described, the use of the QD laser that can be operated at the high temperature environment in addition to its own low power consumption gives a solution for this issue. Under the FIRST program that started in 2011 in Japan, a silicon photonic interposer integrating

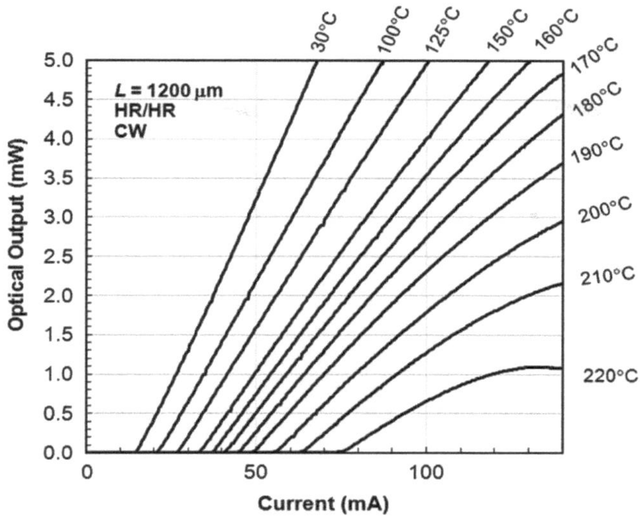

Fig. 3 High temperature operation of quantum dot lasers up to 220 °C (Kageyama et al., 2011).

a QD laser array implemented by hybrid flip-chip bonding was first demonstrated. The silicon optical interposer acheived error-free transmission of 12.5 Gbps and a high transmission band density of 15 Tbps/cm^2 over a wide temperature range of 25–125 °C, without any temperature control on lasers, modulators and photodetectors (Urino et al., 2015). Since the QD lasers can realize a small α parameter of ~0, the QD laser is insensitive to optical feedback effect, which enables to realize silicon photonics chip without any optical isolators (Mizutani et al., 2015). On the basis of these achievements, a venture company named AIO-Core Co., Ltd. was launched in 2017 that delivers 5 mm^2 transmitting/receiving silicon photonics chips integrating the QD lasers. In the following sections, integration of the QD lasers to silicon photonics systems will be discussed.

2.4 Conclusion

In this section, advances in the QD lasers are briefly reviewed, including the first proposal in 1982, superior lasing characteristics theoretically forecasted, the realization of the QD lasers in 1990s and the demonstration of temperature insensitive QD lasers. Furthermore, implementation of the QD lasers in silicon photonic systems is also introduced. The QD lasers has already established itself as the most promising light sources at 1.3 mm for silicon photonics technology.

3. Flip-chip bonding for quantum dot lasers on silicon and its application

In this section, QD laser integration by flip-chip bonding is introduced as a laser diodes (LD) integration method on Si. First, the requirements for LDs which are assembled onto silicon photonics integrated circuits by flip-chip bonding, are mentioned in Section 3.1. Especially, we mention that QD lasers have perfect characteristics for assembling onto silicon photonics integrated circuits. Then, an assembly method of QD lasers onto silicon photonics integrated circuits by flip-chip bonding is introduced in Section 3.2. Finally, applications of this method to optical I/O cores for board-to-board interconnection and to silicon optical interposers for chip-to-chip interconnection are introduced in Section 3.3.

3.1 Requirements for laser diode assembly onto silicon photonics integrated circuits

The common LD package is shown in Fig. 4A. The LD package consists of an LD, two lenses, an isolator, Peltier device and so on. Therefore, LD assembly onto silicon photonics integrated circuits requires that all functions of the LD package should be integrated into a tiny silicon photonics chip. Fig. 4B shows the concept of our LD assembly onto a silicon photonics integrated circuit. The left figure of Fig. 4B shows the functions of the LD package. The right figure shows the method of integrating the LD package's functions. The functions of the thermoelectric controller (TEC) and isolator are replaced with that of the QD laser, as mentioned in Sections 3.1.1 and 3.1.2. The function of a lens is enabled by a spot-size converter (SSC) with a silicon waveguide and LD, as mentioned in Section 3.1.4. Furthermore, the hermetic package is not needed due to bare-chip reliability at high temperature and high humidity, as mentioned in Section 3.1.3.

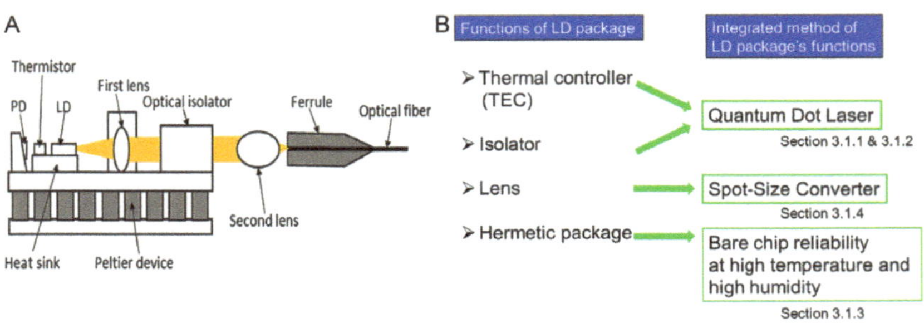

Fig. 4 (A) Example of LD package and (B) concept of LD assembly.

3.1.1 High-temperature stability enabled by quantum dot lasers

As the energy level of QDs is discretized, the injected electrons are efficiently recombined with holes without the electrons radiating to the upper energy level and the light is efficiently generated. Furthermore, a QD laser has high-temperature stability because of this discretized energy level compared to a quantum well (QW) laser (Arakawa and Sakaki, 1982).

The temperature dependence on the LD output power at the injected currents of 200 mA is shown in Fig. 5 for a 1.3-μm QD-Fabry-Perot (FP) laser and a 1.3-μm QW-FP laser (Yashiki et al., 2016). The QD layer is composed of eight layers of InAs/GaAs QDs sandwiched between 1400-nm $Al_{0.4}Ga_{0.6}As$ cladding layers. The QW layer is composed of a strain compensated QW structure consisting of five InGaAsP compressive strained (+1%) wells with six InGaAsP tensile strained (−0.1%) barriers sandwiched between 1.0-μm composition InGaAsP separate confinement heterostructure layers. The cavity length of the QD-FP and QW-FP lasers is 600 μm and the front and rear facet reflectivity is 30% and 90%, respectively. These lasers are mounted on an AlN heatsink for measuring light output power. As shown in Fig. 5, the degradation of light output power on the QD-FP laser at high temperature (1.5 dB) is less than that on the QW-FP laser (8 dB). This means that no TEC is required due to using the QD-FP laser. This result is consistent with the characteristics predicted in 1982 (Arakawa and Sakaki, 1982).

3.1.2 High optical feedback tolerance enabled by quantum dot lasers

For assembling the optical light source onto silicon photonics integrated circuits, it is necessary to suppress the repeated mode competition under operation conditions as well as use a FP laser assembled on these circuits without isolators. When mode competition occurs between several FP modes at high optical power, the relative intensity noise drastically increases.

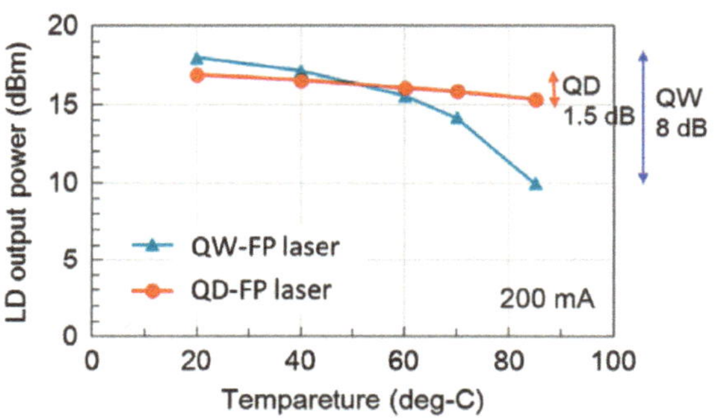

Fig. 5 Temperature dependence of LD output power of QD-FP laser and QW-FP laser (Nakamura et al., 2019).

However, these degradations cannot be suppressed under the worst phase condition and might easily occur due to several near-end reflections in silicon photonics integrated circuits. Several research groups have stated that QD lasers have strong optical feedback tolerance to the far-end reflection due to their small alpha parameter and strongly suppressed relaxation oscillation frequency (Duan et al., 2019; Huyet et al., 2004; Otto et al., 2012). However, for application to silicon photonics integrated circuits, optical feedback tolerance to the near-end reflection should also be taken into consideration. The excited carriers in a QD laser are strongly confined in all directions. Then, the density of states can be expressed like a delta-function in each QD and the energy relaxation is strongly suppressed. Therefore, we believe that the transition between the FP modes of a LD by the relaxation process can be restrained and the mode competition can be avoided. We previously demonstrated strong optical feedback tolerance of a QD laser to the near-end reflection for the first time (Mizutani et al., 2015). Optical feedback tolerance to the near-end reflection was measured using the system shown in Fig. 6A, consisting of a 1×2 Mach-Zehnder interferometer (MZI), photo diode (PD), and LD. One arm in the MZI was connected to an air facet and the other arm was connected to the PD. By shifting the phase of one arm in the MZI, the reflected light intensity from the air facet to the LD could be changed. In this study, we measured and compared the optical feedback tolerance for QD-FP and QW-FP lasers. Fig. 6B shows the estimated worst signal-to-noise ratio (SNR) between 50 and 150 mA in the LD current as a function of the feedback condition ($C_{feedback}$), the equation of feedback condition is indicated in Fig. 6A

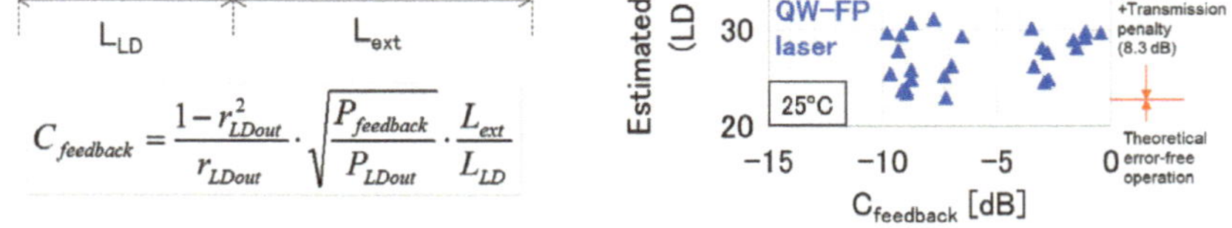

Fig. 6 (A) System for measuring optical feedback tolerance at near-end reflection and the equation of feedback condition and (B) estimated worst SNR versus feedback condition (Nakamura et al., 2019).

(Nakamura et al., 2019). The bit error rate (BER) requirement of data communication application is generally under 10^{-12} for error-free operation. Therefore, an SNR of at least 23 dB is necessary according to the theoretical equation given in a previous study (Yariv, 1997). Furthermore, it is necessary to consider the optical loss and the transmission penalty in the link. The optical loss in the silicon photonics integrated circuit including the coupling loss of both LD and optical fiber is assumed to be about 10 dB. The transmission penalty is generally assumed to be 8.3 dB in data communication (Ganga, 2008). Then, normally, an SNR_{LD} of 41.3 dB is necessary for error-free operation without forward error correction. As shown in Fig. 6B, compared to a QW-FP laser, the estimated worst SNR for a QD-FP laser was better than 40 dB in a wide region of $C_{feedback}$, which was sufficient for 25-Gbps error-free operation. These results indicate that QD-FP lasers have sufficient high optical feedback tolerance to be used in silicon photonics integrated circuits without optical isolators.

Based on Sections 3.1.1 and 3.1.2, QD lasers are suitable for assembly onto silicon photonics integrated circuits because of both the Peltier device and isolator-free ability.

3.1.3 High bare-chip reliability enabled by quantum dot lasers

Generally, each LD is hermetically packaged for reliability. However, if the bare chip of an LD has good reliability at high temperature and high humidity, a hermetic package is no longer required and it contributes to reduced size and cost of silicon photonics integrated circuits. We conducted a reliability test of a QD-FP laser under high temperature and high humidity. Fig. 7 shows the

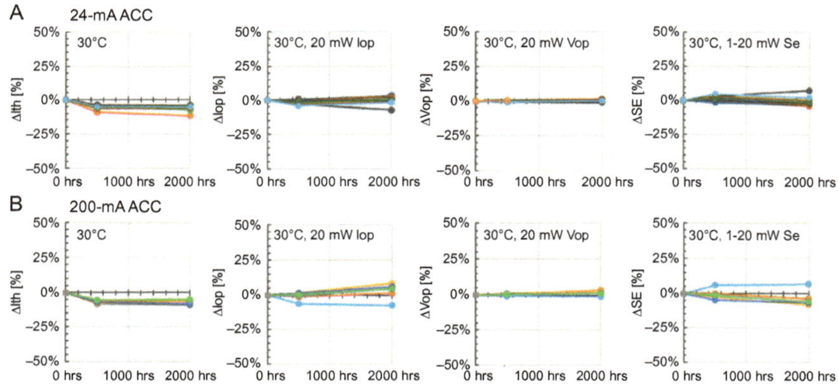

Fig. 7 QD-FP laser bare chip reliability test at 85 °C, 85% humidity and under two auto current control (ACC) conditions of 24 and 200 mA (Nakamura et al., 2019).

results of this QD-FP laser bare chip reliability test at 85 °C, 85% humidity and under two auto current control (ACC) conditions. We performed two ACC conditions. One was the maximum injected current use condition of 200-mA and the other was the above threshold current at 85 °C. As the practical humidity is lower than 85% due to the heat of the LD chip at the maximum injected current condition, the lower current condition was added.

The degradation of threshold currents, operating currents, operating voltages, and slope efficiencies under both conditions were less than ±10% up to 2000 h and QD-FP laser bare chips had good reliability. This result demonstrates that hermetic packaging free is available for QD-FP lasers.

3.1.4 Highly efficient coupling enabled by spot-size converter

Highly optical coupling efficiency is important between an LD and waveguide, such as fiber, Si waveguide and so on. Generally, a lens is used to improve these optical coupling efficiencies. However, in tiny silicon photonics integrated circuits, it is difficult to use lenses from the view point of size and cost. Instead of lenses, a spot-size converter (SSC), by changing the waveguide shape, is generally used in silicon photonics integrated circuits. Then, the optical coupling loss between a QD laser and Si waveguide with a tip tapered SSC along the horizontal and vertical directions was measured, as shown in Fig. 8. The tip width was 160 nm and the gap between the QD laser facet and waveguide facet was 1 μm. The minimum optical coupling loss between the QD laser and Si waveguide was 2.4 dB (Nakamura et al., 2019). Furthermore, if a SSC is also introduced to a QD laser, the minimum coupling loss will improve to 1.4 dB. Therefore, LD assembly onto silicon photonics integrated circuits is available without lenses by using an SSC.

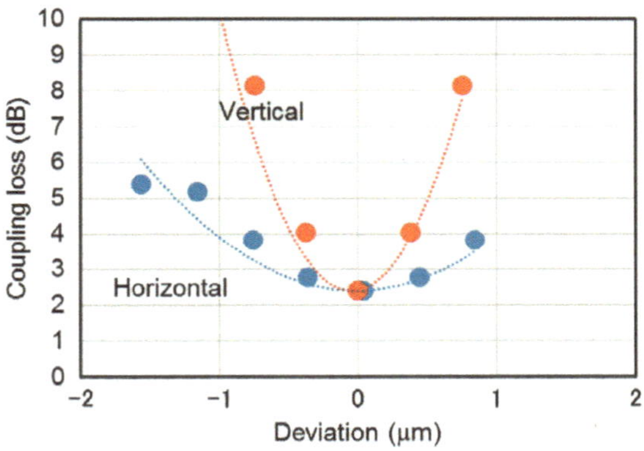

Fig. 8 Optical coupling loss between QD laser and Si waveguide with tip tapered SSC along the horizontal and vertical directions (Nakamura et al., 2019).

3.2 Assembly method of quantum dot lasers onto silicon photonics integrated circuit by flip-chip bonding

There are mainly two LD assembly methods by flip-chip bonding, one is active alignment and the other is passive alignment. By active alignment, we can achieve accurate positioning. However, it takes a long alignment time and is difficult to fix the alignment position. On the other hand, passive alignment is low cost because of the short alignment time. However, it requires high precision alignment. We chose passive alignment method for low-cost silicon photonics integration and have developed high precision passive alignment technique. Next, I will introduce this technique.

It is important to precisely align the LD stripe and Si waveguide for passive alignment of a LD onto a silicon photonics integrated circuit by flip-chip bonding. Fig. 9A shows our LD assembly method with passive alignment method (Shimizu et al., 2014).

After fabricating the silicon photonics integrated circuit, the LD mounting stage with Si alignment marks and Si pedestals was fabricated, as shown in Fig. 9B. First, to form the waveguide facet (Fig. 9(b-1)), both the upper cladding and box layer were removed down to the Si substrate by dry etching (Fig. 9(b-2)). Second, the Si alignment marks and Si pedestals were patterned onto the Si substrate with dry etching (Fig. 9(b-3)). Finally, the electrodes were formed on the Si substrate (Fig. 9(b-4)). The alignment marks were also fabricated on the LD chip. Fig. 10A shows the passive alignment method of the LD chip on the LD mounting stage. As shown in this cross-sectional view, visual alignment was carried out using an infrared (IR) camera image through the substrate transmission light. Fig. 10B is an example of alignment image adjusting between markers on the LD chip and LD mounting stage. The horizontal misalignment between the LD and Si substrate was better than $\pm 0.5\,\mu m$ as shown in Fig. 10C. The vertical positioning was determined from the Si pedestal fabrication process. Therefore, vertical misalignment was better than $\pm 0.1\,\mu m$.

The optical coupling tolerance between a QD laser and Si waveguide with a tip tapered SSC along the horizontal and vertical directions has been shown in Fig. 8 in Section 3.1.4. The minimum optical coupling loss between the QD laser and Si waveguide was 2.4 dB. The alignment tolerances for 1-dB excess loss were $\pm 0.7\,\mu m$ and $\pm 0.3\,\mu m$ in the horizontal and vertical directions, respectively. Using our LD assembly method, <3-dB coupling loss could be obtained, with almost no temperature dependence of the coupling loss up to 100 °C (Shimizu et al., 2013).

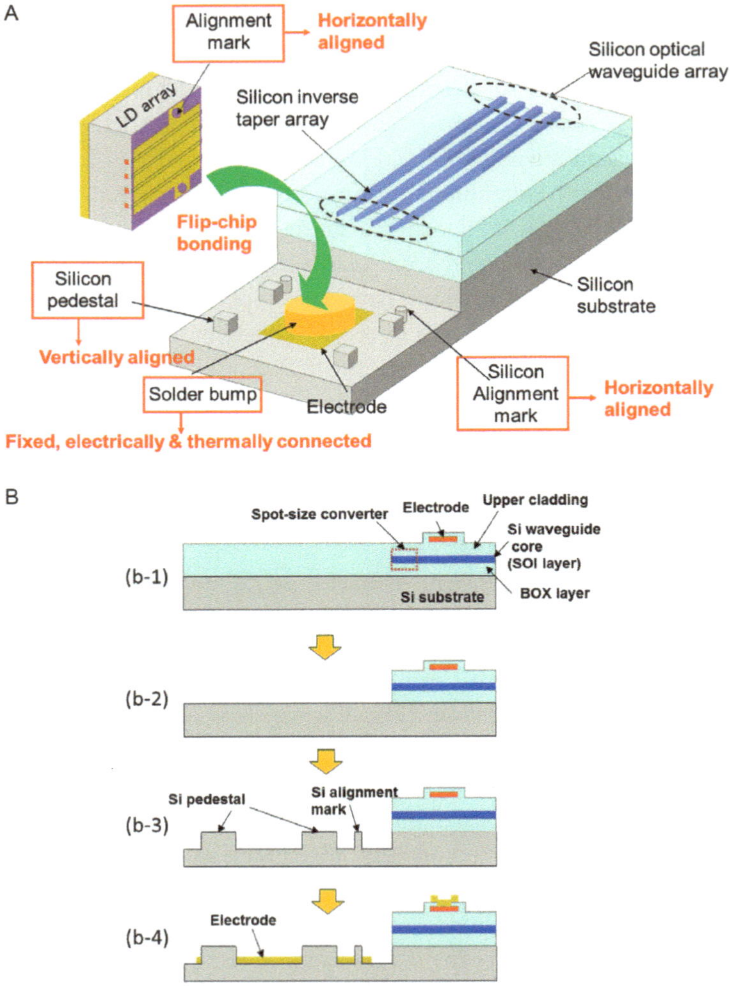

Fig. 9 (A) LD assembly method with passive alignment technique and (B) fabrication process flow of LD mounting stage.

3.3 Application

In this section, the applications of the method of assembling QD lasers onto silicon photonics integrated circuits by flip-chip bonding are introduced. One is an optical I/O core as an embedded optical module for board-to-board optical interconnection and the other is a silicon optical interposer for chip-to-chip optical interconnection.

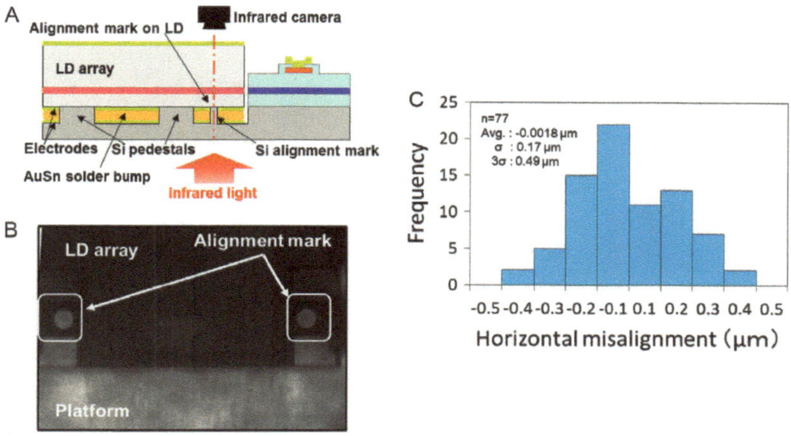

Fig. 10 (A) Passive alignment method of LD chip on the LD mounting stage, (B) example of alignment image adjusting between markers on LD chip and LD mounting stage (Nakamura et al., 2019), and (C) measured horizontal misalignment between LD and Si substrate.

3.3.1 Optical I/O core

We previously developed optical I/O cores as fingertip-size optical modules with high bandwidth density, low power consumption, and high temperature operation. Fig. 11 shows photographs and cross-sectional views of our developed optical I/O cores, transmitter (Tx) and receiver (Rx), respectively (Yashiki et al., 2015). These optical I/O cores have a small footprint of $5 \times 5\,mm^2$ and maximum capacity of 300 Gbps ($25\,Gbps \times 12$ channels). As shown in the cross-sectional views in Fig. 11, the optical I/O cores consist of technologies based on both silicon photonics integration and optical/electrical assembly. More specifically, the Tx consists of a QD laser, silicon photonics integrated circuit including a Si optical waveguide with an SSC, carrier-depletion Si optical modulators based on a symmetric MZI, and grating couplers (GCs), a CMOS driver IC, optical pins made from ultraviolet curable resin, and through glass via (TGV). The Rx consists of optical pins, germanium (Ge) PDs, a CMOS transimpedance amplifier (TIA) IC, and TGV. Especially, the QD laser was assembled onto the optical I/O core in the same manner as described in Section 3.2.

Fig. 12A shows LD injection currents dependence of the BERs and eye patterns of the optical I/O cores at 25 Gbps, 25 °C with pseudo random bit sequence (PRBS) 31, mounted with a QD laser and QW laser. The BERs of the optical I/O core mounted with the QW laser indicated a rapid increase with an increase of the injection current. At the point where BERs increase

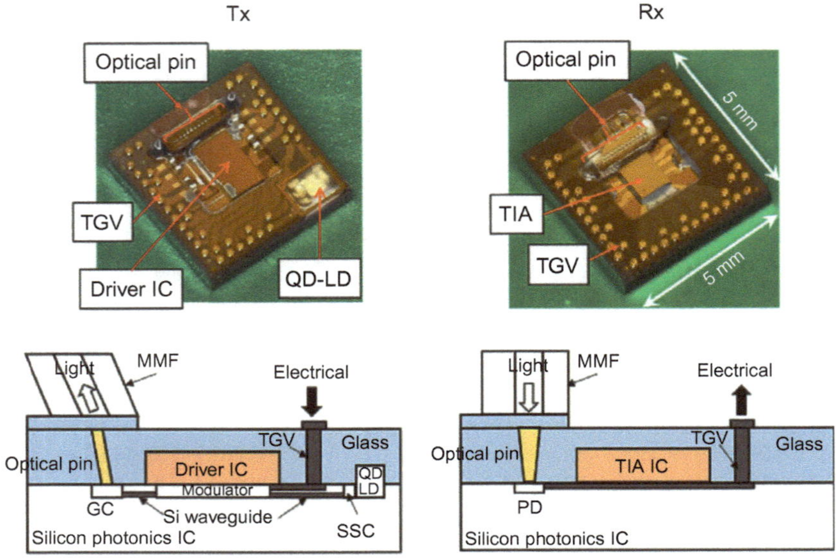

Fig. 11 Photographs and cross-sectional views of optical I/O core Tx and Rx, respectively (Nakamura et al., 2019).

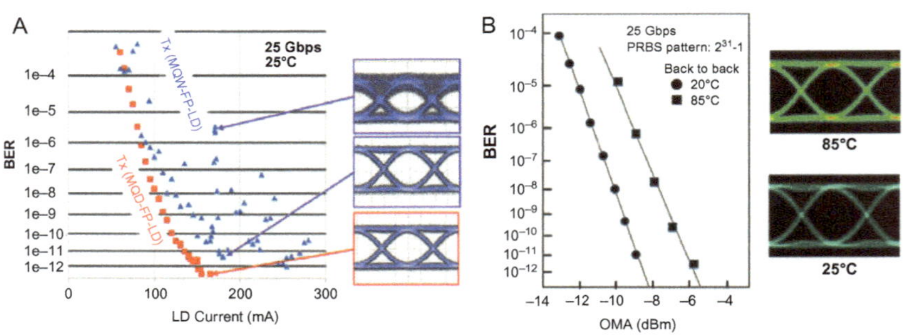

Fig. 12 (A) LD injection currents dependence of BERs and eye patterns of optical I/O cores mounted with QD laser and QW laser, respectively, at 25 Gbps, 25°C, PRBS 31 and (B) temperature dependence of eye patterns and BERs of optical I/O core mounted with QD laser (Nakamura et al., 2019).

rapidly, the noise also increased in the eye patterns. This phenomenon was thought to be caused by the near-end reflection from the optical I/O core. On the other hand, the BERs and eye patterns of that mounted with the QD laser shows normal behavior. This means that a QD laser has strong optical feedback tolerance to the near-end reflection.

Fig. 12B shows the temperature dependence of eye patterns and BERs of the optical I/O core mounted with the QD laser at 25 Gbps with PRBS 31. Clear eye patterns were observed from 25 up to 85 °C. Furthermore, BERs

$<10^{-12}$ were demonstrated at 20 and 85 °C (Yashiki et al., 2016). This isolator-free operation in a wide temperature range is achieved by means of a QD laser.

3.3.2 Silicon optical interposers

For solving the bandwidth bottleneck among LSIs, we developed silicon optical interposers fully integrated with optical components for achieving a high bandwidth density of 30 Tbps/cm^2 at room temperature. For practical applications, interposers should be usable under high temperature or rapid temperature change so that they can cope with the heat generated by the mounted LSIs. Fig. 13 shows our fabricated temperature-independent silicon optical interposers integrated with temperature-insensitive components on a Si substrate. An arrayed LD was flip-chip bonded to the substrate, the assembly method of which is mentioned in Section 3.2. Each LD had QD layers with a 1.3-μm lasing wavelength. The output power was higher than 10 mW per channel up to 100 °C. Si optical modulators and Ge PD arrays were monolithically integrated onto the substrate. The modulators were structured as symmetric MZIs, which were inherently insensitive to temperature and wavelength. The phase shifters composed of p-i-n diodes were stable against temperature with constant bias currents. The photo current of the PDs was also temperature insensitive and the photo-to-dark

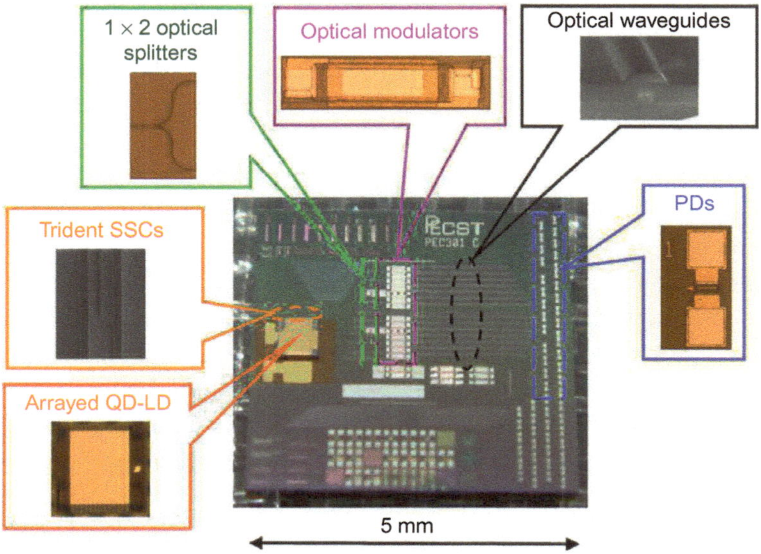

Fig. 13 Photograph of one of our fabricated temperature-independent silicon optical interposer with close-up images of the components (Urino et al., 2014).

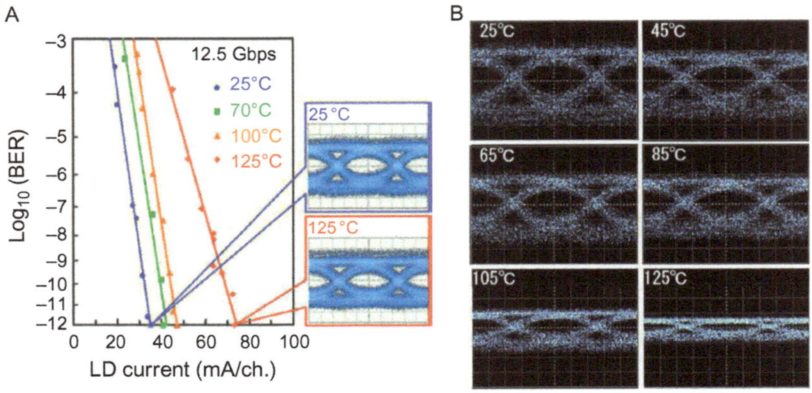

Fig. 14 (A) Measured eye diagrams and BERs of the PD output at 12.5 Gbps and (B) measured eye-diagram snapshots at 12.5 Gbps under the fixed bias and signal conditions for the LDs, modulators, and PDs while temperature was continuously changed from 25 to 125 °C (Urino et al., 2014).

current ratio was higher than 30 dB up to 100 °C. The measured eye diagrams and BERs of the PD outputs at 12.5-Gbps NRZ with PRBS 7 at various temperatures are shown in Fig. 14A. The horizontal axis shows the LD current so that the temperature dependence of the LD output power can be seen. In these measurements, we intentionally changed LD current to change the BER at each temperature, but did not adjust bias current of the modulator, bias voltage of the PD, or RF input signals to the modulators throughout the entire temperature range from 25 to 125 °C. We achieved error-free data links at 12.5 Gbps operating from 25 to 125 °C with the interposers without bias adjustments for the modulators, and PDs. We also observed the eye diagrams at 12.5 Gbps under the fixed bias and signal conditions for the LDs, modulators, and PDs while temperature was continuously changed from 25 to 125 °C. Snapshots at certain temperatures are shown in Fig. 14B. Although the eye amplitude decreased due to the decreased LD output as temperature increased above 100 °C, we confirmed that the eye kept opening across the entire temperature range without any bias and signal adjustments. Since maximum junction temperatures in most LSIs have been below 125 °C and will be in the future, the interposers are tolerant of heat generated by LSIs and suitable for inter-chip interconnects.

3.4 Conclusion

In this section, we introduced QD laser integration by flip-chip bonding as an LD integration method on Si. First, the requirements for LDs, which are assembled onto silicon photonics integrated circuits by flip-chip bonding, are outlined. Especially, we mentioned that QD lasers have perfect characteristics for assembling onto silicon photonics integrated circuits from the viewpoints of wide-temperature-range operation and strong optical feedback tolerance. Then, the assembly method of QD lasers onto silicon photonics circuits by flip-chip bonding was introduced. The passive alignment method was adopted by IR visual alignment between the marks on the LD and those on silicon photonics integrated circuits. We achieved <3 dB coupling loss between the QD laser and silicon photonics integrated circuit. Furthermore, applications of this assembly method to optical I/O cores for board-to-board optical interconnection and silicon optical interposers for chip-to-chip optical interconnection were introduced. The optical I/O core mounted with a QD laser achieved wide-temperature-range operation from 25 to 85 °C at 25 Gbps per channel without a TEC and optical isolator. Furthermore, the silicon optical interposer mounted with a QD laser achieved error-free data links at 12.5 Gbps operating from 25 to 125 °C without bias adjustments for the LDs, modulators, and PDs.

4. Wafer bonding for quantum dot laser on silicon

Wafer bonding is one of the most promising methods for forming InAs/GaAs QD lasers on Si substrates. We adopt metal intermediate bonding and direct bonding as wafer-bonding methods. In this section, we first introduce the process wafer bonding and laser fabrication. We then describe the characteristics of this fabricated laser.

4.1 Metal intermediate bonding

Metal intermediate bonding is a method of bonding two wafers with a metal layer under heat and pressure. This method has attracted much attention because it is not greatly affected by the structure and roughness of the wafer surface and can use intermediate metal as an electrode. There is a disadvantage of light reflection in the metal layer, however there are structural solutions such as partial bonding of metal stripe.

In this section, we introduce the bonding method of QD lasers using various intermediate metals. We also introduce a QD laser integrated on a silicon-on-insulator (SOI) substrate and Si rib structure by using the metal-stripe wafer-bonding method (Tanabe et al., 2010).

For metal wafer bonding, InAs/GaAs QD laser structures on a GaAs substrate were grown by MOCVD or MBE. The area density of the formed

InAs QDs was 4–6 × 10^{10} cm^{-2}. The InAs QD layers were stacked in 5–10 layers. The QD active layer was sandwiched between 1.4-μm $Al_{0.4}Ga_{0.6}As$ layers for light confinement. The center wavelength of the ground state of the QDs was in the range of 1.27–1.30 μm (O-band).

For the first time, we succeeded in wafer bonding using AuGeNi/Au/AuGeNi intermediate metal on QD laser wafers and AuGeNi on Si wafers. This metal-bonding layer improves bonding strength and interface electrical conductivity using the Au-Ge alloy's low eutectic point (~280 °C). The wafers were diced to ~1 cm^2 area.

The two wafers were brought into contact at 300 °C in air for 3 h under a uniaxial pressure of 0.1 MPa. The GaAs substrate was then removed at room temperature by selective etching with H_3PO_4-H_2O_2 (3:7 vol.). The $Al_{0.7}Ga_{0.3}As$ etch-stop layer was then removed using HCl aq. Broad-area FP lasers were then fabricated by applying electrodes at the top and bottom of the structure.

The devices consisted of a double hetero-laser structure with 3.45 μm thickness bonded onto Si substrates through a metal layer with a thickness of 380 nm. The length and width of the laser cavity were 2.4 mm and 100 μm, respectively. High-reflection (HR) coating was not used. The mirror loss α m for this cavity was calculated as 5 cm^{-1}. Electroluminescence measurements were conducted at room temperature under pulsed operation.

Fig. 15 shows a design and a fabricated QD laser structure. Fig. 16 shows the corresponding L-I characteristics. Insets show the electroluminescence spectra at $J = 0.2$ and 1.0 kA/cm^2.

We then developed a metal-stripe wafer-bonding method that involves dissipative bonding to the Si rib waveguide on SOI substrates and fabricated InAs/GaAs QD lasers on these substrates, the basis of Si photonic integrated circuits, using Si-stripe wafer-bonding technology. In our device design, the

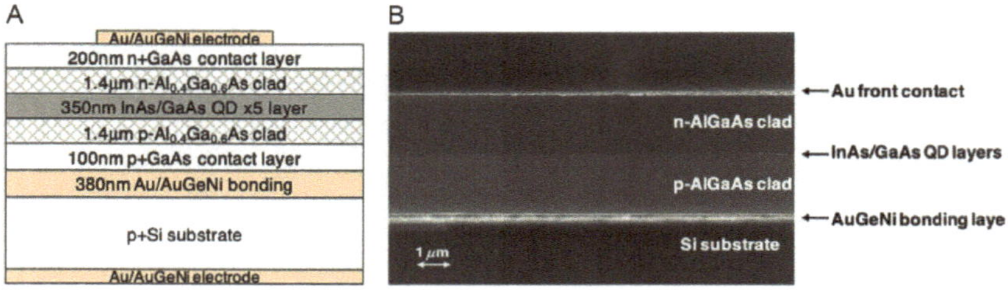

Fig. 15 (A) Cross-sectional schematic of fabricated QD laser metal bonded on Si substrate and (B) cross-sectional SEM image of broad-area QD laser on Si structure (Tanabe et al., 2010).

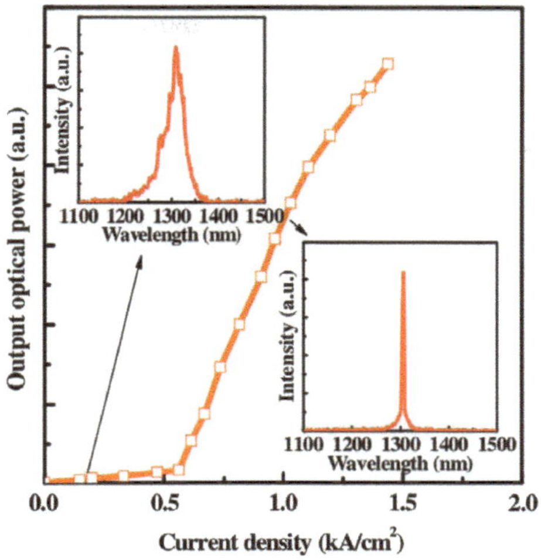

Fig. 16 Light-current (L-I) characteristics of laser. Insets show electroluminescence spectra (Tanabe et al., 2010).

Si rib structure is introduced to mimic the optical waveguide and planned as an intermediate process with the final goal of evanescent coupling. The metal strip we installed in the laser device serves as an electrode that provides ohmic contact to the bonded laser structure as well as the bonding interlayer. Metal-based bonding schemes generally allow higher interfacial electrical conductivity and surface-roughness tolerances to provide mechanically stable bonding (Jhang et al., 2015).

First, the wafer surface is coated with photoresist to protect the surface from particles generated in the dicing process. Quantum dot lasers and SOI wafers are diced to 0.64 and 1 cm^2 dies, respectively, with a 100 nm thick SiO_2 layer formed by sputtering. Next, a Si rib array having a width of 8 μm and a pitch of 500 μm was formed by photolithography and wet etching: at the initial stage, the upper SiO_2 mask was removed at room temperature using a buffered HF solution. The two die pieces are immediately brought into contact with each other and annealed at 400 °C for 3 h under a uniaxial pressure of 0.1 MPa. The GaAs substrate was removed by chemical wet etching.

A broad-area FP laser with a cavity length of 2 mm was formed by depositing a 100-μm-wide Au/AuGeNi electrode strip on top of the structure. For lateral current confinement, a ridge laser structure was fabricated with a top width of 5 μm by photolithography and wet etching. A 100-nm-thick SiO_2 layer was then deposited on top of the laser structure by sputtering to

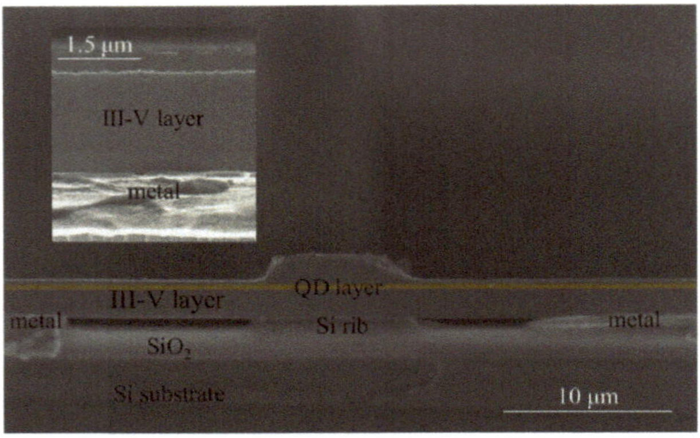

Fig. 17 SEM image of ridge QD laser on SOI substrate with Si rib structure by metal-stripe bonding (Jhang et al., 2015).

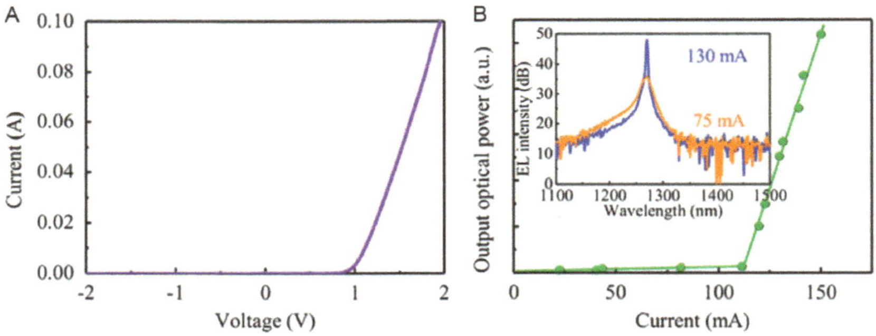

Fig. 18 (A) DC current-voltage characteristics and (B) room-temperature L-I characteristics (Jhang et al., 2015).

leave an open window 2 μm wide for current injection (see Fig. 17). Fig. 18 shows the device characteristics of a broad-area QD laser on an SOI substrate. The current-voltage (I–V) characteristics of the LD shows the ohmic behavior of the GaAs/metal interface, as shown in Fig. 18A, and the device resistance is 6.2 Ω when the resistance is obtained. During the bonding process, the GaAs/metal interface is the same as the laser. The L-I characteristics were measured under the pulsed current injection condition (500 Hz, 1 μs). A clear kink in the L-I curve and a laser threshold current of 1040 mA are shown in Fig. 18B. The inset shows the electroluminescence spectra of the InAs/GaAs QD broad-area laser on the SOI substrate fabricated using our metal-stripe wafer-bonding method. From the above investigation, we demonstrated metal bonding between the QD laser and Si substrate. Metal bonding is less affected by the roughness of the hetero interface. The current is also injected into the single crystal material rather than the hetero-interface to improve performance. Furthermore, opening the window to a part of the metal bonding and partially applying the direct bonding makes it possible to solve the problem that the hetero-interface is opaque and waveguide coupling is impossible.

4.2 Direct bonding

We developed a direct-bonding method to Si wafers and laser wafers without the use of intermediate metals and oxides. With this method, the current path can be obtained through the hetero-interface, and integration of the III-V light source on the Si substrate can be facilitated. All bonding processes can be carried out at 300 °C or lower and be free from problems such as impurity diffusion of Si devices. Such Si photonics light-source-integrated process technology is more suitable for the CMOS process (Tanabe et al., 2012).

The process of this direct-bonding method is as follows, with several modifications in each experiment we conducted for this study.

1. Photoresist coating on laser wafer and Si wafers
2. Dicing the laser and Si wafers
3. Photoresist removing (dipping, acetone)
4. Native oxide removing (wet chemical etching, HF)
5. Ozone ashing for bonding surface
6. Wafer bonding
7. GaAs substrate removing (wet chemical etching, $H_3PO_4 + H_2O_2$)
8. AlGaAs etch-stop-layer removing (wet chemical etching, HF)

For these experiments, the bonding surfaces of GaAs (100) and Si (100) wafers were coated with photoresist to protect these surfaces from particles produced in the scribing process. The wafer was then diced into two each with an area of 1 cm^2. The photoresist was then removed with acetone and the bonding surface was degreased. The native oxides on both GaAs and Si wafers were removed using HF aq. (20 vol%). The two dies were brought into contact with aligned (011) edges useful for cleaving in laser fabrication. The GaAs/Si samples were then annealed at 300 °C in ambient air for 3 h with a uniaxial pressure of 0.1 MPa. The I–V characteristics of the bonded GaAs/Si interface was measured using a DC bias source with an AuGeNi/Au ohmic contact applied to the outer surface of the bonded GaAs/Si samples.

An amorphous layer at the GaAs/Si interface with a thickness of around 2 nm can be seen in the image (see Fig. 19). Even if this interfacial layer is an

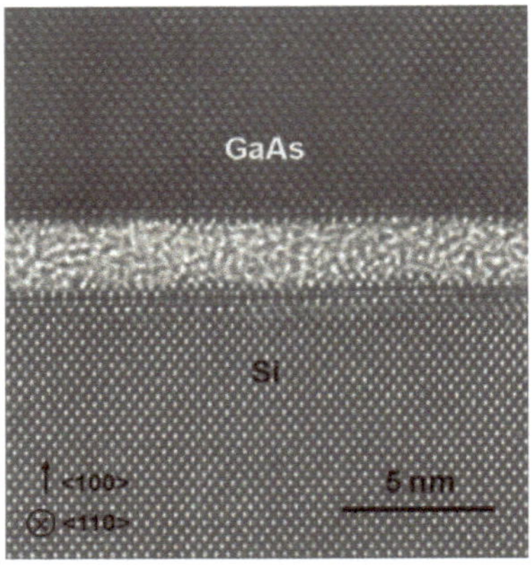

Fig. 19 Cross-sectional TEM image of direct-bonded p-GaAs/p-Si hetero-interface (Tanabe et al., 2012).

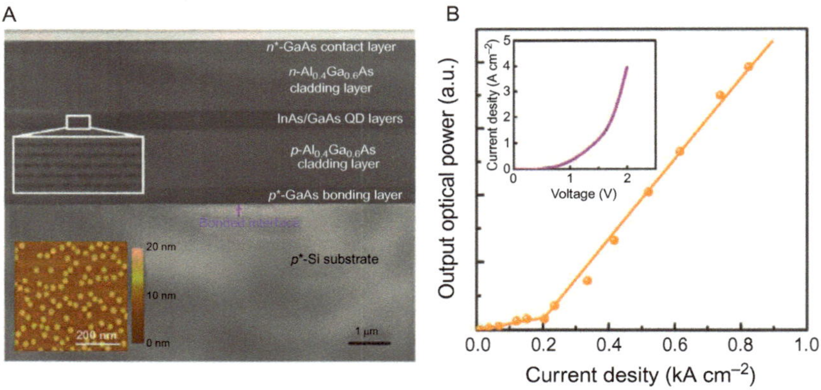

Fig. 20 (A) Cross-sectional TEM image of InAs/GaAs QD laser on Si substrate. (Inset) AFM image of high density InAs/GaAs QD included in laser structure. (B) L-I characteristics and electroluminescence spectra of QD laser (Tanabe et al., 2012).

oxide, this thickness is sufficiently thin to provide ohmic interfacial conductivity by inducing a tunneling current or oxide breakdown by the applied voltage. The final device includes a 3.9-mm-thick III-V semiconductor double-hetero laser structure on top of a Si substrate, as shown in Fig. 20A. The clear kink in the L-I curve indicates the lasing turn-on with a threshold current density of 205 A/cm^2, the lowest threshold current density. The inset of Fig. 20B shows the I–V characteristics of the laser. The resistivity in the linear I–V region at higher voltages is around 0.1 V/cm^2.

These results indicate that an electrically conductive and optically transparent hetero-interface can be formed at a CMOS compatible temperature

($<300\,°C$) by using our wafer-bonding methods between GaAs and Si. By using these methods, it is possible to supply current through the Si-substrate side, which increases the ease of integration. The possibility of evanescent coupling to a Si waveguide was also demonstrated.

After demonstrating our direct-bonding method between QD lasers on GaAs and Si substrates, we developed technology to improve the performance of laser devices and evaluated their dynamic characteristics. We confirmed the direct modulation performance of a QD laser on a Si device, which is considered important for practical application (Jhang et al., 2016).

For this evaluation, the InAs/GaAs core layer was sandwiched between p-type and n-type $Al_{0.4}Ga_{0.6}As$ layers. An $Al_{0.7}Ga_{0.3}As$ etch-stop layer having a thickness of 1 μm was grown between the GaAs substrate and lower cladding layer. The laser and Si wafers were coated with photoresist to protect the bonding surfaces from contaminants generated from the scribing process. The laser and Si wafers were then diced to $0.8 \times 0.8\,cm^2$ and $1 \times 1\,cm^2$, respectively. The photoresist and organic contaminants on the wafer surfaces were removed using acetone. The native oxide on the laser and Si dies were etched using a HF solution. Prior to the bonding process, the laser and Si dies were treated with ultraviolet ozone for 5 min to hydrophilize the bonding surface and improve bonding strength. The two dies were immediately brought into contact and annealed at 300 °C in ambient air for 3 h at a uniaxial pressure of 0.1 MPa. After bonding, the GaAs substrate was removed by wet chemical etching with H_3PO_4-H_2O_2 followed by citric acid-H_2O_2 solution. Finally, the etch-stop AlGaAs layer was removed at room temperature using HF solution.

Next, the ridge laser structure was fabricated with a top width of 5 μm using photolithography and wet etching using a mixture of citric acid-H_2O_2. A 100-nm-thick SiO_2 passivation layer was deposited on top of the ridge structure by sputtering, leaving a 2-μm-wide opening for current injection. The lower electrode was deposited using electron beam deposition and consisted of Ti (20 nm)/Pt (30 nm)/Au (500 nm). A top electrode with a width of 100 μm was deposited using a similar method but using a thicker (1800 nm) Au layer prepared for the chip assembly for the submount. The cross-sectional scanning electron microscopy (SEM) image shown in Fig. 21 (right) indicates that the III-V laser structure is tightly bonded to the Si substrate without any clear defects. The bonded structure was finished by forming an ohmic contact by annealing at 350 °C in ambient air for 30 min. Manual cutting was used to form an FB cavity with a length of 500 μm. The sample was then diced further with a $500 \times 500\,\mu m^2$ laser chip

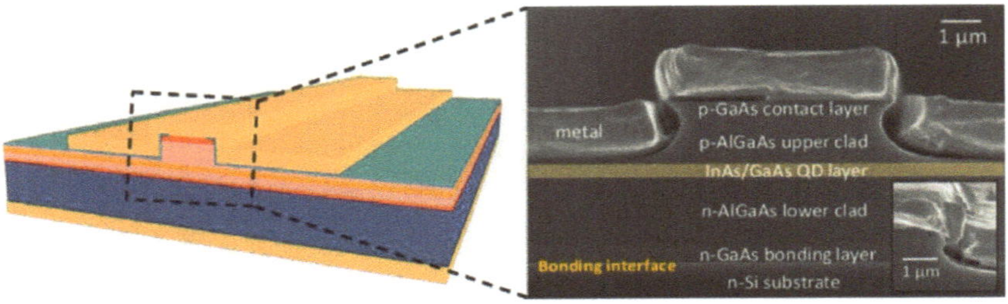

Fig. 21 Schematic of structure (left) and corresponding cross-sectional SEM image (right) of direct-bonded QD laser on Si substrate. (Inset) Cross-sectional image indicating continuous top metal layer (Jhang et al., 2016).

then assembled onto the submount using die bonding and wire bonding for the next modulation experiment.

Fast direct modulation of the bonded laser on the Si die was evaluated based on the bit rate performance of the laser. A ground signal contacting submount with 50 Ω-termination ground probe was connected to a DC source and a pulse pattern generator using a bias tee. The output signal was collected using a detector connected to an oscilloscope for data analysis. Large signal modulation measurements were conducted using a signal analyzer and an NRZ PRBS pattern. The bonded laser was driven above the threshold with a bias current of 100 mA applied to obtain an optimized eye diagram under a driving electrical signal of 2.5-V swing at the voltage. Fig. 22 shows a measured eye diagram from 2.5 to 10 Gbps modulation at room temperature. The eye pattern indicates that the combined laser can modulate up to 10 Gbps with an extinction ratio of 1.9 dB. We further investigated the modulation performance at various temperatures. Fig. 23 shows 6-Gbps modulation of the Si-bonded laser at 40 and 60 °C under the same operating conditions without current or voltage regulation. The 60 °C eye diagram shows an extinction ratio of 4.5 dB. We demonstrated continuous wave (CW)-pumped direct-bonded InAs/GaAs QD lasers directly on a Si [wafer?] and conducted the first modulation experiment with the device. Our devices were directly modulated at a bit rate of 10 Gbps at room temperature, and 6-Gbps modulation was achieved up to 60 °C without current or voltage regulation.

We developed wafer-bonding technology and bonded a QD laser onto a Si waveguide. We also fabricated Si hybrid QD lasers by fabricating a DBR resonator on a Si waveguide and forming a tapered structure. The following section describes how we fabricated this hybrid Si InAs/GaAs QD laser on a Si waveguide (Jang et al., 2016).

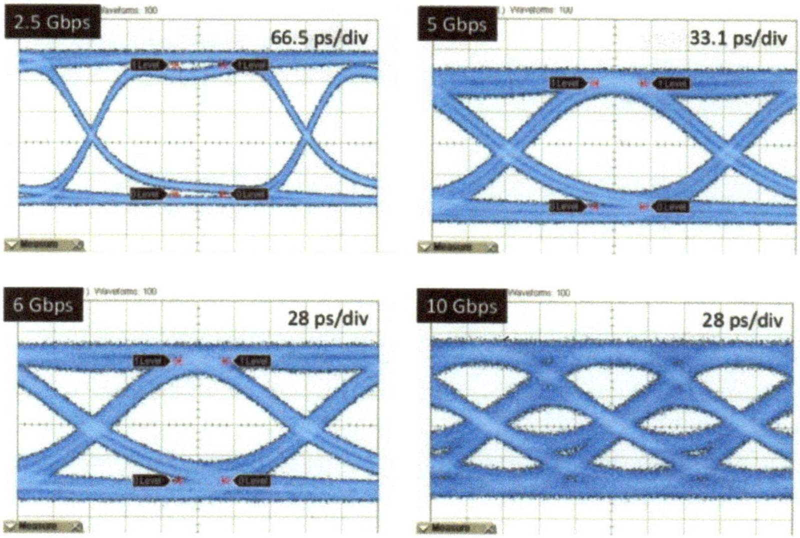

Fig. 22 Eye diagrams of optical signals for direct-bonded QD laser on Si substrate at modulation speeds from 2.5 to 10 Gbps at room temperature (Jhang et al., 2016).

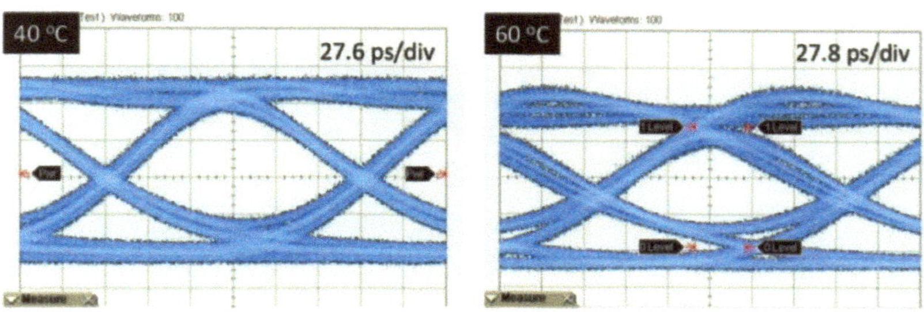

Fig. 23 Eye diagrams of optical signal for 6-Gbps modulation of direct-bonded QD laser on Si substrate at 40 °C (left), and 60 °C (right) (Jhang et al., 2016).

We needed an optical design for the coupling of a Si waveguide including a DBR and QD laser structure with only top side AlGaAs clad. First, a cavity was formed by two DBRs on both sides of the bonding interface of the Si waveguide. The DBR consisted of 100 pairs of side wall grids. The center wavelength of the stop band of the DBR was 1255 nm, the bandwidth of the stop band was measured at 20 nm, and the reflectance was designed to have a reflectance of 90% or more. The one-sided clad QD laser for evanescent coupling was designed with an epitaxial structure containing a thin $Al_{0.7}Ga_{0.3}As/GaAs$ layer over the InAs/GaAs QD active layer.

The QD laser was a 20-μm-wide, 5-mm striped mesa structure. The light generated in the QD active layer was coupled to the Si waveguide by a taper. It was then reflected from the DBR then returned to the QD active layer to induce the stimulated emission.

Silicon rib waveguides and DBRs were fabricated on 4-in. SOI wafers using electron beam lithography and dry etching with an etch depth of 300 nm. The InAs/GaAs QD-laser and SOI wafers were coated with photoresist to protect the bonding surfaces from particles and contaminants created in the cleaving process. After cleaving, acetone was used to remove the photoresist and clean the bonding surfaces. Before direct-wafer bonding, the native oxide on the laser and SOI dies were removed by dipping both in HF for 30 s. The two dies were then brought into contact with and annealed at 300 °C in ambient air for 3 h under a uniaxial pressure of 0.1 MPa to finish the bonding process. For selective chemical wet etching of the GaAs substrate, H_3PO_4-H_2O_2 and citric acid-H_2O_2 were used. The $Al_{0.7}Ga_{0.3}As$ etch-stop layer was removed using HF solution at room temperature. The ridge-type laser structures were fabricated by typical photolithography and wet etching. Fig. 24A shows an optical microscopy [image?] of the fabricated hybrid laser. Fig. 24B and C shows SEM images

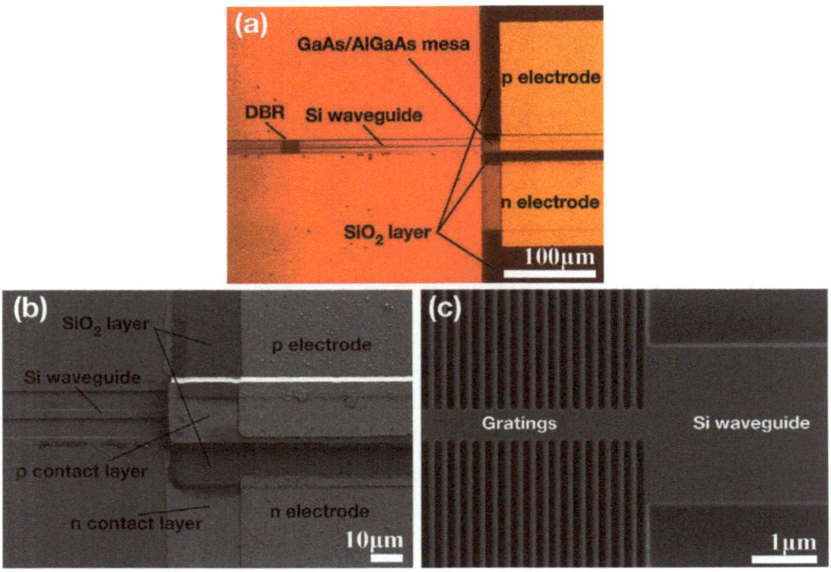

Fig. 24 (A) Top-view optical microscopy image of edge of hybrid Si evanescent QD laser. (B) SEM image of GaAs-AlGaAs mesa edge. (C) SEM image of side-wall gratings on Si waveguide after laser fabrication (Jang et al., 2016).

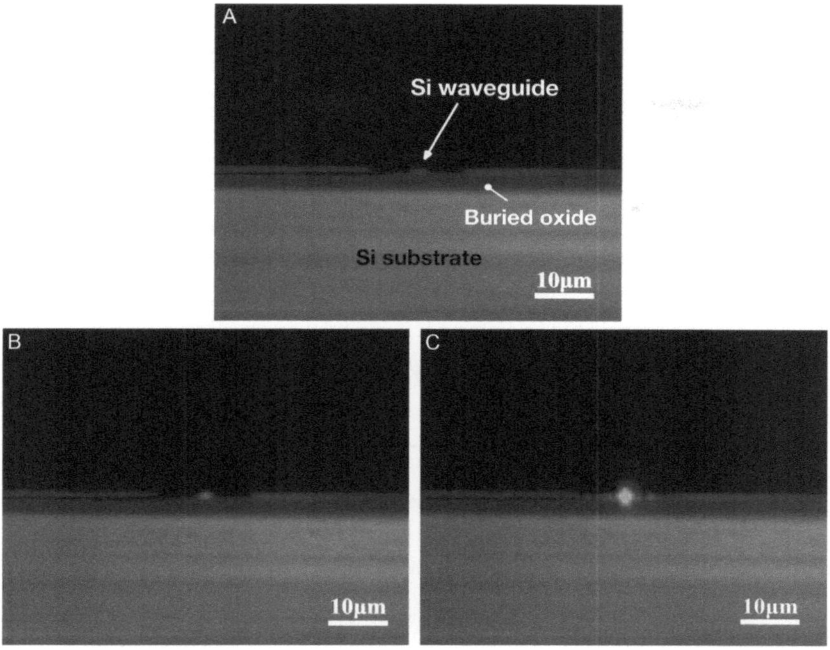

Fig. 25 Near-field images of Si waveguide facet under current injections of (A) 0 mA, (B) 250 mA, and (C) 350 mA (Jang et al., 2016).

of the hybrid laser structure edge and a DBR, respectively. It is clear that the GaAs/AlGaAs mesa structure is aligned along the Si waveguide, and there is no damage to the sidewall gratings.

To verify the gain-waveguide optical coupling and laser emission in the Si waveguide, cross-section observation was carried out with an IR camera focused on the device edge (Si waveguide side). Fig. 25A shows a laser facet without current injection; therefore, no light emission. The spontaneous- and lasing-emission patterns are shown in Fig. 25B and C, respectively. Complementary optical measurements using polarizing filters confirmed TE (transverse electric) polarization for spontaneous and laser emission.

The improvement of the QD laser on hybrid silicon was achieved by reducing the width of the GaAs/AlGaAs mesa to 8 μm, and CW operation was achieved with a threshold current of 100 mA as shown in Fig. 26. We have calculated the critical current density of this device to be 250 A/cm^2, in our best knowledge, the lowest value of the hybrid silicon laser we have reported so far (Chang et al., 2007). The low-threshold current density device achieved in lasers is based in part on the QD gain medium

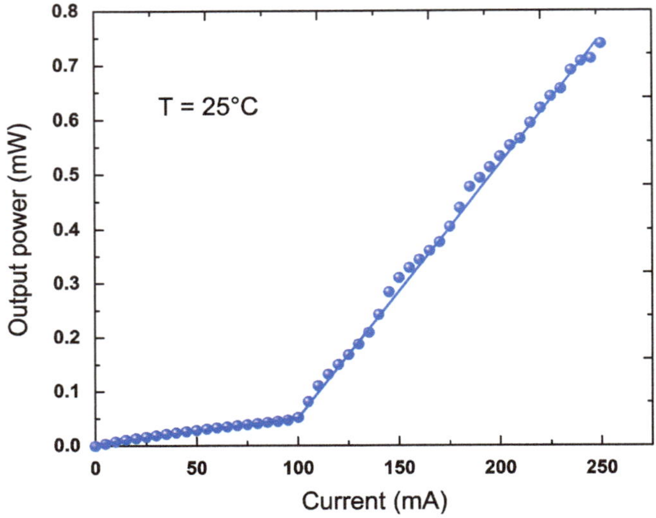

Fig. 26 CW L-I curve of hybrid Si evanescent QD laser with GaAs-AlGaAs mesa width of 8 μm. Threshold current was reduced to 100 mA.

(Asada et al., 1986; Freisem et al., 2008). In addition, high temperature operation up to 115 °C was observed. Additionally, characteristic temperature (T_0) of 60 K at 100 °C and 303 K at near room temperature was achieved.

The hybrid QD laser on Si was improved by reducing the width of the GaAs/AlGaAs mesa to 8 μm, and CW operation was achieved with a threshold current of 100 mA, as shown in Fig. 26. We calculated the critical current density of this device to be 250 A/cm^2, the lowest value of a hybrid Si laser we have reported thus far (Chang et al., 2007). The low-threshold current density device achieved in lasers is based in part on the QD gain medium (Asada et al., 1986; Freisem et al., 2008). High-temperature operation up to 115 °C was observed, and characteristic temperatures (T_0) of 60 K at 100 °C and 303 K at near room temperature were achieved.

We then developed a QD laser on a Si distributed-feedback (DFB) structure. The distributed-feedback structure is fabricated by corrugations on the side of the Si waveguide, which is an effective structure for the fabrication process. This laser device was fabricated by adhesive bonding and general laser processing.

First, an SOI wafer was processed to form a Si waveguide having a grating and taper structure by electron beam lithography and dry etching. Fig. 27 shows an SEM image of the top view of a λ/4-phase shift region

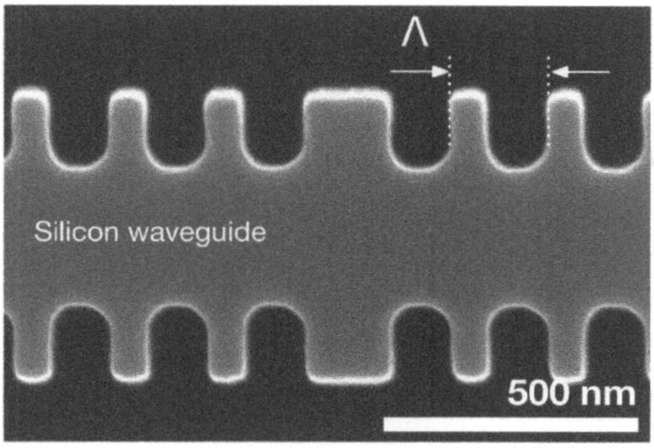

Fig. 27 SEM image of sidewall grating presenting λ/4-phase shift in middle of grating.

in the middle of the grating. The duty cycle of the grating can be seen to be around 40%. The InAs/GaAs QD and SOI Si waveguide wafers were then cleaned by dicing for spin-on-glass (SOG) bonding. The SOG solution was spin-coated on a GaAs wafer at 4000 rpm for 60 s then soft-baked at 100 °C for 1 min to prevent void formation. The GaAs and SOI dices were brought into contact and annealed at 300 °C for 3 h under a uniaxial pressure of 0.1 MPa. The thickness of the SOG after the wafer bonding process was measured to be 40 nm. Subsequent GaAs substrate removal was carried out to complete the layer-transfer process with selective wet chemical etching. The III-V mesa and tapered mesa structures were defined using photolithography and wet chemical etching. A SiO_2 layer was then deposited as a protective layer by sputtering. To form p- and n-contacts, the contact windows were etched using HF solution and the metal contacts were formed by electron beam evaporation. Fig. 28A shows an optical micrograph of the fabricated laser device. QD lasers were then fabricated on a Si waveguide chip. It is clear that the tapered III-V mesa structure is aligned on the Si waveguide and the tip width is about 1 μm in accordance with the device design (see Fig. 28B). Since these III-V tapered mesa structures were formed by wet chemical etching using a photoresist mask, they have a width of about 1 μm, unlike an ideal sharp tapered structure. Finally, both sides of the Si waveguide were cleaved and an antireflective coating was formed. Fig. 29 shows the measured L-I-voltage (L-I-V) curves of the DFB QD laser under CW operation at room temperature. The series resistance of the LD is 3.8 Ω, and the threshold current is 58 mA corresponding to a Jth of 1.16 kA/cm^2. This Jth was compared to the minimum Jth of 1.05 kA/cm^2 of previously reported hybrid DFB lasers. The lower critical current density of QD lasers than that of multi quantum well (MQW) lasers

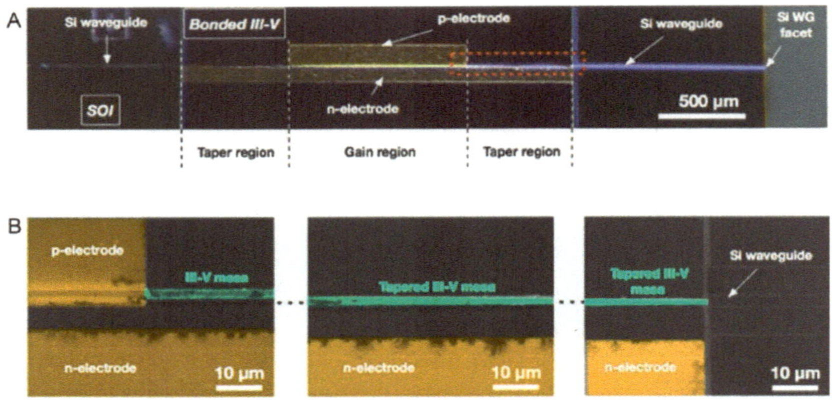

Fig. 28 (A) Optical microscopy image of hybrid Si DFB QD laser and (B) enlarged SEM image of red dotted rectangle in (A).

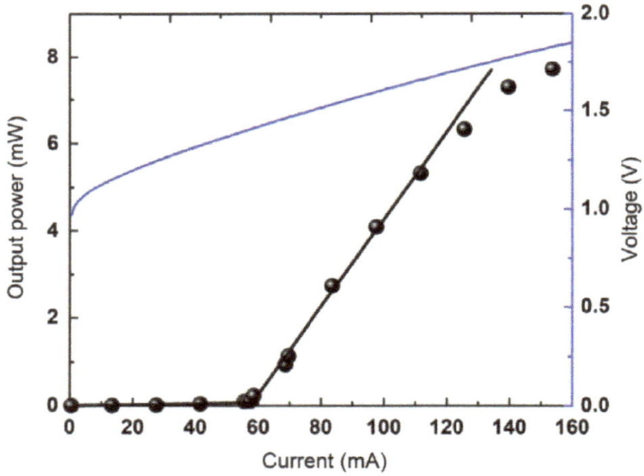

Fig. 29 L-I-V curve at room temperature under CW operation.

is one of the outstanding characteristics of QD lasers, as presented in various reports (Duprez et al., 2016; Fang et al., 2008; Stankovic et al., 2012). For our fabricated DFB QD laser, the reason is the absorption region of the non-electrical injection region. The emission spectrum of our fabricated DFB QD laser at a 100-mA drive current is shown in Fig. 30. A single wavelength emission can be seen in the laser emission spectrum. The peak wavelength is about 1266 nm and the side-mode suppression ratio (SMSR) is >40 dB. The emission wavelength is consistent with the device design. A 30-nm span spectrum (inset in Fig. 30) also shows that this laser clearly operates in single wavelength emission.

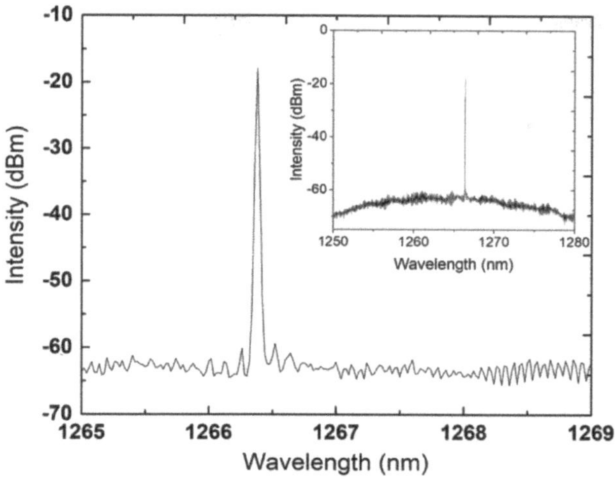

Fig. 30 Lasing spectrum at 100 mA CW current injection with over 40-dB SMSR. (Inset) Lasing spectrum.

The demonstration of single-mode QD lasers on Si substrate using a DBR and DFB, as described above, suggests not only their operation as hybrid laser devices but also the possibility of multichannel operation for wavelength-division multiplexing applications.

5. Direct epitaxial growth for quantum dot laser on silicon

Direct growth is one of the most promising methods for forming InAs/GaAs QD lasers on Si substrates. We conducted an experiment involving growing InAs/GaAs QD laser structures on Si substrates by using only MBE. The results indicate that a single epitaxy growth on a CMOS-compatible Si (001) on-axis substrate can form a QD laser structure. In this section, we first introduce the process flow of crystal growth and laser fabrication. We then describe the characteristics of the fabricated laser.

5.1 Crystal growth

Prior to crystal growth, we prepared 3-in. n-type on-axis Si (001) substrates for the experiment. The off angle of these substrates is <0.2° forward [110], meaning that the substrates are CMOS compatible (Volz et al., 2011). The as-received substrates had a 2-nm native oxide on their surfaces, which was introduced into the MBE chamber immediately after removing the oxide film by wet etching using 1% diluted HF. The surfaces of the Si substrates subjected

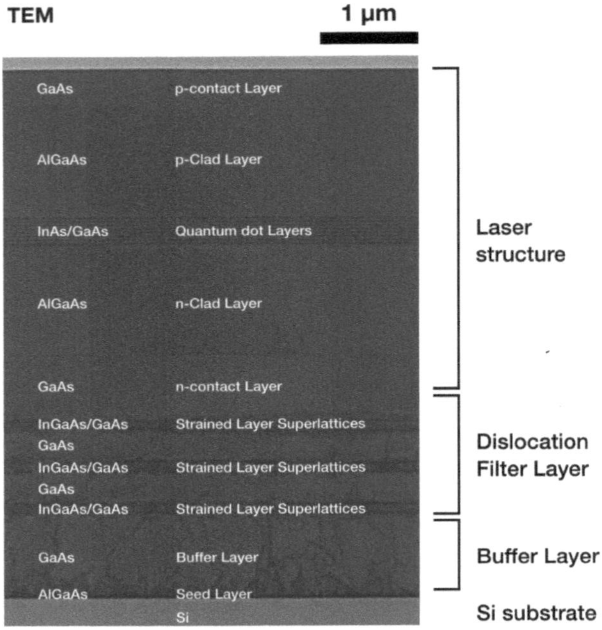

Fig. 31 Epitaxial structure of QD laser directly grown on on-axis Si(001) substrate.

to wet etching were assumed to be hydrophobic and H-terminated. These substrates underwent outgassing for 1 h at 450 °C in the preparation chamber. They were then transferred to a deposition chamber where the heater was thermally cleaned at a temperature of 950 °C for 10 min. We lowered the temperature of the heat-cleaned Si substrate with As4 irradiation.

Direct growth of a QD laser on a Si substrate is largely composed of three structures (see Fig. 31). The first structure is the growth of the buffer layer including the seed layer on the Si substrate. The second structure is a dislocation filter layer for lowering the high dislocation density between III-V and Si. The third structure is the growth of the same QD laser layer as that on the GaAs substrate.

For our experiment, we called the III-V layer first grown on a Si substrate the seed layer. A hetero-epitaxial interface was formed between this layer and the Si surface. This interface is important because it is the origin of the majority of defects of the subsequently deposited III-V layer (Kwoen et al., 2018, 2019).

An antiphase boundary (APB) is a boundary formed by Ga–Ga and As–As bonds when GaAs, which is a polar material, is grown on Si, which is a nonpolar material. Since the injected carriers extinct at the local charge of an APB, the electrical characteristics of the sample containing an APB deteriorate. High-density defects are also formed because the strain is concentrated in APBs. The main approach to GaAs-on-Si growth to date is to suppress the generation of APBs. Since an APB is generated from a single-step layer on the Si (001) surface, the formation of the APB can be suppressed by changing the surface structure from a single-step layer to a double-step layer or limiting the growth area. Silicon (001) substrates with off-axis angles of 4–6°, annealed at high temperatures, form double-step layers on the surface (Bringans et al., 1986; Nakayama et al., 1987). Vicinal substrates have been widely used in the field of III-V direct growth on Si, even in recent studies (Chen et al., 2016; Liu et al., 2015, 2016). It has recently been reported that double-step layers are generated by annealing an Si substrate having a very small inclination angle (<0.15°) in a H2 environment (Alcotte et al., 2016; Brückner et al., 2013; Martin et al., 2016). In particular, these methods, which involve MOCVD with in situ measurement, are believed to have significant industrial value because APB-free GaAs films can be obtained over the entire 12-in. Si substrate (Alcotte et al., 2016). To fabricate template substrate such as an etched v-groove oxide fin structure, oxide masking is another effective approach for suppressing APBs (Li and Lau, 2017; Li et al., 2016; Wan et al., 2017). However, these methods have not yet been used with MBE because of several hurdles such as high cost of the H_2 supply system and high adsorption coefficient of group III adatoms. Nevertheless, MBE has several advantages for growing light-emitting structures, such as the ability to grow a good crystalline AlGaAs layer at low temperature, which can be used as a clad layer, and to produce an InAs QD with high quality. From these aspects, forming single-domain GaAs without APBs on a CMOS-compatible Si (001) substrate by MBE is promising for manufacturing monolithic and high-performance light sources by a single process of crystal growth.

The AlGaAs nucleation layers were grown at a relatively high growth rate of 1.1 μm/h and high substrate temperature of 500 °C. Thanks to the high growth temperature, the surface temperature could be measured using pyrometers. Forty-nm-thick $Al_xGa_{1-x}As$ samples with Al compositions x; (x=0 (sample A), 0.3 (sample B), 0.5 (sample C), 0.7 (sample D)) were grown on Si.

After etching with NH_4OH-based etchant (NH_4OH:H_2O_2:H_2O= 3:100:9897) for 10 s, cross-sectional SEM was used for observing the elimination of APBs.

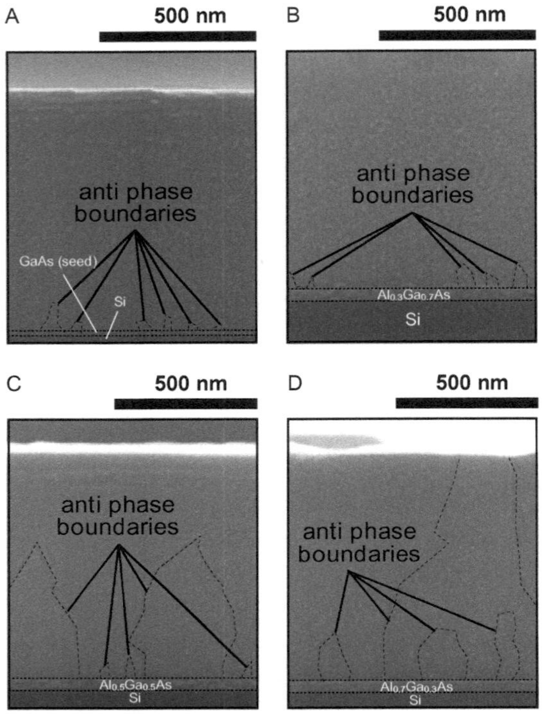

Fig. 32 Cross-sectional SEM images of GaAs layer on (A) GaAs nucleation layer (Sample A), (B) $Al_{0.3}Ga_{0.7}As$ nucleation layer (Sample B), (C) $Al_{0.5}Ga_{0.5}As$ nucleation layer (Sample C), and (D) $Al_{0.7}Ga_{0.3}As$ nucleation layer (Sample D) grown on Si (001) substrate (Kwoen et al., 2019).

Fig. 32 shows a cross-sectional SEM image of all samples. For Sample B, the APB was self-extinguishing in the region near the nucleation layer. However, in Sample D, the APB extended to the GaAs surface. These results are consistent with surface atomic force microscopy (AFM) observations and indicate the origin of the APBs observed on the GaAs surface. The interface between the AlGaAs nucleation layer and GaAs buffer layer is also remarkable. Sample B had a rough interface due to the island structure formed in the nucleation layer.

Fig. 33 shows the dependence of Al composition of the nucleation layer on APB-annihilated thickness, Root mean squared (RMS) roughness, and XRD ω-2θ full-width at half-maximum (FWHM). The RMS roughness of the GaAs surface was almost the same with Al composition from 0 to 0.5 (see Fig. 3A). This suggests that the APBs were annihilated from the 800-nm thick GaAs at Al fractions below 0.5 (see Fig. 3B). Therefore, we found that GaAs on the $Al_{0.3}Ga_{0.7}As$ nucleation layer has good crystal quality.

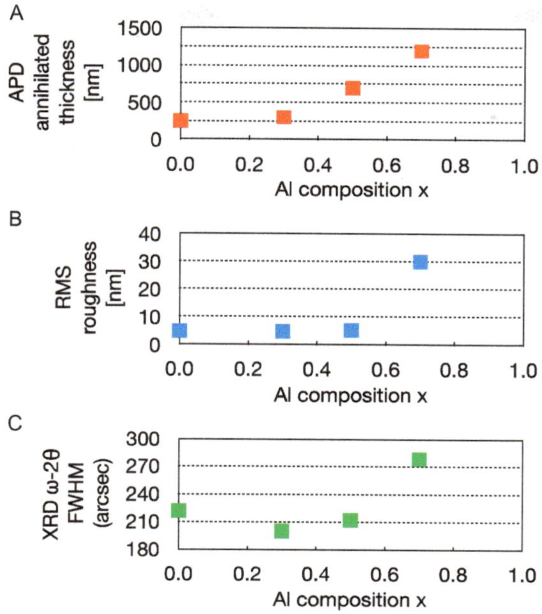

Fig. 33 Dependence of Al composition x of nucleation layer and (A) APB-annihilated thickness, (B) RMS roughness, and (C) XRD ω-2θ FWHM (Kwoen et al., 2019).

Fig. 34A shows a cross sectional bright-field TEM image of the buffer layer structure on a GaAs/Si (001) substrate. The density of dislocations at the interface between the Si substrate and GaAs epi-layer decreased through the three groups of InGaAs/GaAs strained layer superlattice (SLS). Plane-view TEM was used to accurately estimate treading dislocation densities (TDDs) in the sample below $1 \times 10^8/\mathrm{cm}^2$. Fig. 34B shows a plane-view bright-field TEM image of the GaAs layer on an n-clad layer. This is the layer that immediately precedes that on which the InAs QD layers are grown. The white arrows in Fig. 34B indicate threading dislocations. The TDD was $4.7 \times 10^7/\mathrm{cm}^2$, which is higher than the lowest reported TDD, but still corresponds to only one threading location per 1000 QDs. Since a buffer layer was not applied to reduce the number of dislocations via a technique such as thermal cycling annealing, it is suggested that incorporating this technique can reduce the TDD to below $10^7/\mathrm{cm}^2$ (Jung et al., 2017, 2018b; Li et al., 2018). Stacking faults were not observed in the plane-view TEM images.

We also grew the InAs/GaAs QDs on GaAs and GaAs/Si (001) substrates by adopting the same growth method and evaluated their sizes and densities. Fig. 35 shows AFM images of the uncapped InAs/GaAs QDs grown on

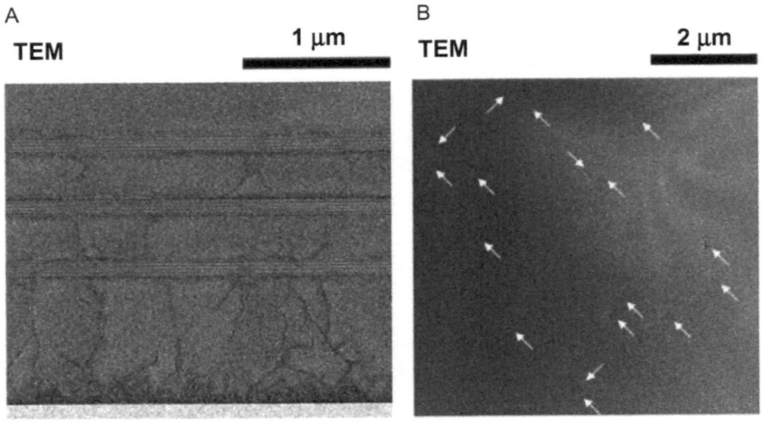

Fig. 34 (A) Cross-sectional TEM image of buffer layer and dislocation filter layer. Purple areas indicate InGaAs/GaAs strained layer superlattice. (B) Plane-view TEM image of GaAs layer directly under InAs QD layer (Kwoen et al., 2019).

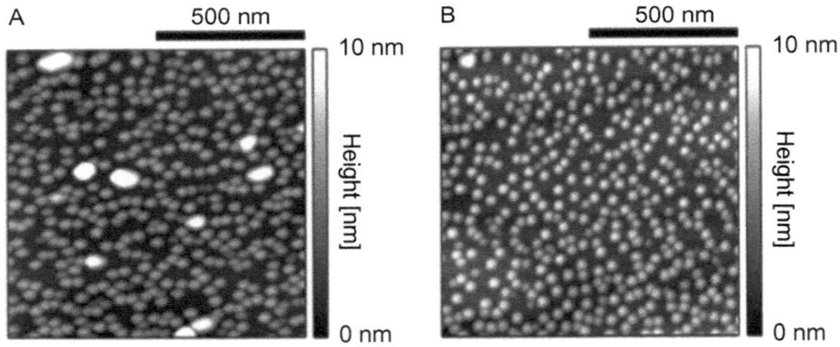

Fig. 35 $1 \times 1\,\mu m^2$ AFM images of uncapped InAs/GaAs QDs grown on (A) GaAs and (B) GaAs/Si (001) substrates, respectively (Kwoen et al., 2018).

GaAs and GaAs/Si (001) substrates, respectively. From the AFM results, we observed the formation of InAs/GaAs QDs of almost the same size and density on both substrates. The average lateral sizes of the formed QDs were 30.1 and 30.2 nm, respectively, and the density was about $5.1 \times 10^{10}\,cm^{-2}$ and $5.2 \times 10^{10}\,cm^{-2}$, respectively. In addition, we found lower giant QD densities on GaAs/Si samples than that on GaAs samples. Since the surface of GaAs/Si substrate is rougher than that of GaAs substrate, it is thought that there is lower local concentration of stress; and thus, less likely to cause the formation of giant InAs QDs is less likely.

Fig. 36 shows the room-temperature photoluminescence (PL) spectra from a single QD layer of the grown structures on Si (001) and GaAs

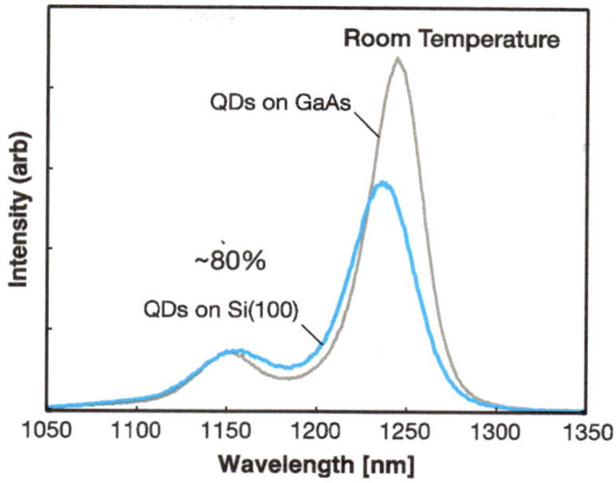

Fig. 36 PL comparison of QD sample grown on GaAs/Si (001) to reference sample grown on GaAs substrate at room temperature under same pump conditions (Kwoen et al., 2018).

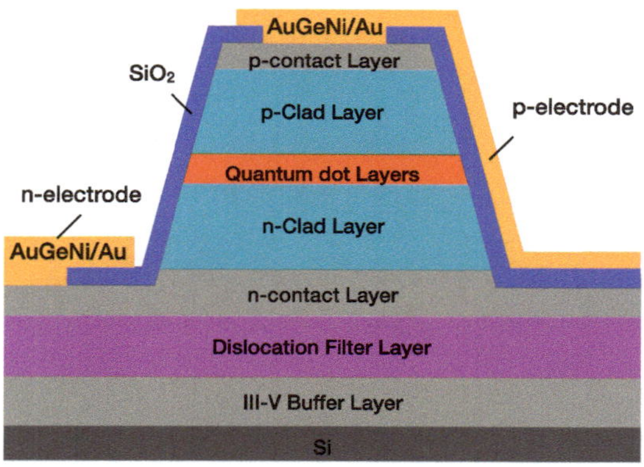

Fig. 37 Schematic of QD laser structure on on-axis Si (001) substrate (Kwoen et al., 2019).

substrates. The integrated PL intensity of QDs on the Si (001) sample was ~80% that on the GaAs sample. At room-temperature, we observed a PL emission of the ground state at ~1250 nm with a FWHM of 31 meV. Additionally, the emission wavelength of excited level was at ~1150 nm. Due to the narrow emission linewidth and the large energy difference between the ground and excited levels (86 meV), the peaks could be clearly distinguished.

5.2 Fabrication process

Fig. 37 is a schematic of the fabricated InAs/GaAs QD laser structures. Laser mesas with a 7-μm width were fabricated using photolithography and wet etching. The AuGeNi/Au was deposited onto the p- and n-contact layers

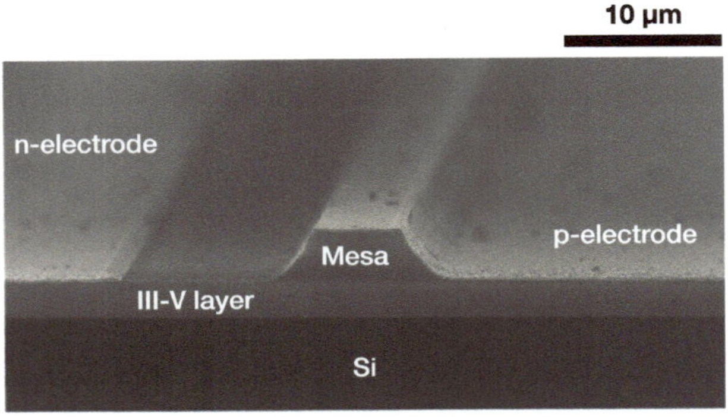

Fig. 38 Cross-sectional view of a scanning electron microscope image of an InAs/GaAs QD laser device fabricated on a Si substrate (Kwoen et al., 2019).

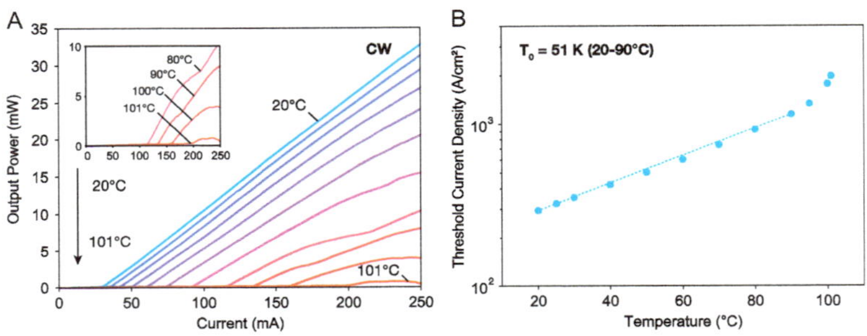

Fig. 39 (A) Temperature dependence of L-I characteristics under CW operation on QD laser directly grown on on-axis Si (001) substrate. (B) Temperature dependence of threshold current density on QD laser directly grown on Si (001) substrate (Kwoen et al., 2019).

(see Fig. 38). After lapping the backside of the Si substrate to 100 μm, the structures were cleaved to make 1.1-mm-long devices. Note that no high-reflection coatings were applied to the cleaved facets.

5.3 Quantum-dot laser characteristics

Fig. 39A shows the temperature dependence of the L-I characteristics in CW operation. The threshold current at room temperature was 27.6 mA, and the corresponding Jth was as low as 370 A/cm^2. CW operation was also observed at up to 101 °C in the fabricated devices. The slope efficiencies were 53.2 mW/A at 20 °C, 49.5 mW/A at 50 °C, 38.5 mW/A at 80 °C, and 26.3 mW/A at 100 °C. Fig. 39B shows the temperature dependence of Jth. The T_0 was nearly constant at approximately 50 K between operating temperatures of 20 and 90 °C. Fig. 40A shows the dependence of the

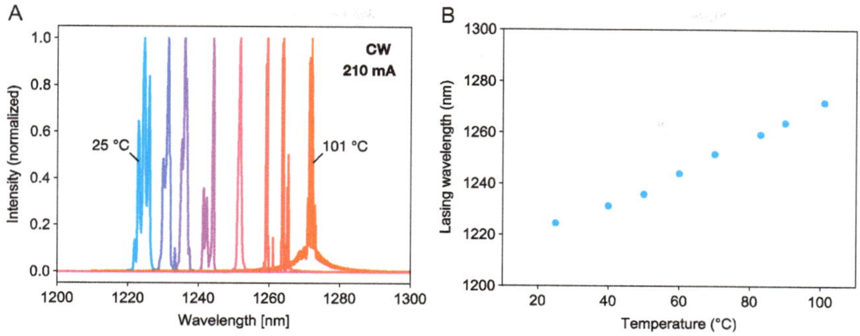

Fig. 40 (A) Temperature dependence of emission spectra on QD laser directly grown on on-axis Si (001) substrate and (B) peak wavelength shifts due to heat-sink temperature change (Kwoen et al., 2019).

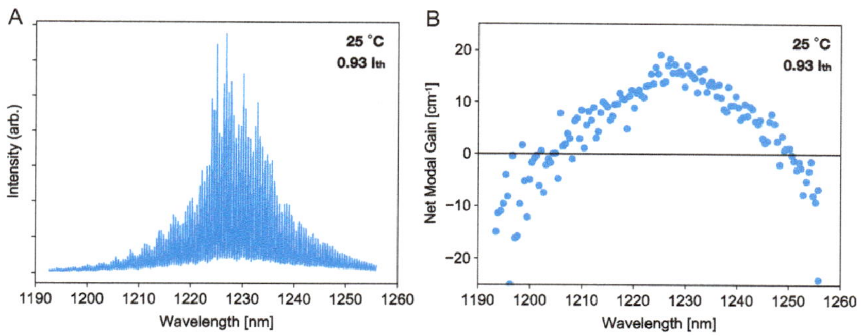

Fig. 41 (A) Amplified spontaneous emission spectrum of QD laser on Si substrate and (B) gain characteristics of QD laser on Si substrate by using Hakki-Paoli method (Kwoen et al., 2019).

ground-state lasing spectrum on the heat-sink temperature. Under the bias current of 210 mA, the ground-state lasing wavelengths at 25 and 101 °C were 1224.6 and 1272.0 nm, respectively, and the wavelength shift was 0.62 nm/K (see Fig. 40B). We measured the sub-threshold electro-luminescence spectrum and evaluated the net modal gain of QD lasers by using the Hakki-Paoli method at 25 °C using a 450-μm-long FP laser (see Fig. 41). The maximum net modal gain was 20 cm^{-1} at around 1225 nm. This gain is smaller than those on QD lasers on GaAs substrates (Kageyama et al., 2011; Salhi et al., 2008), but it is higher than those on QD lasers on on-Si substrates (Liu et al., 2015).

By narrowing the mesa width, we were able to achieve room-temperature and high-temperature CW operation, which was not achieved in our group's previous study (Kwoen et al., 2018). This is because narrow mesa structures result in more efficient heat dissipation and current constriction. Additionally, the maximum operating temperature and T_0 of our QD lasers were lower than those for previously reported QD lasers on-GaAs without p-doping (Liu et al., 2005). This is believed due to the increased non-radiative recombination associated with the relatively high TDD caused by the surface roughness-induced uniformity of the QD structure. However, implementing various approaches to reduce the TDD of GaAs on Si may have improved the thermal properties via optimization of the GaAs buffer layer's crystal quality.

Our QD lasers also demonstrated a buffer-layer TDD value that is approximately eight times higher than that for previously reported QD lasers on on-axis Si (001) substrates (Jung et al., 2018a,b), although we did not apply the modulation p-doping commonly used to improve the high-temperature characteristics. Conversely, the maximum operation temperature of our QD lasers was comparable to that of previously reported QD lasers. It is believed that the adoption of a GaAs capping structure and the high gain of the eight-layer stacking QD structure improve the thermal characteristics of the lasers fabricated for this study. Although it was not achieved in this study, we believe that a lower TDD in the buffer layer and modulation p-doping could improve the thermal characteristics of the lasers in this study.

We successfully demonstrated the high characteristic temperature and high-temperature CW operation of InAs/GaAs QD lasers directly grown on on-axis Si (001) substrates by using only MBE. Despite the relatively high TDD of the buffer layer and the absence of modulation p-doping, high-temperature operation was achieved with the fabricated lasers. This suggests that QD lasers are realistic candidates for integration into Si CMOS technology, which requires high-temperature operation.

6. Conclusion

In this chapter, we discussed recent advances in QD lasers for silicon photonics and demonstrated that QD lasers are indispensable laser sources for silicon photonics. After giving a brief overview of the progress in QD lasers, integration of QD lasers on silicon substrates by using flip-chip bonding was introduced as a semiconductor laser-integration method. We also improved the passive assembling method of the QD lasers via IR visual alignment between the marks on QD lasers and those on silicon photonics integrated circuits, demonstrating less than 3-dB coupling loss between QD lasers and silicon photonics integrated circuits. Furthermore, applications of this assembling method to optical I/O cores for board-to-board optical interconnection and silicon optical interposers for chip-to-chip optical interconnection were discussed. The silicon optical interposers mounted with the QD lasers achieved error-free data links at 12.5 Gbps operating from 25 to 125 °C without bias adjustments for the QD lasers, modulators and photodetectors.

Next we developed wafer bonding method, with which a hybrid Si III-V QD laser evanescently coupled to a silicon waveguide was fabricated. The InAs/GaAs QD laser structures with thin AlGaAs lower cladding layers were transferred onto silicon waveguides defining cavities with adiabatic taper structures and distributed Bragg reflectors. The laser operated at temperatures up to 115 °C under pulsed current conditions, with a T_0 of 303 K near room temperature. A hybrid DFB QD laser on silicon was also fabricated. Finally, we discussed an all-MBE grown InAs/GaAs QD laser on on-axis Si (001) substrates, demonstrating that highly efficient QD lasers can be fabricated using only MBE without MOCVD or substrate patterning. The fabricated QD laser was successfully operated at room temperature. This direct-growth method makes it possible to fabricate QD lasers monolithically on silicon (001) substrates using just a simple growth procedure.

The above three methods have paved the way toward integrating the QD lasers into silicon-based platforms as indispensable III-V laser devices. The QD lasers have established its position as the most promising light sources at 1.3 μm for silicon photonics technologies.

Acknowledgment

This chapter is based on results obtained from a project commissioned by the New Energy and Industrial Technology Development Organization (NEDO).

References

Alcotte, R., Martin, M., Moeyaert, J., Cipro, R., David, S., Bassani, F., Ducroquet, F., Bogumilowicz, Y., Sanchez, E., Ye, Z., Bao, X.Y., Pin, J.B., Baron, T., 2016. Epitaxial growth of antiphase boundary free GaAs layer on 300 mm Si(001) substrate by metalorganic chemical vapour deposition with high mobility. APL Mater. 4, 046101. https://doi.org/10.1063/1.4945586.

Arakawa, Y., Sakaki, H., 1982. Multidimensional quantum well laser and temperature dependence of its threshold current. Appl. Phys. Lett. 40, 939–941. https://doi.org/10.1063/1.92959.

Arakawa, Y., Yariv, A., 1986. Quantum well lasers—gain, spectra, dynamics. IEEE J. Quantum Electron. 22, 1887–1899. https://doi.org/10.1109/JQE.1986.1073185.

Arakawa, Y., Sakaki, H., Nishioka, M., Okamoto, H., Miura, N., 1983. Spontaneous emission characteristics of quantum well lasers in strong magnetic fields—an approach to quantum-well-box light source—. Jpn. J. Appl. Phys. 22, L804–L806. https://doi.org/10.1143/JJAP.22.L804.

Arakawa, Y., Vahala, K., Yariv, A., 1984. Quantum noise and dynamics in quantum well and quantum wire lasers. Appl. Phys. Lett. 45, 950–952. https://doi.org/10.1063/1.95453.

Arakawa, Y., Nakamura, T., Urino, Y., Fujita, T., 2013. Silicon photonics for next generation system integration platform. IEEE Commun. Mag. 51, 72. https://doi.org/10.1109/MCOM.2013.6476868.

Asada, M., Miyamoto, Y., Suematsu, Y., 1986. Gain and the threshold of three-dimensional quantum-box lasers. IEEE J. Quantum Electron. 22, 1915–1921. https://doi.org/10.1109/JQE.1986.1073149.

Bringans, R.D., Uhrberg, R.I.G., Olmstead, M.A., Bachrach, R.Z., 1986. Surface bands for single-domain 2×1 reconstructed Si(100) and Si(100):As. Photoemission results for off-axis crystals. Phys. Rev. B 34, 7447–7450. https://doi.org/10.1103/PhysRevB.34.7447.

Brückner, S., Kleinschmidt, P., Supplie, O., Döscher, H., Hannappel, T., 2013. Domain-sensitive *in situ* observation of layer-by-layer removal at Si(100) in H_2 ambient. New J. Phys. 15, 113049. https://doi.org/10.1088/1367-2630/15/11/113049.

Chang, H.-H., Fang, A.W., Sysak, M.N., Park, H., Cohen, O., Raday, O., Paniccia, M.J., Bowers, J.E., 2007. 1310nm silicon evanescent laser. Opt. Exp. 15, 11466–11471. https://doi.org/10.1364/OE.15.011466.

Chen, S., Li, W., Wu, J., Jiang, Q., Tang, M., Shutts, S., Elliott, S.N., Sobiesierski, A., Seeds, A.J., Ross, I., Smowton, P.M., Liu, H., 2016. Electrically pumped continuous-wave III–V quantum dot lasers on silicon. Nat. Photonics 10, 307–311. https://doi.org/10.1038/nphoton.2016.21.

Duan, J., Huang, H., Dong, B., Jung, D., Norman, J., Bowers, J., Grillot, F., 2019. 1.3-μm reflection insensitive InAs/GaAs quantum dot lasers directly grown on silicon. IEEE Photon. Tech. Lett., 31, 345–348. https://doi.org/10.1109/LPT.2019.2895049.

Duperez, H., Jany, C., Seassal, C., Bakir, B.B., 2016. Highly tunable heterogeneously integrated III-V on silicon sampled-grating distributed Bragg reflector lasers operating in the O-band. Opt. Exp. 24, 20895–20903. https://doi.org/10.1364/OE.24.020895.

Esaki, L., Tsu, R., 1970. Superlattice and negative differential conductivity in semiconductors. IBM J. Res. Dev. 14, 61–65.

Fang, A.W., Lively, E., Kuo, Y., Liang, D., Bowers, J.E., 2008. A distributed feedback silicon evanescent laser. Opt. Exp. 16, 4413–4419. https://doi.org/10.1364/OE.16.004413.

Freisem, S., Ozgur, G., Shavritranuruk, K., Chen, H., Deppe, D.G., 2008. Very-low-threshold current density continuous-wave quantum-dot laser diode. Electron. Lett. 44, 679. https://doi.org/10.1049/el:20080656.

Ganga, I., 2008. Chief Editor's Report. Presented at the IEEE P802.3ba Task Force Meeting.

Glas, G., Guille, G., Henoc, P., Houzay, F., 1987. TEM Study of the Molecular Beam Epitaxy Island Growth of InAs on GaAs. In: Institute of Physics Conference Series 87, p. 71.

Goldstein, L., Glas, F., Marzin, J.Y., Charasse, M.N., Le Roux, G., 1985. Growth by molecular beam epitaxy and characterization of InAs/GaAs strained-layer superlattices. Appl. Phys. Lett. 47, 1099–1101. https://doi.org/10.1063/1.96342.

Guha, S., Madhukar, A., Rajkumar, K.C., 1990. Onset of incoherency and defect introduction in the initial stages of molecular beam epitaxical growth of highly strained $In_x Ga_{1-x} As$ on GaAs(100). Appl. Phys. Lett. 57, 2110–2112. https://doi.org/10.1063/1.103914.

Huyet, G., O'Brien, D., Hegarty, S.P., McInerney, J.G., Uskov, A.V., Bimberg, D., Ribbat, C., Ustinov, V.M., Zhukov, A.E., Mikhrin, S.S., Kovsh, A.R., White, J.K., Hinzer, K., SpringThorpe, A.J., 2004. Quantum dot semiconductor lasers with optical feedback. Phys. Status Solidi (A) 201, 345–352. https://doi.org/10.1002/pssa.200303971.

Jang, B., Tanabe, K., Kako, S., Iwamoto, S., Tsuchizawa, T., Nishi, H., Hatori, N., Noguchi, M., Nakamura, T., Takemasa, K., Sugawara, M., Arakawa, Y., 2016. A hybrid silicon evanescent quantum dot laser. Appl. Phys. Express 9, 092102. https://doi.org/10.7567/APEX.9.092102.

Jhang, Y.-H., Tanabe, K., Iwamoto, S., Arakawa, Y., 2015. InAs/GaAs quantum dot lasers on silicon-on-insulator substrates by metal-stripe wafer bonding. IEEE Photon. Technol. Lett. 27, 875–878. https://doi.org/10.1109/LPT.2015.2398465.

Jhang, Y.-H., Mochida, R., Tanabe, K., Takemasa, K., Sugawara, M., Iwamoto, S., Arakawa, Y., 2016. Direct modulation of 1.3 μm quantum dot lasers on silicon at 60 °C. Opt. Express 24, 18428. https://doi.org/10.1364/OE.24.018428.

Jung, D., Callahan, P.G., Shin, B., Mukherjee, K., Gossard, A.C., Bowers, J.E., 2017. Low threading dislocation density GaAs growth on on-axis GaP/Si (001). J. Appl. Phys. 122, 225703. https://doi.org/10.1063/1.5001360.

Jung, D., Herrick, R., Norman, J., Turnlund, K., Jan, C., Feng, K., Gossard, A.C., Bowers, J.E., 2018a. Impact of threading dislocation density on the lifetime of InAs quantum dot lasers on Si. Appl. Phys. Lett. 112, 153507. https://doi.org/10.1063/1.5026147.

Jung, D., Zhang, Z., Norman, J., Herrick, R., Kennedy, M.J., Patel, P., Turnlund, K., Jan, C., Wan, Y., Gossard, A.C., Bowers, J.E., 2018b. Highly reliable low-threshold InAs quantum dot lasers on on-Axis (001) Si with 87% injection efficiency. ACS Photon. 5, 1094–1100. https://doi.org/10.1021/acsphotonics.7b01387.

Kageyama, T., Nishi, K., Yamaguchi, M., Mochida, R., Maeda, Y., Takemasa, K., Tanaka, Y., Yamamoto, T., Sugawara, M., Arakawa, Y., 2011. Extremely high temperature (220°C) continuous-wave operation of 1300-nm-range quantum-dot lasers. 2011 Conference on Lasers and Electro-Optics Europe and 12th European Quantum Electronics Conference (CLEO EUROPE/EQEC). https://doi.org/10.1109/CLEOE.2011.5943701.

Kirstaedter, N., Ledentsov, N.N., Grundmann, M., Bimberg, D., Ustinov, V.M., Ruvimov, S.S., Maximov, M.V., Kop'ev, P.S., Alferov, Z.I., Richter, U., Werner, P., Gösele, U., Heydenreich, J., 1994. Low threshold, large T0 injection laser emission from (InGa)As quantum dots. Electron. Lett. 30, 1416–1417. https://doi.org/10.1049/el:19940939.

Kwoen, J., Jang, B., Lee, J., Kageyama, T., Watanabe, K., Arakawa, Y., 2018. All MBE grown InAs/GaAs quantum dot lasers on on-axis Si (001). Opt. Express 26, 11568. https://doi.org/10.1364/OE.26.011568.

Kwoen, J., Lee, J., Watanabe, K., Arakawa, Y., 2019. Elimination of anti-phase boundaries in a GaAs layer directly-grown on an on-axis Si(001) substrate by optimizing an AlGaAs nucleation layer. Jpn. J. Appl. Phys. 58. https://doi.org/10.7567/1347-4065/aaffc2. SBBE07.

Leonard, D., Krishnamurthy, M., Reaves, C.M., Denbaars, S.P., Petroff, P.M., 1993. Direct formation of quantum-sized dots from uniform coherent islands of InGaAs on GaAs surfaces. Appl. Phys. Lett. 63, 3203–3205. https://doi.org/10.1063/1.110199.

Li, Q., Lau, K.M., 2017. Epitaxial growth of highly mismatched III-V materials on (001) silicon for electronics and optoelectronics. Prog. Cryst. Growth Charact. Mater. 63, 105–120. https://doi.org/10.1016/j.pcrysgrow.2017.10.001.

Li, Q., Wan, Y., Liu, A.Y., Gossard, A.C., Bowers, J.E., Hu, E.L., Lau, K.M., 2016. 1.3-μm InAs quantum-dot micro-disk lasers on V-groove patterned and unpatterned (001) silicon. Opt. Express 24, 21038. https://doi.org/10.1364/OE.24.021038.

Li, W., Chen, S., Tang, M., Wu, J., Hogg, R., Seeds, A., Liu, H., Ross, I., 2018. Effect of rapid thermal annealing on threading dislocation density in III-V epilayers monolithically grown on silicon. J. Appl. Phys. 123, 215303. https://doi.org/10.1063/1.5011161.

Liu, H.Y., Childs, D.T., Badcock, T.J., Groom, K.M., Sellers, I.R., Hopkinson, M., Hogg, R.A., Robbins, D.J., Mowbray, D.J., Skolnick, M.S., 2005. High-performance three-layer 1.3-μm InAs-GaAs quantum-dot lasers with very low continuous-wave room-temperature threshold currents. IEEE Photon. Technol. Lett. 17, 1139–1141. https://doi.org/10.1109/LPT.2005.846948.

Liu, A.Y., Srinivasan, S., Norman, J., Gossard, A.C., Bowers, J.E., 2015. Quantum dot lasers for silicon photonics [invited]. Photon. Res. 3, B1. https://doi.org/10.1364/PRJ.3.0000B1.

Liu, H., Wang, Q., Chen, J., Liu, K., Ren, X., 2016. MOCVD growth and characterization of multi-stacked InAs/GaAs quantum dots on misoriented Si(100) emitting near 1.3 μm. J. Cryst. Growth 455, 168–171. https://doi.org/10.1016/j.jcrysgro.2016.09.033.

Martin, M., Caliste, D., Cipro, R., Alcotte, R., Moeyaert, J., David, S., Bassani, F., Cerba, T., Bogumilowicz, Y., Sanchez, E., Ye, Z., Bao, X.Y., Pin, J.B., Baron, T., Pochet, P., 2016. Toward the III–V/Si co-integration by controlling the biatomic steps on hydrogenated Si(001). Appl. Phys. Lett. 109, 253103. https://doi.org/10.1063/1.4972394.

Mizutani, K., Yashiki, K., Kurihara, M., Suzuki, Y., Hagihara, Y., Hatori, N., Shimizu, T., Urino, Y., Nakamura, T., Kurata, K., Arakawa, Y., 2015. Optical I/O core transmitter with high tolerance to optical feedback using quantum dot laser. In: 2015 European Conference on Optical Communication (ECOC). Presented at the 2015 European Conference on Optical Communication (ECOC), IEEE, Valencia, Spain, pp. 1–3. https://doi.org/10.1109/ECOC.2015.7341763.

Nakamura, T., Yashiki, K., Mizutani, K., Nedachi, T., Fujikata, J., Tokushima, M., Ushida, J., Noguchi, M., Okamoto, D., Suzuki, Y., Shimizu, T., Takemura, K., Ukita, A., Ibusuki, Y., Kurihara, M., Kinoshita, K., Horikawa, T., Yamaguchi, H., Tsuchida, J., Hagihara, Y., Kurata, K., 2019. Fingertip-size optical module, "optical I/O core", and its application in FPGA. IEICE Trans. Electron. E102.C, 333–339.

Nakayama, T., Tanishiro, Y., Takayanagi, K., 1987. Biatomic layer-high steps on Si(001)2 × 1 surface. Jpn. J. Appl. Phys. 26, L280–L282. https://doi.org/10.1143/JJAP.26.L280.

Oshinowo, J., Nishioka, M., Ishida, S., Arakawa, Y., 1994. Highly uniform InGaAs/GaAs quantum dots (~15 nm) by metalorganic chemical vapor deposition. Appl. Phys. Lett. 65, 1421–1423. https://doi.org/10.1063/1.112070.

Otsubo, K., Hatori, N., Ishida, M., Okumura, S., Akiyama, T., Nakata, Y., Ebe, H., Sugawara, M., Arakawa, Y., 2004. Temperature-insensitive eye-opening under 10-Gb/s modulation of 1.3-μm P-doped quantum-dot lasers without current adjustments. Jpn. J. Appl. Phys. 43, L1124–L1126. https://doi.org/10.1143/JJAP.43.L1124.

Otto, C., Globisch, B., Lüdge, K., Schöll, E., Erneux, T., 2012. Compex dynamics of semiconductor quantum dot lasers subject to delayed optical feedback. Int. J. Bifurc. Chaos 22, 1250246. https://doi.org/10.1142/S021812741250246X.

Salhi, A., Rainò, G., Fortunato, L., Tasco, V., Martiradonna, L., Todaro, M.T., De Giorgi, M., Cingolani, R., Passaseo, A., Luna, E., Trampert, A., De Vittorio, M., 2008. Linear increase of the modal gain in 1.3 μm InAs/GaAs quantum dot lasers containing up to seven-stacked QD layers. Nanotechnology 19, 275401. https://doi.org/10.1088/0957-4484/19/27/275401.

Shimizu, T., Hatori, N., Ishizaka, M., Urino, Y., Yamamoto, T., Nakamura, T., Okano, M., Mori, M., Arakawa, Y., 2013. Hybrid integration technology of laser source with laser diode arrays on silicon optical waveguide platform by flip-chip bonding for silicon photonics. In: 2013 3rd IEEE CPMT Symposium Japan. Presented at the 2013 IEEE CPMT Symposium Japan (Formerly VLSI Packaging Workshop of Japan), IEEE, Kyoto, Japan, pp. 1–4. https://doi.org/10.1109/ICSJ.2013.6756091.

Shimizu, T., Ishizaka, M., Hatori, N., Okano, M., Yamamoto, T., Mori, M., Urino, Y., Nakamura, T., Arakawa, Y., 2014. Multi-channel hybrid integrated light source for ultra-high-bandwidth optical interconnections and its structural optimization for low power consumption by considering thermal interference between LD arrays. Trans. Jpn. Inst. Electron. Packag. 7, 94–103. https://doi.org/10.5104/jiepeng.7.94.

Shoji, H., Mukai, K., Ohtsuka, N., Sugawara, M., Uchida, T., Ishikawa, H., 1995. Lasing at three-dimensionally quantum-confined sublevel of self-organized in/sub 0.5/Ga/sub 0.5/As quantum dots by current injection. IEEE Photon. Technol. Lett. 7, 1385–1387. https://doi.org/10.1109/68.477257.

Snyder, C.W., Orr, B.G., Kessler, D., Sander, L.M., 1991. Effect of strain on surface morphology in highly strained InGaAs films. Phys. Rev. Lett. 66, 3032–3035. https://doi.org/10.1103/PhysRevLett.66.3032.

Stankovic, S., Jones, R., Sysak, M.N., Heck, J.M., Roelkens, G., Van Thourhout, D., 2012. Hybrid III–V/Si distributed-feedback laser based on adhesive bonding. IEEE Photon. Tech. Lett. 24, 2155–2158. https://doi.org/10.1109/LPT.2012.2223666.

Tabuchi, M., Noda, S., Sasaki, A., 1992. Mesoscopic structure in lattice-mismatched heteroepitaxial interface layers. In: Namba, S., Hamaguchi, C., Ando, T. (Eds.), Science and Technology of Mesoscopic Structures. Springer Japan, Tokyo, pp. 379–384. https://doi.org/10.1007/978-4-431-66922-7_40.

Tanabe, K., Guimard, D., Bordel, D., Iwamoto, S., Arakawa, Y., 2010. Electrically pumped 1.3 μm room-temperature InAs/GaAs quantum dot lasers on Si substrates by metal-mediated wafer bonding and layer transfer. Opt. Express 18, 10604. https://doi.org/10.1364/OE.18.010604.

Tanabe, K., Watanabe, K., Arakawa, Y., 2012. III-V/Si hybrid photonic devices by direct fusion bonding. Sci. Rep. 2. https://doi.org/10.1038/srep00349.

Urino, Y., Akiyama, S., Akagawa, T., Baba, T., Usuki, T., Okamoto, D., Miura, M., Fujikata, J., Shimizu, T., Okano, M., Hatori, N., Ishizaka, M., Yamamoto, T., Takahashi, H., Noguchi, Y., Noguchi, M., Imai, M., Yamagishi, M., Saitou, S., Hirayama, N., Takahashi, M., Saito, E., Shimura, D., Okayama, H., Onawa, Y., Yaegashi, H., Nishi, H., Fukuda, H., Yamada, K., Mori, M., Horikawa, T., Nakamura, T., Arakawa, Y., 2013. Demonstration of 30-TBPS/cm2 bandwidth density by silicon optical interposers fully integrated with optical components. In: 39th European Conference and Exhibition on Optical Communication (ECOC 2013) Mo.4.B.2. https://doi.org/10.1049/cp.2013.1305.

Urino, Y., Usuki, T., Fujikata, J., Ishizaka, M., Yamada, K., Horikawa, T., Nakamura, T., Arakawa, Y., 2014. High-density and wide-bandwidth optical interconnects with silicon optical interposers. Photon. Res. 2, A1–A7. https://doi.org/10.1364/PRJ.2.0000A1.

Urino, Y., Hatori, N., Mizutani, K., Usuki, T., Fujikata, J., Yamada, K., Horikawa, T., Nakamura, T., Arakawa, Y., 2015. First demonstration of athermal silicon optical interposers with quantum dot lasers operating up to 125°C. J. Lightwave Technol. 33, 1223–1229. https://doi.org/10.1109/JLT.2014.2380811.

Volz, K., Beyer, A., Witte, W., Ohlmann, J., Németh, I., Kunert, B., Stolz, W., 2011. GaP-nucleation on exact Si (001) substrates for III/V device integration. J. Cryst. Growth 315, 37–47. https://doi.org/10.1016/j.jcrysgro.2010.10.036.

Wan, Y., Jung, D., Norman, J., Shang, C., MacFarlane, I., Li, Q., Kennedy, M.J., Gossard, A.C., Lau, K.M., Bowers, J.E., 2017. O-band electrically injected quantum dot micro-ring lasers on on-axis (001) GaP/Si and V-groove Si. Opt. Express 25, 26853. https://doi.org/10.1364/OE.25.026853.

Yariv, A., 1997. Optical electronics in modern communications. In: The Oxford Series in Electrical and Computer Engineering, fifth ed. Oxford University Press, New York, p. 533.

Yashiki, K., Suzuki, Y., Hagihara, Y., Kurihara, M., Tokushima, M., Fujikata, J., Ukita, A., Takemura, K., Shimizu, T., Okamoto, D., Ushida, J., Takahashi, S., Uemura, T., Okano, M., Tsuchida, J., Nedachi, T., Fushimi, M., Ogura, I., Inasaka, J., Kurata, K., 2015. 5 mW/Gbps hybrid-integrated Si-photonics-based optical I/O cores and their

25-Gbps/ch error-free operation with over 300-m MMF. In: Optical Fiber Communication Conference. Presented at the Optical Fiber Communication Conference, OSA, Los Angeles, California. p.Th1G.1. https://doi.org/10.1364/OFC.2015.Th1G.1.

Yashiki, K., Mizutani, K., Ushida, J., Suzuki, Y., Kurihara, M., Tokushima, M., Fujikata, J., Hagihara, Y., Kurata, K., 2016. 25-Gbps error-free operation of chip-scale Si-photonics optical transmitter over 70 °C with integrated quantum dot laser 3.

Further reading

Nishi, K., Takemasa, K., Sugawara, M., Arakawa, Y., 2017. Development of quantum dot lasers for data-com and silicon photonics applications. IEEE J. Sel. Top. Quant. Electron. 23, 1. https://doi.org/10.1109/JSTQE.2017.2699787.

CHAPTER FOUR

Microchannel epitaxy of III–V layers on Si substrates

Shigeya Naritsuka*
Department of Materials Science and Engineering, Faculty of Science and Technology, Meijo University, Nagoya, Japan
*Corresponding author: e-mail address: narit@meijo-u.ac.jp

Contents

1. Introduction

Dislocations nucleated during highly-mismatched heteroepitaxy are a challenging problem during the fabrication of various devices, and reducing the dislocation density is required to ensure suitable levels of performance and reliability. Microchannel epitaxy (MCE) is a specialized technique that effectively eliminates dislocations in active areas (Nishinaga, 2002; Nishinaga et al., 1988). In MCE, a narrow opening, termed a "microchannel," is cut into a mask and used as a seed for subsequent growth. Consequently, dislocations in the buffer layer are eliminated from the top or the side surfaces of the grown layer, as shown in Fig. 1. This occurs because dislocations tend to propagate in straight lines as this minimizes their energy of formation. The dislocation reduction strategy in MCE is simple but quite effective; only the crystal information is transferred to the newly grown layer, without transference of dislocations. In the case of horizontal MCE (H-MCE), dislocations are eliminated on the top surface of the grown layer, as in Fig. 1A and, as a result, the laterally-grown areas are dislocation-free and are suitable

Semiconductors and Semimetals, Volume 101
ISSN 0080-8784
https://doi.org/10.1016/bs.semsem.2019.05.001

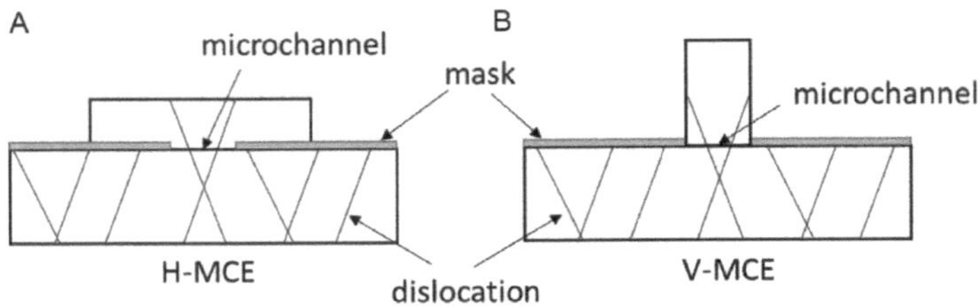

Fig. 1 Two-types of microchannel epitaxy: horizontal microchannel epitaxy (H-MCE) (A) and vertical microchannel epitaxy (V-MCE) (B). *From Naritsuka, S., 2016. Microchannel epitaxy. Prog. Cryst. Growth Charact. Mater. 62, 302–316; Copyright (2016), with permission from Elsevier.*

Strategy for excellent MCE

- Selective growth
  - Mask selection
    - Weak or no-interaction to growth species
  - Low supersaturation
    - Suppression of 2D and 3D nucleations
  - Thermal re-evaporation of adatoms
- Directional growth
  - Facet formation
    - LPE, MBE
  - Shadow effect
    - MBE (V-MCE, BEILE)
  - Low angle incidence of molecular beam
    - MBE (LAIMCE)

Fig. 2 Strategy for large aspect ratio in MCE.

for device fabrication. In contrast, dislocations are terminated on the side surfaces when applying vertical-MCE (V-MCE), as in Fig. 1B. The active regions of such devices are fabricated on the top of the grown layer.

Increasing the area suitable for device fabrication will require larger dislocation-free zones and reduced defective areas, and this can be realized by the growth of a thin yet wide H-MCE layer, which therefore has a large width (W) to thickness (T) ratio. The strategy for the excellent MCE is summarized in Fig. 2. Though directional growth is the key issue in the strategy, selective growth is also important and indispensable to accomplishing superior directional growth. Mask selection with weak or without interaction to growth species and setting of low supersaturation for growth are very important to secure selective growth. After accomplishing selective growth, direct growth is realized by the use of the several techniques, such as facet formation, shadow effect and low angle incidence of molecular beams. A nano-wire is a well-known example of directional growth, but is one-dimensional (Tomioka and Fukui, 2015), whereas MCE requires two-dimensional directional growth. The formation of facets is used to meet this goal in liquid phase epitaxy (LPE), while a shadow mask is employed in molecular beam epitaxy (MBE). On the other hand, perfect selective growth is readily achieved when using equilibrium methods, such as LPE, but is difficult in the case of non-equilibrium methods. The strategy of the non-equilibrium methods also demonstrates in Fig. 2 that both supersaturation control and mask selection are important to ensure the complete re-evaporative growth of species from the growing surface so as to obtain perfect selectivity. Full removal of the adatoms is necessary to suppressing three-dimensional nucleation on the mask and realizing superior two-dimensional directional growth.

In LPE, the growth rate largely depends on the crystal orientation. This occurs because the growth proceeds in a step flow mode, where the forward movement of the surface steps produces vertical growth via the formation of mono-atomic layers. Therefore, the step sources control the growth rate. If the degree of supersaturation is less than the critical value required for two-dimensional nucleation, which is one of the most strong step sources, the growth rate for low-index planes becomes very slow. Conversely, atomically-rough surfaces will exhibit a high growth rate even at a low supersaturation because all growth species will readily attach to the surface. Therefore, it is evident that the aspect ratio of the directional growth can be maximized by the appropriate choice of the microchannel direction and the substrate plane. This scenario will generate a rough surface on the sides and a low-index plane on the top surface.

In the case of non-equilibrium growth methods such as MBE, several techniques have been proposed to realize superior directional growth, as shown in Fig. 3. One is cavity growth, in which a cavity with a mask is used to generate vertical growth by MBE (Matsunaga et al., 2002). This process represents a type of shadow mask growth, where the mask is used to restrict the growth area to a narrow opening to ensure solely vertical growth. In beam-induced lateral epitaxy (BILE) (Suzuki et al., 2003), the shadow of the neighboring mesa is used to restrict the growth area to the top part of the mesa's side. A molecular beam having a low angle of incidence will be partly interrupted by the top of the neighboring mesa, thus reducing the intensity of the beam at the bottom of the next mesa. This effect produces both selective growth and eaves-like overhanging lateral growth on the side of the adjacent mesa.

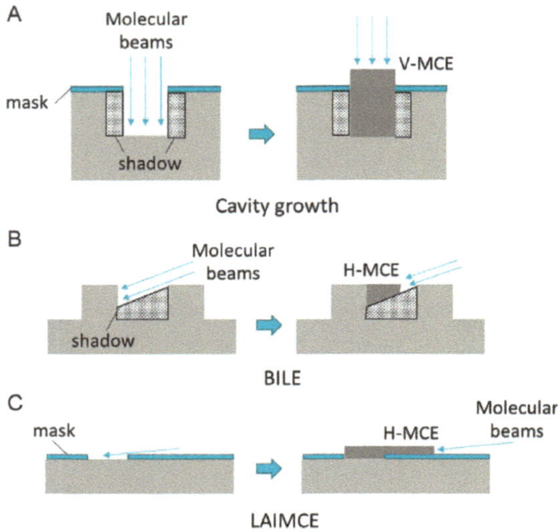

Fig. 3 Way of directional growth in MCE by MBE: (A) cavity growth, (B) beam-induced lateral epitaxy (BILE), and (C) low angle incidence microchannel epitaxy (LAIMCE).

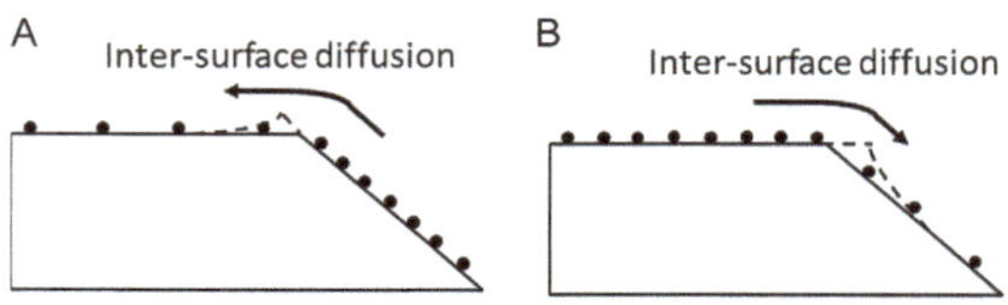

Fig. 4 Illustration for inter-surface diffusion: (A) from side to top surfaces and (B) from top to side surfaces.

A low molecular beam angle of incidence is itself useful in terms of achieving lateral growth even without the assistance of a mesa structure. This technique is known as low angle incidence microchannel epitaxy (LAIMCE) (Bacchin and Nishinaga, 2000). Using this approach, lateral growth is effectively enhanced by supplying the molecular beam at a low angle. The inter-surface diffusion of adatoms between adjacent crystal planes is also an important factor in the exceptional lateral growth obtained with both BILE and LAIMCE (Nishinaga et al., 1996; Shitara and Nishinaga, 1989; Yamashiki et al., 1997), driven by variations in adatom concentrations. As an example, when the adatom concentration on the sides is greater than on the top plane, adatom diffusion occurs from the sides to the top along the planar boundaries, as shown in Fig. 4A. Consequently, the side zones expand to form a facet. Conversely, if the adatom concentration is greater on the top plane than on the side planes, the adatoms move from the top to the sides and the growth rate is increased in the vicinity of the top plane boundaries, and the top plane gradually expands, as shown in Fig. 4B. Although the range of inter-surface diffusion is limited to the surface diffusion length, this technique is effective in conjunction with MCE when micro-scale phenomena are important. The adatom concentration is determined by the balance between the supply and consumption of adatoms on the crystal surface (i.e., the balance between the rates at which the raw material is input and at which it is incorporated into the crystal), and can be controlled by the growth conditions. The effective exploitation of inter-surface diffusion during H-MCE also enables the migratory diffusion of top-plane adatoms to the growth side plane. This technique both reduces the growth rate in the vertical direction and accelerates the growth rate in the lateral direction.

The following sections describe experimental demonstrations obtained using various MCE techniques. Dislocation reduction mechanisms are also discussed and categorized as either equilibrium or non-equilibrium methods.

2. Equilibrium method

2.1 LPE

2.1.1 GaAs H-MCE/Si

GaAs MCE was carried out on a GaAs-coated Si substrate using LPE (Chang et al., 1998), and Fig. 5A presents an optical microphotograph of the resulting wide MCE layer after KOH etching. The LPE conditions consisted of a start temperature, initial supersaturation, cooling rate, and growth time of 530 °C, 1.0 °C, 0.1 °C/min, and 3 h, respectively. A high density of etch pits are clearly evident in the region above the microchannel. These dislocations on the MCE surface had propagated from the surface of the GaAs-coated Si substrate through the narrow microchannel, as illustrated in Fig. 5B. However, no etch pits are observed in the laterally overgrown region, because the propagation of dislocations in the buffer layer was terminated by the SiO_2 mask. The dislocations in the microchannel moved through the MCE layer to reach the top surface of the layer. The straightforward progression of these dislocations ensures that they are confined to the center region of the layer. The width of the etch-pit-free region is approximately 43 μm, which is sufficient for the fabrication of devices such as Vertical-Cavity Surface Emitting Lasers (VCSELs).

Here, it is helpful to briefly describe the H-MCE growth mechanism (Chang et al., 1997). Fig. 5C presents a plot summarizing the typical dependence of the growth rate on supersaturation, in which three general growth modes are evident. When the growth surface is atomically rough, the growth rate increases linearly with supersaturation, as in curve A. In this mode, the rate at which atoms are incorporated is very fast because all the atoms arriving at

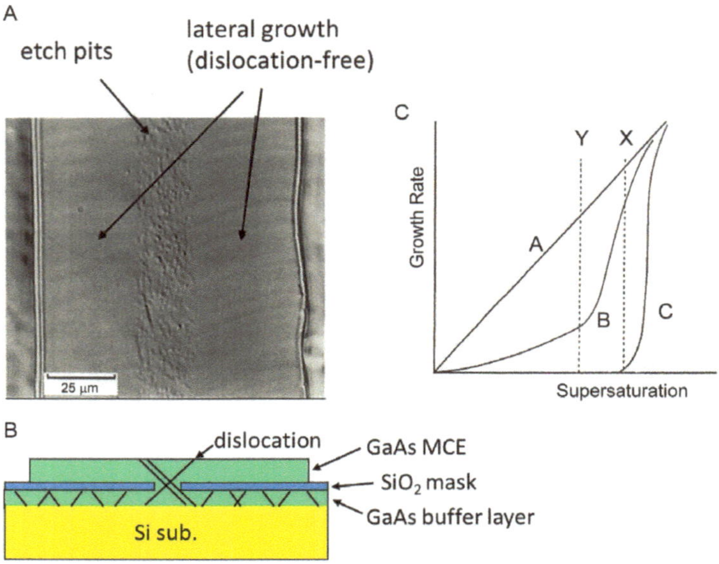

Fig. 5 GaAs MCE on GaAs-coated (001) Si substrate: Optical microscope image of H-MCE layer after KOH etching (A), illustration of GaAs H-MCE on GaAs-coated Si substrate (B), and growth rate dependences of typical growth modes on supersaturation in LPE (C). *Panel (A): From Chang, Y.S., Naritsuka, S., Nishinaga, T., 1998. Optimization of growth condition for wide dislocation-free GaAs on Si substrate by microchannel epitaxy. J. Crystal Growth 192, 18–22; Copyright (1998), with permission from Elsevier; Panel (C): From Chang, Y.S., Naritsuka, S., Nishinaga, T., 1997. Effect of growth temperature on epitaxial lateral overgrowth of GaAs on Si substrate. J. Cryst. Growth 174, 630–634; Copyright (1997), with permission from Elsevier.*

the rough surface are incorporated immediately. The growth denoted by curve B results from a constant supply of steps based on screw dislocations originating from the substrate. Growth by the third mechanism (curve C) is induced by two-dimensional nucleation and can be prevented by maintaining a low degree of supersaturation. In the present MCE process, the growth in the vertical direction proceeds via mode B as screw dislocations from the MBE-grown buffer layer supply growth steps. In contrast, the lateral growth occurs via mode A since the side surfaces of the MCE layer are atomically rough based on the appropriate choice of the microchannel direction. From Fig. 5C, it is apparent that decreasing the supersaturation from X to Y produces a greater difference in the lateral and vertical growth rates. Thus, by optimizing the supersaturation, this difference can be maximized so as to obtain wide lateral growth with a large W/T ratio.

Fig. 6A presents a transmission electron microscopy (TEM) image of the microchannels (Nishinaga, 2002), in which a large number of dislocations can be seen in the buffer layer. The microchannel sections show that these

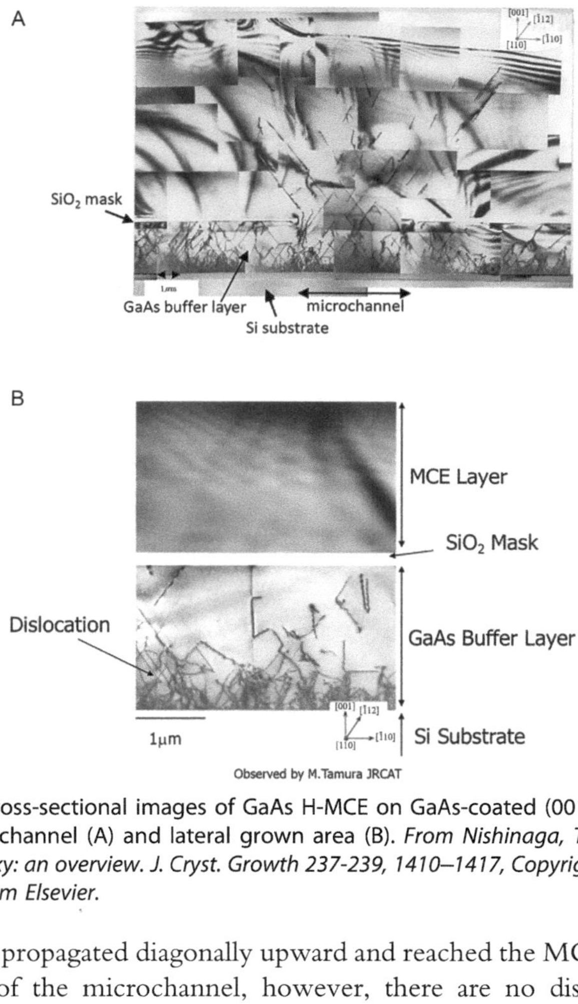

Fig. 6 TEM cross-sectional images of GaAs H-MCE on GaAs-coated (001) Si substrate: around microchannel (A) and lateral grown area (B). *From Nishinaga, T., 2002. Microchannel epitaxy: an overview. J. Cryst. Growth 237-239, 1410–1417, Copyright (2002); with permission from Elsevier.*

dislocations propagated diagonally upward and reached the MCE layer. On both sides of the microchannel, however, there are no dislocations, as dislocations present in the buffer layer are blocked by the mask and do not propagate to the lateral growth regions further from the microchannels (Fig. 6B). Consequently, dislocation-free areas are obtained. These results demonstrate that MCE is extremely effective at reducing dislocations. The study of growth mechanism of MCE was further proceeded and the relation between the W/T ratio and the dislocation density in the substrate was already reported (Chang et al., 2010).

2.1.2 AlGaAs-based laser array fabrication on Si

The following describes the fabrication of one of the key devices for opto-electronic integrated circuits (OEICs), that is, an AlGaAs laser array

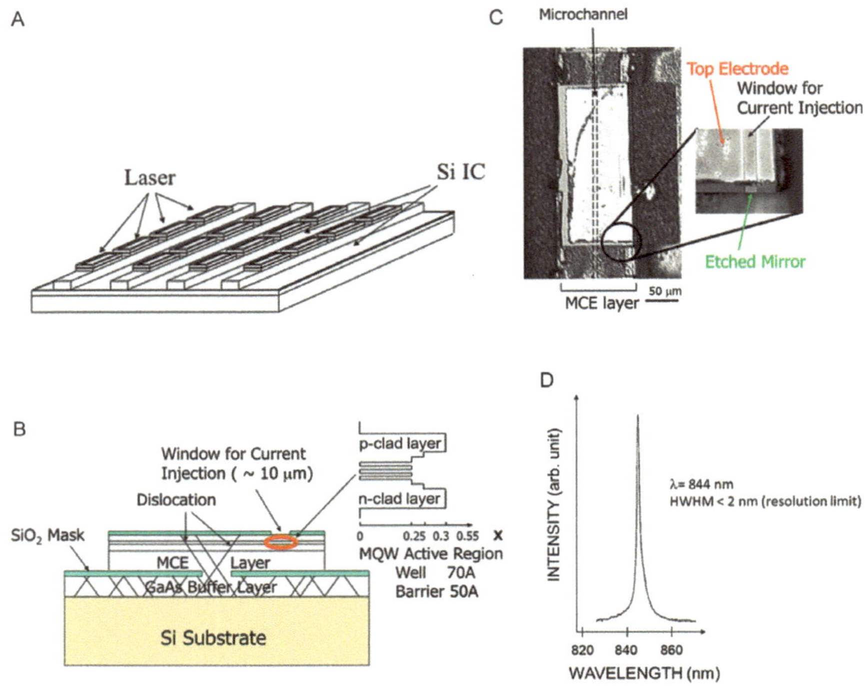

Fig. 7 Illustration of laser array fabricated on GaAs MCE layer on Si substrate (A), cross-sectional illustration of laser structure fabricated on GaAs MCE layer on Si substrate with layer structure of laser (B), optical microscope image of individual laser in laser array on GaAs MCE on Si (C), and light output vs. pulsed-current characteristic of laser (D). *From Naritsuka, S., 2016. Microchannel epitaxy. Prog. Cryst. Growth Charact. Mater. 62, 302–316; Copyright (2016), with permission from Elsevier.*

on Si substrate. The lasers were fabricated on GaAs dislocation-free areas grown by MCE (Naritsuka, 2016; Naritsuka et al., 2001). An illustration of the laser array is provided in Fig. 7A. It shows the array of lasers lined up in a series for the MCE layers arranged in parallel. These lasers can be fabricated entirely by the planar process including edge-emitting mirror formation through reactive ion etching (RIE), which is compatible with the processes used in the fabrication of normal ICs.

The lasers have an edge-emitting type structure with mirrors on both ends, and active layers (shown by ovals) on dislocation-free areas produced by MCE run parallel to the microchannels, as shown in the cross-sectional illustration of Fig. 7B. Mirrors were formed by RIE etching at both ends of the active layers to form optical resonators. The laser's active layer is multi quantum well (MQW), which consists of three GaAs quantum wells and four $Al_{0.25}Ga_{0.75}As$ barriers. The MQW is sandwiched on both sides by

$Al_{0.27}Ga_{0.73}As$ side layers to increase the injection efficiency of the carriers. A p-type cladding layer exists above the active layer and an n-type cladding layer on the bottom. An n-electrode is fabricated on the back of the substrate, while the p-electrode on the top. The optical confinement is carried out by a gain-guided structure, which is formed by the striped openings with 10-μm widths on the SiO_2 mask for injection of electric current into p-type cladding layer.

A micrograph of the upper plane of the fabricated laser structure is shown in Fig. 7C. Compared to the central microchannel, the striped opening for electric-current injection exists slightly on the right side for escaping the dislocation area. The mirrors are formed above and below of the laser by RIE. The enlarged view in the figure presents the top gold electrode, the opening established for the SiO_2 mask, and the edge-emitting mirror.

A spectra of laser oscillation at room-temperature by pulsed current are shown in Fig. 7D. Laser oscillation occurred at a threshold current of approximately 135 mA and an oscillation wavelength of 844 nm, with the cavity length of 250 μm.

2.1.3 InP H-MCE/Si

Fig. 8A shows a cross-sectional micrograph of an H-MCE InP layer on an InP-coated Si substrate (Naritsuka and Nishinaga, 1995a). As seen in the figure, a wide InP H-MCE layer (having a width of approximately 90 μm) with a relatively minimal thickness (about 6 μm) was obtained.

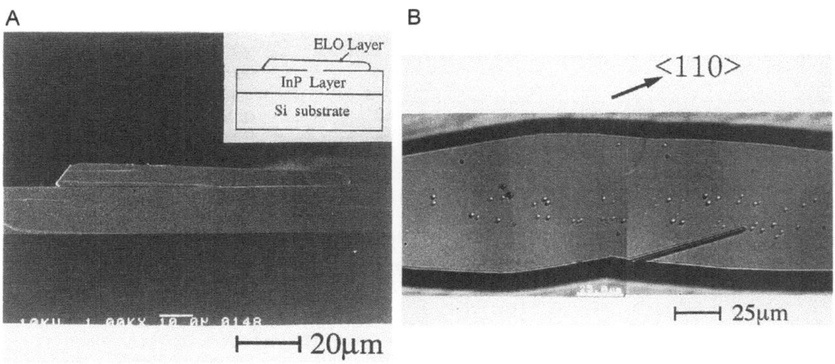

Fig. 8 Cross-sectional SEM image of InP H-MCE on InP-coated (001) Si substrate (A), and optical microscope image of InP H-MCE layer after etching (B). *From Naritsuka, S., Nishinaga, T., 1995a. InP layer grown on (001) silicon substrate by epitaxial lateral overgrowth. Jpn. J. Appl. Phys. 34, L1432–L1435; Copyright (1995) The Japan Society of Applied Physics.*

The aspect ratio of this sample was 15, which is almost equal to values obtained in the case of homoepitaxial specimens (Naritsuka and Nishinaga, 1995b). This result confirms that a good InP MCE layer can be grown on an InP-coated Si substrate if the appropriate H-MCE growth conditions are selected. Moreover, since the H-MCE layer shows no evidence of facet formation on the front edge of the H-MCE, it is likely that an even wider H-MCE could be obtained simply by extending the growth time.

InP H-MCE layers grown on InP-coated Si substrates were chemically etched to ascertain the dislocation densities; an H-MCE layer after etching is shown in Fig. 8B. It can be seen that the etch pits align along two parallel lines, which may correspond to the two sides of the line seed. A few etch pits are also observed in the lateral overgrowth regions, along with the presence of a stacking fault on an inclined {111}B plane. In contrast, no stacking faults on inclined {111}A planes were observed in any of the etched samples. To date, GaAs layers grown on Si substrates have been reported to have several dislocation types, including 30°, 60° and screw type (Tamura et al., 1991). All these dislocations, except the screw type, glide on {111} planes. Based on the similarity in the dislocations in the InP and GaAs, the etch pits aligned in two lines in Fig. 8B appear to indicate the propagation of 60° type dislocations on {111} planes through the line seed. The fact that there are rather few dislocations appearing in the area just above the line seed opening also suggests that the densities of screw and 30° type dislocations were rather low.

Fig. 8B also shows the appearance of dislocation-free InP areas, having widths of approximately 30 μm, in the outer regions of the dislocation areas. This finding indicates that the SiO_2 film between the lateral overgrowth layer and the substrate effectively prevented the propagation of the dislocations from the substrate. Thus, dislocation-free InP areas were obtained on Si substrates, and such specimens are expected to have applications in various devices.

The spatially resolved photoluminescence (SRPL) spectrum of an InP MCE layer is provided in Fig. 9. This figure also includes the SRPL spectra of an InP-coated Si substrate and an InP reference sample grown directly by LPE on an InP substrate under similar growth conditions. The peak wavelength of the H-MCE layer was 880 nm, which was almost the same as that of the reference sample, confirming minimal stress in the MCE layer. That is, the H-MCE process was able to decrease the stress caused by both the lattice mismatch and the difference in thermal expansion coefficients. The strong SRPL intensity relative to that of the reference sample and the narrow full

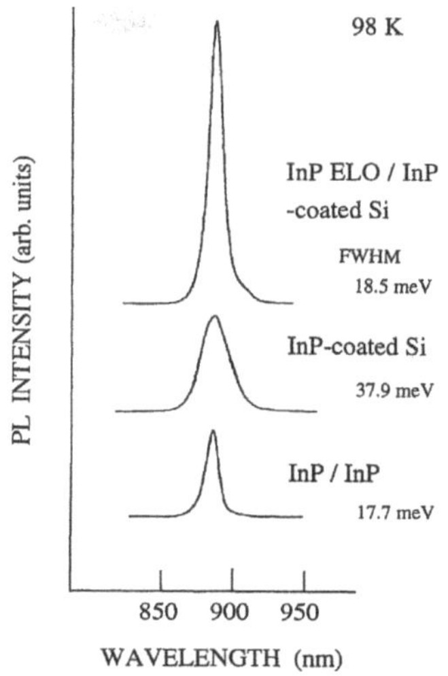

Fig. 9 Spatially resolved photoluminescence (SRPL) spectra of InP MCE on InP-coated (001) Si substrate, InP-coated Si substrate itself, and InP homoepitaxial layer on (001) InP substrate. *From Naritsuka, S., Nishinaga, T., 1995a. InP layer grown on (001) silicon substrate by epitaxial lateral overgrowth. Jpn. J. Appl. Phys. 34, L1432–L1435; Copyright (1995) The Japan Society of Applied Physics.*

width at half maximum (FWHM) (18.5 meV) demonstrate that the optical quality of this MCE layer was almost equal to that of the reference sample. Thus, the MCE process is beneficial in terms of reducing the dislocation density and relieving stress, which results in MCE layers with good optical quality.

To obtain flat, wide MCE layers, neighboring mesas are typically combined (McClelland et al., 1980; Vennegues et al., 2000). However, the area where two adjacent mesas have coalesced generally contains newly nucleated dislocations. Therefore, a new method was investigated to suppress the nucleation of dislocations in the coalesced region (Yan et al., 2000). In the course of this study, we identified two modes of coalescence, which we term one-zipper (Fig. 10A) and two-zipper (Fig. 10B). Once coalescence is initiated at a specific point, it proceeds outward as shown by the arrows in Fig. 10A. As a result, the open region between the two approaching islands closes via simultaneous zipper motions. In the case of this growth mode, it is likely that dislocations will not be introduced into the coalesced region. We have termed this "one-zipper" growth. In contrast, when the coalescence starts from two points, a closed-off open area is expected to appear. As such, the lateral growth occurs inwardly as shown

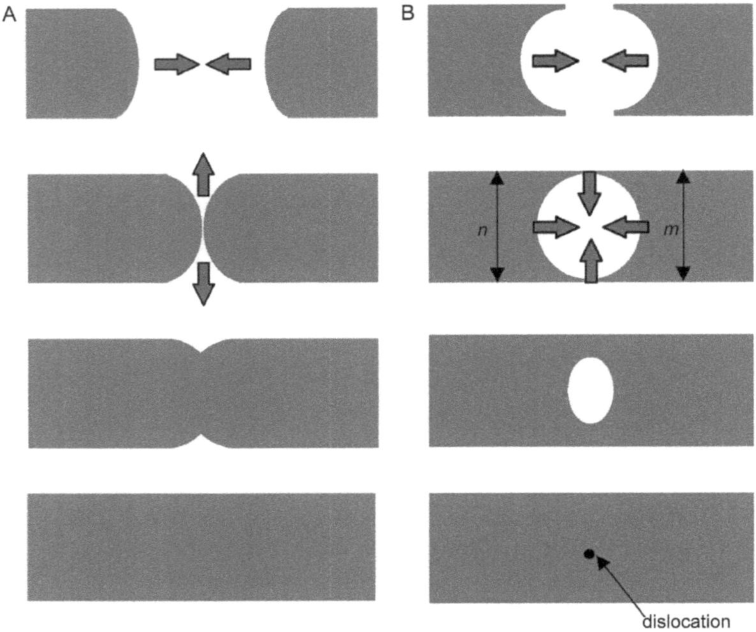

Fig. 10 Schematic diagrams of "one-zipper" mode and "two-zipper" mode growth. (A) "One-zipper" mode, where the coalescence starts from one point, (B) "two-zipper" mode where the coalescence starts from two points. Arrows show the directions of the lateral growth. n and m denote the number of lattice points between the two connecting positions of the coalescence. *From Yan, Z., Hamaoka, Y., Naritsuka, S., Nishinaga, T., 2000. Coalescence in microchannel epitaxy of InP. J. Cryst. Growth 212, 1–10; Copyright (2000), with permission from Elsevier.*

in Fig. 10B and the zipper motion proceeds in two opposite directions. In Fig. 10B, n and m denote the number of lattice points between the two connecting positions. Because the substrate is not an ideal crystal, dislocations may exist in the substrate region beneath the enclosed open area, which indicates that $n \neq m$. Therefore, misfit dislocations will appear as the open area is closed by the lateral growth process. Another possible scenario is that the two coalescing islands will not grow in exactly the same plane, and so dislocations with a screw component will also be generated. We have named this growth mode "two-zipper" growth. New microchannel patterns were designed to realize the one-zipper (Fig. 11A) and two-zipper (Fig. 11C) growth modes. Subsequently, the coalesced InP H-MCE islands were chemically etched and it was found that dislocations were generated in the coalesced region when employing the two-zipper mode (Fig. 11D),

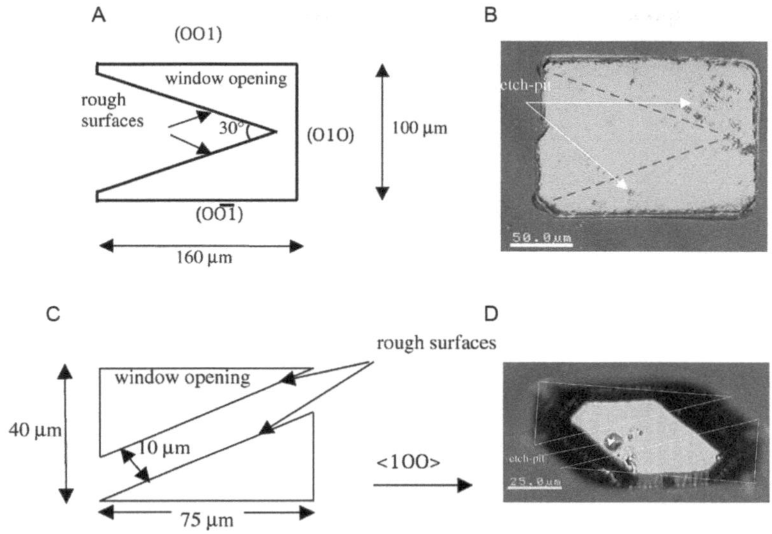

Fig. 11 Microchannel designed for "one-zipper" mode growth (A), etched InP H-MCE island grown by "one-zipper" mode (B). The dotted line shows the original position of microchannel edge. Microchannel designed for "two-zipper" mode growth (C), and etched InP H-MCE island grown by "two-zipper" mode (D). The etch pits are indicated by arrows and the solid lines show the original position of the microchannel edges. *From Yan, Z., Hamaoka, Y., Naritsuka, S., Nishinaga, T., 2000. Coalescence in microchannel epitaxy of InP. J. Cryst. Growth 212, 1–10; Copyright (2000), with permission from Elsevier.*

while the one-zipper mode did not result in dislocations (Fig. 11B). It is concluded that conducting the growth in the one-zipper mode avoids the formation of dislocations in the coalesced region.

3. Non-equilibrium method

3.1 MBE

3.1.1 V-MCE/Si

Employing a non-equilibrium method, not only directional growth but also selective growth must be addressed. During MBE, selective growth is obtained via either the thermal re-evaporation of surface adatoms or the shadow effect, which reduces the supply of the molecular beam. This method also involves the surface migration of adatoms from the mask region to the growth area to improve the selective growth. In V-MCE, vertical growth is achieved with the aid of a cavity (Matsunaga et al., 2002). In

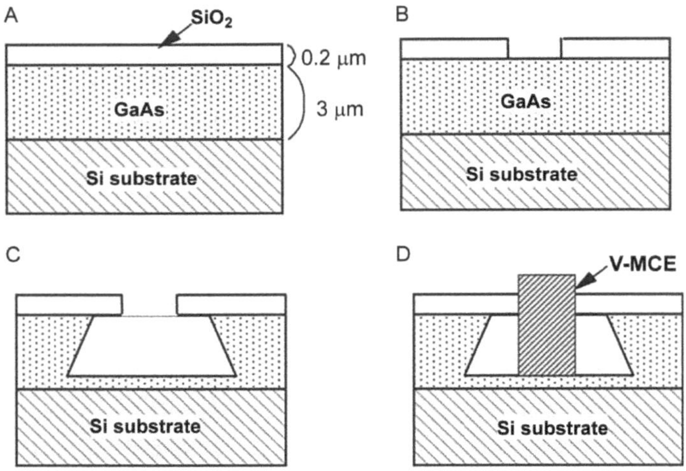

Fig. 12 Schematic illustrations of fabrication process of shadow mask; deposition of SiO_2 mask (A), patterning of SiO_2 mask (B), cavity structure fabricated by anisotropic etching of GaAs buffer layer (C), and typical V-MCE structure grown through shadow mask (D). *From Matsunaga, Y., Naritsuka, S., Nishinaga, T., 2002. A new way to achieve dislocation-free heteroepitaxial growth by molecular beam epitaxy: vertical microchannel epitaxy. J. Crystal Growth 237–239, 1460–1465; Copyright (2002), with permission from Elsevier.*

the present work, GaAs-coated Si was employed as the substrate (Fig. 12A) and microchannel openings were fabricated on an SiO_2 mask (Fig. 12B). The GaAs layers were subsequently etched using this mask to fabricate a cavity structure, as shown in Fig. 12C. The selective supply of Ga and As molecular beams as source materials from above the substrate to a limited region of the substrate through the mask opening was accomplished using the shadow effect of the SiO_2 mask, which extended partly over the cavity, similar to the eaves of a house. Vertical growth occurred inside the cavity with no source material supplied from the sides. Together with microchannel fabrication in the [01 − 1] direction, this enabled favorable vertical growth with (011) side planes (Fig. 12D). Even when the growth layer extended above the mask, the (011) side plane formation enabled uninterrupted continuation of the V-MCE process.

The cross-section of a GaAs V-MCE layer can be seen in the TEM in Fig. 13A, from which it is evident that a flat leading edge growth layer was achieved. Accordingly, the dislocations in the buffer layers did not propagate to the upper part of the growth layer. A typical photoluminescence (PL) spectrum generated by a GaAs V-MCE layer on Si is presented in

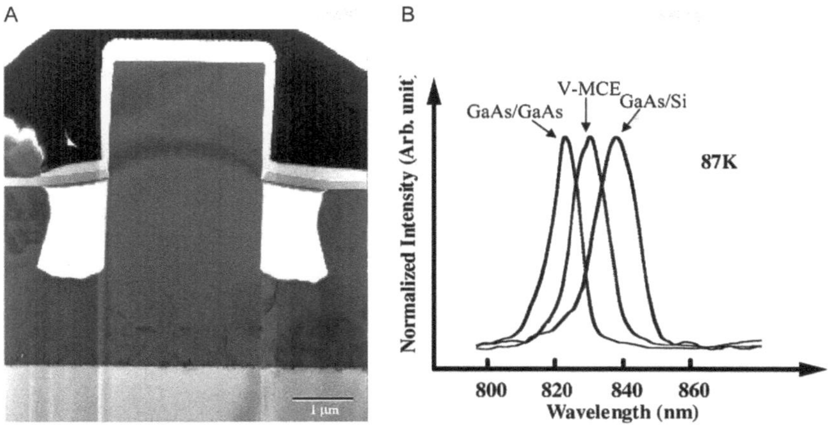

Fig. 13 (A) TEM image of V-MCE on Si with a height of 3.5 μm grown through a linear window opening of 2.5 μm. (B) Typical PL spectrum of GaAs V-MCE on Si. Spectra of GaAs/Si grown without mask and GaAs/GaAs sample are also shown for references. *From Matsunaga, Y., Naritsuka, S., Nishinaga, T., 2002. A new way to achieve dislocation-free heteroepitaxial growth by molecular beam epitaxy: vertical microchannel epitaxy. J. Crystal Growth 237–239, 1460–1465; Copyright (2002), with permission from Elsevier.*

Fig. 13B, along with PL spectra acquired from GaAs grown on Si without a mask and GaAs on homoepitaxial GaAs, as a reference. Note that the peak intensities have been normalized to make all of uniform heights in Fig. 13B. The peak from the V-MCE layer is located between the peaks originating from the conventional GaAs on Si and the GaAs on GaAs, demonstrating that the residual stress in the V-MCE layer was reduced compared with that of GaAs deposited on Si without a mask.

In the case of V-MCE, the stress vertical to the microchannel is easily released by deformation of the grown layer itself, while the stress parallel to the microchannel is difficult to reduce. This is because the resulting structure is longer in the direction of the microchannel and it is difficult to release stress through deformation. The stress in a GaAs V-MCE layer on Si was simulated using a finite element method (Naritsuka et al., 2004a), with the mesh structure in Fig. 14A. The results showed that the stress in the V-MCE was decreased with increasing distance from the hetero-interface and also disappeared in response to a unique set of conditions. In the simulation, tensile stress resulted from differences in the thermal contraction of the GaAs and Si, and Fig. 14B plots stress at the center of the top surface of the V-MCE layer as a function of thickness (H) for various Si substrate

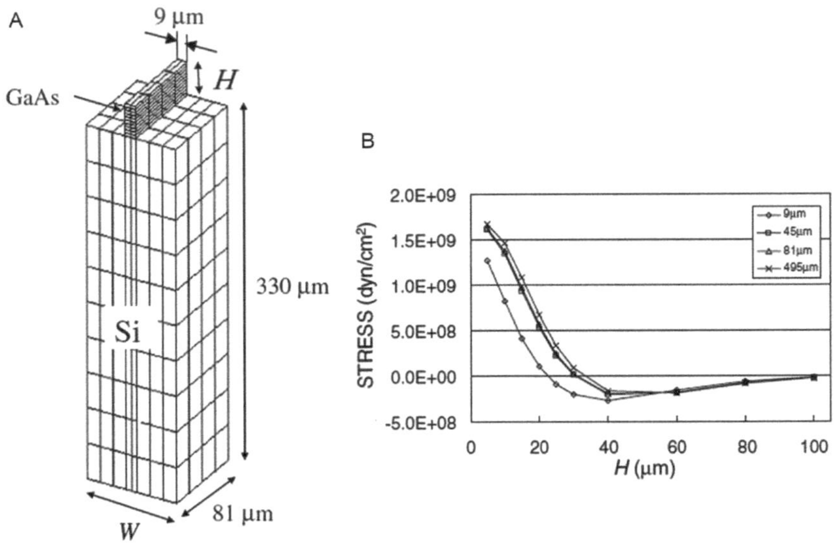

Fig. 14 (A) Simulated mesh structures of GaAs V-MCE layer on Si substrate. The height of V-MCE layer and the width of Si substrate are denoted by H and W, respectively. (B) Stress at the center of the top surface of GaAs V-MCE layers. The parameter W varies from 9 to 495 μm. *From Naritsuka, S., Okada, M., Maruyama, T., 2004a. Simulation of Three-Dimensional Stress in GaAs Microchannel Epitaxy Layer on Si Substrates. Jpn. J. Appl. Phys. 43, 3289–3292; Copyright (2004) The Japan Society of Applied Physics.*

widths (W) between 9 and 495 μm. For all W, the stress decreased with increasing thickness of the V-MCE layer. In addition, the stress was canceled when the V-MCE layer reached a critical thickness at which the tensile stress was equal but opposite to the compressive stress produced by bowing of the substrate. If this scenario can be achieved in the active layer of a device, V-MCE could offer not only a reduction in dislocations but also stress release in devices fabricated based on highly-mismatched heteroepitaxy.

3.1.2 BILE/Si

In BILE, selective growth results from the shadow effect of neighboring mesas in conjunction with a molecular beam at a low angle of incidence (Naritsuka et al., 2004b, 2005; Suzuki et al., 2003). Fig. 15A shows the manner in which the beam strikes the truncated ridges and the expected shape of the lateral growth from the ridge is presented in Fig. 15B. The dashed oval line in Fig. 15A indicates the area without irradiation from the beams for which growth is not expected. The shadowing effect from the neighboring ridge is

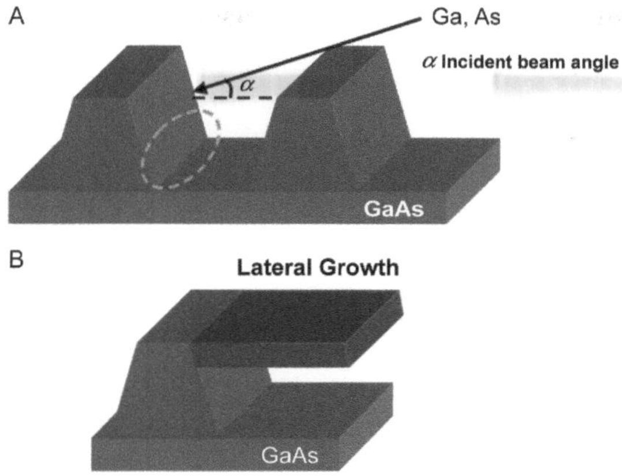

Fig. 15 (A) and (B): Basic features of beam induced lateral epitaxy (BILE). *From Naritsuka, S., Saitoh, K., Suzuki, T., Maruyama, T., 2004b. Beam induced lateral epitaxy: a new way to lateral growth in molecular beam epitaxy. Mat. Res. Soc. Symp. Proc. 799, Z2.4.1-6; Copyright (2004), with permission from Cambridge University Press.*

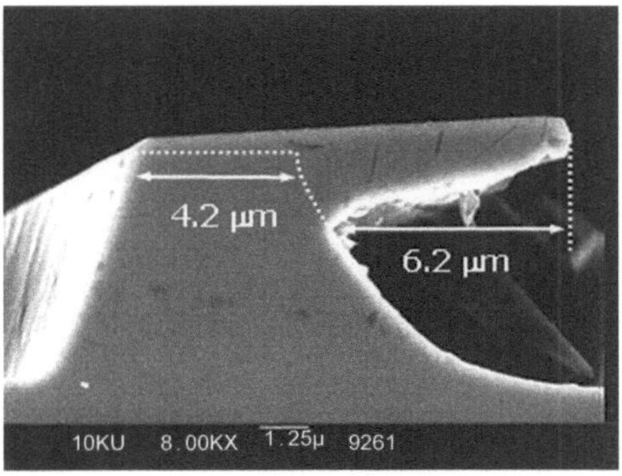

Fig. 16 SEM cross-sectional image of lateral growth on (111) B GaAs substrate for initial truncated ridges aligned parallel to [1 − 10]. Ga incident angle was 12°. *From Naritsuka, S., Saitoh, K., Suzuki, T., Maruyama, T., 2004b. Beam induced lateral epitaxy: a new way to lateral growth in molecular beam epitaxy. Mat. Res. Soc. Symp. Proc. 799, Z2.4.1-6; Copyright (2004), with permission from Cambridge University Press.*

expected to lead to a thin layer of lateral growth on each ridge. Here, the ridge width only affects the area of direct growth while the separation of the ridges primarily affects the shape of the lateral growth by changing the shadow area.

Fig. 16 shows a cross-sectional SEM image of a BILE layer grown on a (111)B GaAs substrate with a truncated ridge aligned parallel to the [1 − 10] direction (Naritsuka et al., 2004b). While producing this sample, the molecular beam was directed such that the projection on the (111)B surface was

[2 − 1 − 1] because this direction was expected to result in wide lateral growth and because exposure in the opposite direction would cause rapid termination of the lateral growth by the formation of a (111)A facet on the growth front. Fig. 16 demonstrates that the top surface of the layer was extremely flat and smooth. The inclination angle of the surface was misaligned relative to the substrate by 4°, indicating the formation of (111)B. Therefore, BILE has been accomplished over a wide area to produce a flat upper surface by using a (111)B substrate.

The BILE process was subsequently applied to grow a GaAs layer on a GaAs-coated Si substrate (Naritsuka et al., 2005). Although the surface morphology of the GaAs-coated Si substrate appeared to have a mirror finish on visual inspection, the surface of the KOH-etched sample actually contained a large quantity of etching pits and stacking faults (Fig. 17A), at a density of approximately 5×10^8 cm^{-2}. Fig. 17B shows an SEM image

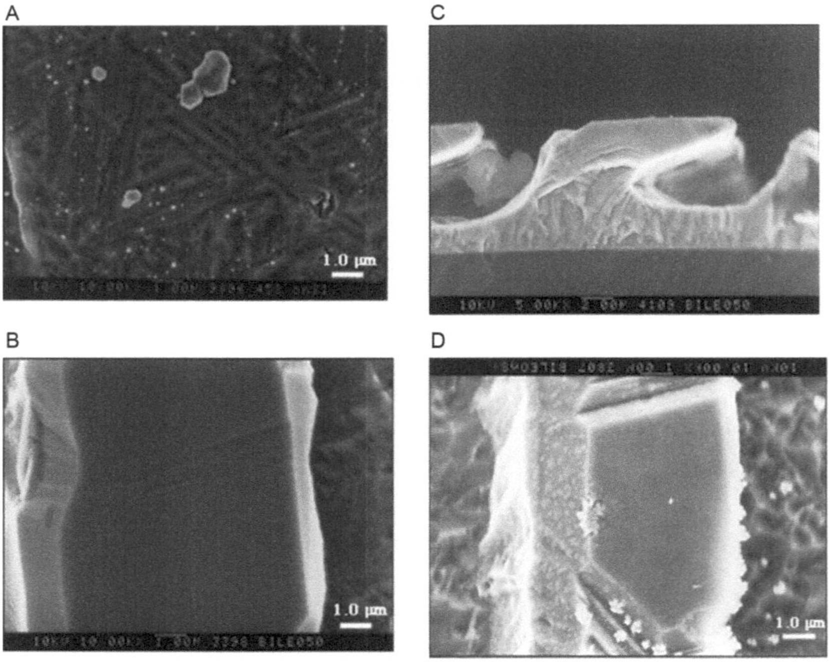

Fig. 17 (A) SEM image of distribution of etch pits on GaAs/Si template after chemical etching. GaAs BILE layer on GaAs/Si template, (B) surface and (C) cross-sectional SEM images. The Ga incident angle was 12°. (D) SEM image showing the distribution of etch pits on the BILE layer. The white particles are residual products from the chemical etching. *From Naritsuka, S., Saitoh, K., Kondo, T., Maruyama, T., 2005. Beam induced lateral epitaxy of GaAs on a GaAs/Si template. Mat. Res. Soc. Symp. Proc. 829, B9.30-1-6; Copyright (2005), with permission from Cambridge University Press.*

of a flat GaAs BILE layer with a width of approximately 10 μm grown on the GaAs-coated Si substrate. The predominant formation of (111) facets resulted in a wide, flat top-surface with a width as much as 7.5 μm, with small micro-facets on both sides. The image in Fig. 17B also shows transverse lines with bumps on their sides. These are likely stacking faults, which would degrade the flatness of the surface. The cross-sectional SEM image in Fig. 17C demonstrates the overgrown region of the GaAs that grew laterally from the side of the truncated ridge, with a width as large as 6.5 μm. The shape of the underside of the layer was primarily determined by the shadow effect of the neighboring ridge, and an SEM image of this layer after KOH etching is provided in Fig. 17D. The triangular grooves in the upper and lower parts of this image indicate stacking faults. However, the lateral growth generated almost no pits due to dislocations. Although stacking faults are detrimental with regard to device applications, this result confirms that BILE is a useful means of reducing the dislocation density in the overgrown region.

3.1.3 LAIMCE/Si

LAIMCE is another unique technique for obtaining lateral growth by MBE, in which lateral growth is performed solely using a low angle of incidence molecular beam (Bacchin and Nishinaga, 2000). Fig. 18 illustrates the basic concept, and Fig. 18A shows the relevant angles in the geometrical arrangement. Here, we define the angles between the beams (Ga and As_4) and the line seed direction in the substrate plane (α_{Ga}, α_{As4}), the angle of incidence of the beams (β_{Ga}, β_{As4}) and the angle between the line seed direction and the crystallographic direction [−110] (γ). In Fig. 18B, we show a cross section of a sample during growth with a low angle between the substrate and the directions of the Ga and As_4 beams.

In these experiments, the Ga and As_4 angles were, respectively, 11.2° (β_{Ga}) and 22.8° (β_{As4}). If both selective growth (that is, no formation of GaAs polycrystalline islands on the mask) and lateral growth occur, epilayers are expected to grow as shown in Fig. 18C. In this figure, d is the window (or microchannel) width, while W is the width of the lateral growth part of the epilayer and T is its thickness. Figures of merit for this epilayer are the W and W/T ratio. The basic assumption of LAIMCE is that the lateral growth rate will be significantly higher than the vertical growth rate due to the reduced angle of incidence of the beams. Furthermore, the difference in the net incident As_4 fluxes on the top and sidewall of the epilayer allows significant diffusion of Ga adatoms from the top surface to the side surface under optimized conditions. The intersurface diffusion of the adatoms helps not only to enlarge the lateral growth but also to obtain a flat top surface.

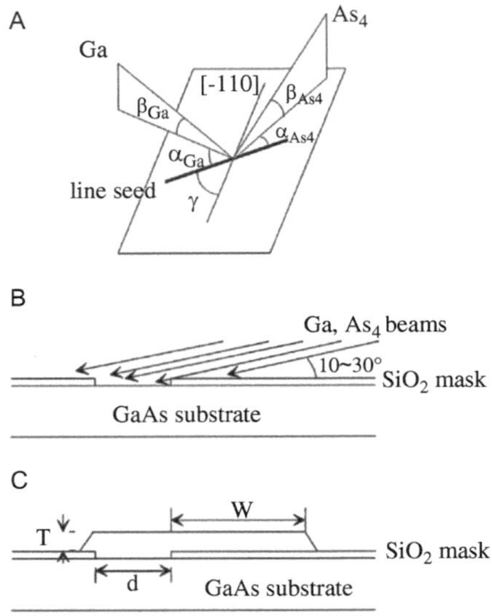

Fig. 18 Schematic of the low angle incidence microchannel epitaxy. (A) relevant angles in the geometrical arrangement of LAIMCE; (B) cross section of the sample structure and beam directions; (C) cross section of the sample structure and of the epilayer after selective and lateral growth. *From Bacchin, G., Nishinaga, T., 2000. A new way to achieve both selective and lateral growth by molecular beam epitaxy: low angle incidence microchannel epitaxy. J. Cryst. Growth 208, 1–10; Copyright (2000), with permission from Elsevier.*

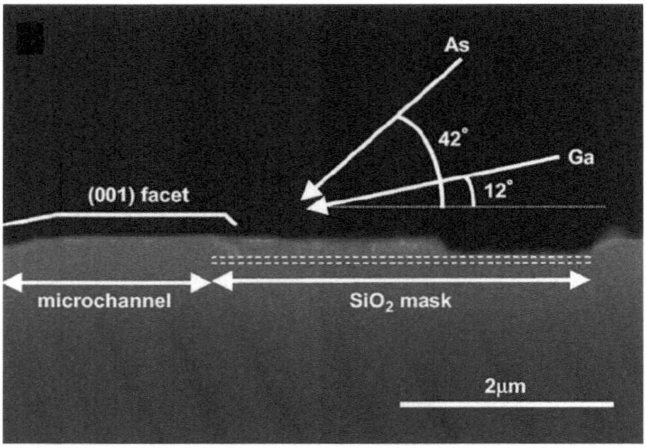

Fig. 19 Cross-sectional SEM images of GaAs LAIMCE on (001) GaAs substrates with microchannels aligned in 20° from [−110] direction. *From Naritsuka, S., Matsuoka, S., Yamashita, Y., Maruyama, T., 2008. Optimization of initial growth in low-angle incidence microchannel epitaxy of GaAs on (001) GaAs substrates. J. Crystal Growth 310, 1571–1575; Copyright (2008), with permission from Elsevier.*

A cross-sectional SEM image of the LAIMCE sample in Fig. 19 confirms the lateral growth of the GaAs layer over the SiO_2 mask (Naritsuka et al., 2008). The flat, smooth surface of the layer also demonstrates that the inter-surface diffusion of the Ga adatoms proceeded from the top to the side surfaces. In addition, Fig. 19 shows the excellent selectivity of the growth. Although the length of the lateral growth region was only 0.25 μm, there is clear evidence of overgrowth. The extremely thin (170 nm) layer suggests that wider lateral growth could be obtained simply by increasing the growth time.

4. Conclusion

In this chapter, the principle and the experimental results of MCE are explained. MCE can be classified into two groups, i.e., H-MCE and V-MCE from the direction of its growth. Selective growth and directional growth are necessary to realize MCE. In the equilibrium method, directional growth is performed by controlling facet formation. On the other hand, in non-equilibrium method, it is difficult to fulfill both directional growth and selective growth. Several techniques of MCE, such as cavity growth, BILE and LAIMCE are proposed to perform the directional growth. The shadow effect accomplishes not only directional growth but also selective growth in cavity growth and BILE. In LAIMCE, low incidence of molecular beams is used to laterally grow the layer. Moreover, it is useful to enhance re-evaporation of adatoms and to obtain an excellent selective growth because the low incidence of molecular beams decreases the adatom concentration on the surface from its small vertical component. These experimental results demonstrate the dramatic reduction of dislocations in highly-mismatched heteroepitaxial growth, such as GaAs/Si or InP/Si. Consequently, dislocation-free areas, whose width are sufficient for the device fabrication such as VCSELs, are successfully obtained on Si substrates.

Acknowledgment

This work was partly supported by JSPS KAKENHI Grant; Numbers 2660089, 15H03558, 26105002, 25000011 and others. The author greatly thanks for long-term encouragement and heartful instruction of emeritus Prof. Tatau Nishinaga of the University of Tokyo, who is my teacher and also the inventor of this outstanding technology of "microchannel epitaxy". The author would like to thank Prof. M. Tanaka of the University of Tokyo, Prof. T. Maruyama of Meijo University, Prof. Y.S. Chang of National Kaohsiung Institute of Marine Technology in Taiwan, and Dr. Y. Matsunaga of Toshiba co., Dr. G. Bacchin and other coworkers for their help in carrying out the present work. He also expresses gratitude to many graduate and undergraduate students at Meijo University, especially Dr. C. Lin, Messrs. T. Suzuki, K. Saito, M. Tomita, D. Kambayashi for their collaboration.

References

Bacchin, G., Nishinaga, T., 2000. A new way to achieve both selective and lateral growth by molecular beam epitaxy: low angle incidence microchannel epitaxy. J. Cryst. Growth 208, 1–10.
Chang, Y.S., Naritsuka, S., Nishinaga, T., 1997. Effect of growth temperature on epitaxial lateral overgrowth of GaAs on Si substrate. J. Cryst. Growth 174, 630–634.
Chang, Y.S., Naritsuka, S., Nishinaga, T., 1998. Optimization of growth condition for wide dislocation-free GaAs on Si substrate by microchannel epitaxy. J. Cryst. Growth 192, 18–22.
Chang, Y.S., Naritsuka, S., Nishinaga, T., 2010. Effect of dislocation density on microchannel epitaxy of GaAs on GaAs/Si substrate. J. Cryst. Growth 312, 629–634.
Matsunaga, Y., Naritsuka, S., Nishinaga, T., 2002. A new way to achieve dislocation-free heteroepitaxial growth by molecular beam epitaxy: vertical microchannel epitaxy. J. Cryst. Growth 237–239, 1460–1465.
McClelland, R.W., Bozler, C.O., Fan, J.C.C., 1980. A technique for producing epitaxial films on reuseable substrates. Appl. Phys. Lett. 37, 560–562.
Naritsuka, S., 2016. Microchannel epitaxy. Prog. Cryst. Growth Charact. Mater. 62, 302–316.
Naritsuka, S., Nishinaga, T., 1995a. InP layer grown on (001) silicon substrate by epitaxial lateral overgrowth. Jpn. J. Appl. Phys. 34, L1432–L1435.
Naritsuka, S., Nishinaga, T., 1995b. Epitaxial lateral overgrowth of InP by liquid phase epitaxy. J. Cryst. Growth 146, 314–318.
Naritsuka, S., Mochizuki, Y., Motodohi, K., Ohya, S., Ikeda, N., Sugimoto, Y., Asakawa, K., Huang, W.D., Mitsuda, K., Kobayashi, O., Nishinaga, T., 2001. Fabrication of laser array on microchannel epitaxy GaAs/Si toward the realization of optoelectronic integrated circuits (OEICs). In: Device Research Conference (DRC), Notre Dame, Indiana, USA, June 25–27.
Naritsuka, S., Okada, M., Maruyama, T., 2004a. Simulation of three-dimensional stress in GaAs microchannel epitaxy layer on Si substrates. Jpn. J. Appl. Phys. 43, 3289–3292.
Naritsuka, S., Saitoh, K., Suzuki, T., Maruyama, T., 2004b. Beam induced lateral epitaxy: a new way to lateral growth in molecular beam epitaxy. Mater. Res. Soc. Symp. Proc. 799. Z2.4, 1-6.
Naritsuka, S., Saitoh, K., Kondo, T., Maruyama, T., 2005. Beam induced lateral epitaxy of GaAs on a GaAs/Si template. Mater. Res. Soc. Symp. Proc. 829. B9.30, 1-6.
Naritsuka, S., Matsuoka, S., Yamashita, Y., Maruyama, T., 2008. Optimization of initial growth in low-angle incidence microchannel epitaxy of GaAs on (001) GaAs substrates. J. Cryst. Growth 310, 1571–1575.
Nishinaga, T., 2002. Microchannel epitaxy: an overview. J. Cryst. Growth 237-239, 1410–1417.
Nishinaga, T., Nakano, T., Zhang, S., 1988. Epitaxial lateral overgrowth of GaAs by LPE. Jpn. J. Appl. Phys. 27, L964–L967.
Nishinaga, T., Shen, X.Q., Kishimoto, D., 1996. Surface diffusion length of cation incorporation studied by microprobe-RHEED/SEM MBE. J. Cryst. Growth 163, 60–66.
Shitara, T., Nishinaga, T., 1989. Surface diffusion length of gallium during MBE growth on the various misoriented GaAs (001) substrates. Jpn. J. Appl. Phys. 28, 1212–1216.
Suzuki, T., Naritsuka, S., Maruyama, T., Nishinaga, T., 2003. Beam induced lateral epitaxy: a new approach for lateral growth in molecular beam epitaxy. Cryst. Res. Technol. 38, 614–618.

Tamura, M., Hashimoto, A., Sugiyama, N., 1991. Threading dislocations in in-doped GaAs/Si. J. Appl. Phys. 70, 4770–4778.
Tomioka, K., Fukui, T., 2015. Formation of III-V nanowires. In: Nishinaga, T. (Ed.), Handbook of Crystal Growth, second ed. In: vol. 1 (Pt. B). Elsevier, Amsterdam, pp. 772–788.
Vennegues, P., Beaumont, B., Bousquet, V., Vaille, M., Gibart, P., 2000. Reduction mechanisms for defect densities in GaN using one- or two-step epitaxial lateral overgrowth methods. J. Appl. Phys. 87, 4175–4181.
Yamashiki, Y., Shen, X.Q., Nishinaga, T., 1997. Arsenic pressure dependence of pure two-face inter-surface diffusion between (001) and (111)B in molecular beam epitaxy of GaAs. J. Cryst. Growth 174, 539–543.
Yan, Z., Hamaoka, Y., Naritsuka, S., Nishinaga, T., 2000. Coalescence in microchannel epitaxy of InP. J. Cryst. Growth 212, 1–10.

CHAPTER FIVE

Epitaxial lateral overgrowth of III-V semiconductors on Si for photonic integration

Yan-Ting Sun*, Sebastian Lourdudoss
Department of Applied Physics, KTH-Royal Institute of Technology, Stockholm, Sweden
*Corresponding author: e-mail address: yasun@kth.se

Contents

1. Introduction

1.1 Heteroepitaxy of III-Vs on Si

Development of III-V/Si integration has attracted attention for many years because of the possibility to integrate III-V semiconductor based optical functions with electronic integrated circuits based on Si to realize cost effective Si photonic integrated circuits (PICs) and optoelectronic integrated circuits (OEICs) with multi-functionalities at chip level. Heteroepitaxial

Semiconductors and Semimetals, Volume 101
ISSN 0080-8784
https://doi.org/10.1016/bs.semsem.2019.07.004

growth of III-V semiconductors on Si offers the feasibility of implementing monolithic integration on full size Si wafers directly. Thus, no expensive III-V substrates are used in the monolithic process flow and cost effectiveness of III-V/Si integration can be realized by exploiting the wide availability of Si wafers and the maturity of CMOS processing technology. However, two major challenges need to be addressed in order to implement monolithic III-V/Si integration by heteroepitaxy for realizing synergetic photonic/electronic functionalities: (1) First, the mismatch of materials properties of III-V semiconductors and Si can cause severe crystalline defects in the heteroepitaxial structure. High density threading dislocations (TDs) can be generated in heteroepitaxial III-V semiconductor layers on Si because of the mismatch of lattice constants between the III-V semiconductors and Si. Such threading dislocations can propagate to the surface of the epitaxial layer with growth and degrade the performance and reliability of III-V photonic devices built on top. In addition, residual thermal strain was often observed in heteroepitaxial III-V layers on Si introduced during the cool-down process (after growth at a higher temperature) because of the mismatch of thermal expansion coefficients in the heterostructure. Whereas a small strain in the heteroepitaxial III-V layer can affect the bandgap of semiconductors, which however does not always lead to adverse effects, a large strain can introduce defects in the optical components during operation. The heteroepitaxial III-V layers on Si are normally under tensile thermal strain arising from the (generally) larger thermal expansion coefficients of III-V semiconductors with respect to that of Si which can result in cracking in thick III-V film. (2) Second, seamless integration of complementary functionalities in III-V semiconductors and Si is essential in the realization of Si PIC and OEIC devices. The coupling loss of optical wave between the III-V laser diode and the Si passive waveguide circuits is critical in Si PIC, which will affect the performance of circuits and the functionality scaling. In electrically injected laser diodes and photodiodes on Si, the current injection and external bias applied from the back contacts on Si via III-V/Si heterojunction will simplify the process and device integration. Therefore, the charge carrier transport through the III-V/Si interface with low electrical loss will play an important role in realizing energy efficient active III-V photonic devices on Si.

To address the challenges presented above, epitaxial lateral overgrowth (ELOG) (Sun and Lourdudoss, 2003) and corrugated epitaxial lateral overgrowth (CELOG) (Sun et al., 2014) technologies have been developed to reduce the threading dislocations in III-V heteroepitaxial layer on Si and achieve synergetic functionalities simultaneously. This chapter deals with these two methods. After a brief description of ELOG and CELOG in Section 1.2 along with appropriate literature and their exploitation potential, Section 2 deals with ELOG of InP/Si where optimization, morphology, strain, crystalline quality along with multi-quantum well (MQW) growth on InP/Si are treated. Section 3 treats CELOG III-V/Si and the crystalline quality of the grown material. We demonstrate not only InP but also GaInP can be grown on Si from InP seed with this technique. Section 4 describes the photonic devices attempted so far on Si by exploiting ELOG and CELOG techniques. Section 5 ends this chapter with summary and conclusions.

1.2 Description of ELOG and CELOG

The concept of ELOG for dislocation filtering is shown in Fig. 1A. ELOG has been investigated in the scheme of microchannel epitaxy (MCE) for GaAs growth on Si by liquid phase epitaxy (LPE) (Chang et al., 1998) and molecular beam epitaxy (MBE) (Matsunaga et al., 2002). The details of MCE for III-V/Si photonic integration is presented in the chapter "Microchannel epitaxy of III–V layers on Si substrates" by S. Naritsuka in this volume. In the ELOG scheme, first a III-V seed layer is deposited on Si by metal-organic vapor phase epitaxy (MOVPE) or MBE. Normally such a seed layer contains a high density of threading dislocations. Then a dielectric thin film mask layer, such as SiO_2 is deposited on the surface of the III-V seed layer on Si. Next, openings are formed on the SiO_2 mask to expose the surface of the III-V seed layer from which the epitaxial growth of III-V semiconductors is initiated by a near-equilibrium process such as hydride vapor phase epitaxy (HVPE) to avoid nucleation on the dielectric

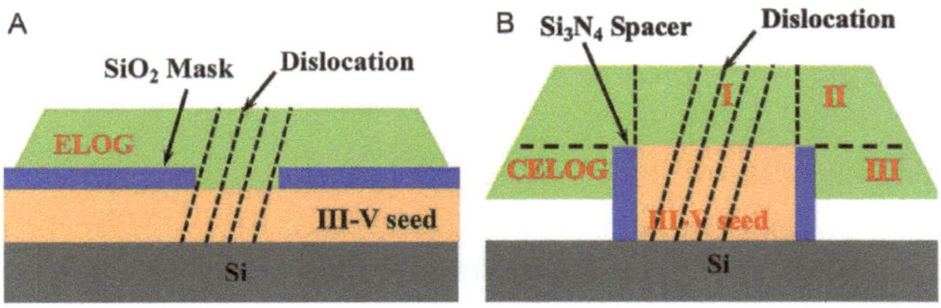

Fig. 1 Concepts of (A) ELOG and (B) CELOG for threading dislocation filtration. *Sun, Y.-T., Kataria, H., Metaferia W., Lourdudoss, S., 2014. Realization of an atomically abrupt InP/Si heterojunction via corrugated epitaxial lateral overgrowth. CrstEngComm 16, 7889–7893; Copyright (2014) Royal Society of Chemistry.*

mask. This growth proceeds vertically from the opening and then laterally over the mask after reaching the brim of the opening. The dislocations in the seed layer will extend to the surface of the epitaxial layer with growth in the region above the opening but not that above the mask. The SiO_2 mask will block the propagation of dislocations in the lateral overgrowth part, which is free of dislocations and can be used as a substrate for device fabrication. By choosing a dielectric material (mask) with a thermal expansion coefficient close to that of III-V heteroepitaxial layer, the thermal strain in ELOG III-V/Si arising from the mismatch of thermal expansion coefficient can be engineered and reduced. In addition, by adjusting the width of the opening to be smaller than the thickness of the mask, dislocations through the opening can also be suppressed by taking advantage of the so-called necking effect (Langdo et al., 2000).

In ELOG, the high crystalline quality III-V lateral overgrowth is separated from the Si substrate by the defective III-V seed layer and the dielectric mask layer, see Fig. 1A. Thus no direct III-V/Si heterojunction exists which hinders the feasibility of charge carrier transport through direct III-V/Si heterojunction. This issue is addressed in the approach of CELOG shown in Fig. 1B. The III-V seed layer is processed to mesas with appropriate dimensions by suitable lithographic and dry etching techniques. The sidewalls of the III-V seed mesas are protected by dielectric spacer layers, such as Si_3N_4, by either photolithography or self-aligned process enabled by anisotropic dry etching. Selective growth of III-V semiconductors gets initiated on the surface of III-V seed mesa and then proceeds laterally. As soon as the lateral growth exceeds the edge of spacers on the sidewalls, the downward $(00\overline{1})$ oriented growth front is developed toward the surface of Si. The high density threading dislocations existing in III-V seed mesas will propagate to the top surface of the grown layer in the vertical growth, as shown in region I in Fig. 1B and depending on the dimension of the opening and vertical layer thickness some dislocations can escape to the lateral overgrowth region II in Fig. 1B. However, these dislocations intersecting the surface cannot bend downward to the surface of Si, which leaves region III free of threading dislocations. In the meantime, the downward growth in region III will approach the surface of Si to form direct III-V/Si heterojunction with low dislocation density. By removing the regions I and II by chemical mechanical polishing (CMP) process, a high crystalline quality III-V semiconductor layer in region III can be exploited for fabricating III-V photonic devices on the top.

Prior to the treatment on ELOG and CELOG we would like to highlight the potential of these two techniques in integration with silicon useful for silicon photonics.

In order to explore the full potential of lateral overgrowth based III-V/Si integration platform, innovative designs of optical and electrical coupling need to be devised. For example, a monolithic evanescently coupled silicon laser (MECSL) based on ELOG platform was proposed in Wang et al. (2012) by incorporating the waveguide structure into the mask in ELOG that filters out the dislocations. As shown in Fig. 2A, the waveguide structure is comprised of SiO_2 cladding layers and Si waveguide core. Narrow openings are etched through the waveguide structure to allow the formation of ELOG InP on the seed layer and block the propagation of threading dislocations. The ELOG InP from adjacent openings will coalesce to form a uniform layer on the top of the Si waveguide, which is separated from the Si waveguide core by SiO_2 cladding layer with thickness of h_s. Quantum well laser device structure can be grown on ELOG InP with thickness of h_i. The optical mode distribution in MECSL is calculated by finite difference mode solver and the quasi-TE mode profile is shown in Fig. 2B. The optical mode is split into cavity/silicon waveguide section and quantum well section. The optical confinement in the quantum wells will determine the optical gain and threshold current density. The ELOG layer will increase the effective refractive index of quantum wells and result in a low confinement in Si waveguide. By adjusting the thickness of ELOG InP layer, h_i, and the height of rib waveguide, h_r, optimal confinement factors for quantum wells and Si waveguide can be achieved simultaneously.

Si based photodiode for telecom wavelength detection has been investigated in Ge/Si (Michel et al., 2010) and InGaAs/Si (Campbell, 2007) materials systems. The operation of photodetectors involves three processes: light absorption, carrier separation, and carrier collection and the properties of Si based heterojunction play an important role to achieve high efficiency photon detection. Wafer bonded InGaAs/Si avalanche photodiode has been developed in which photo-excited electron-hole pairs generated in the InGaAs absorption layer transported across the heterointerface to Si for low noise electron impact ionization to achieve a high gain-bandwidth product up to 315 GHz (Hawkins et al., 1996). When selected III-V materials form heterojunction with Si substrate, the band discontinuity could have type II staggered alignment, in which the electrons and holes will be separated naturally to different materials by the built-in electric field, which requires an atomically abrupt III-V/Si interface with low optical and

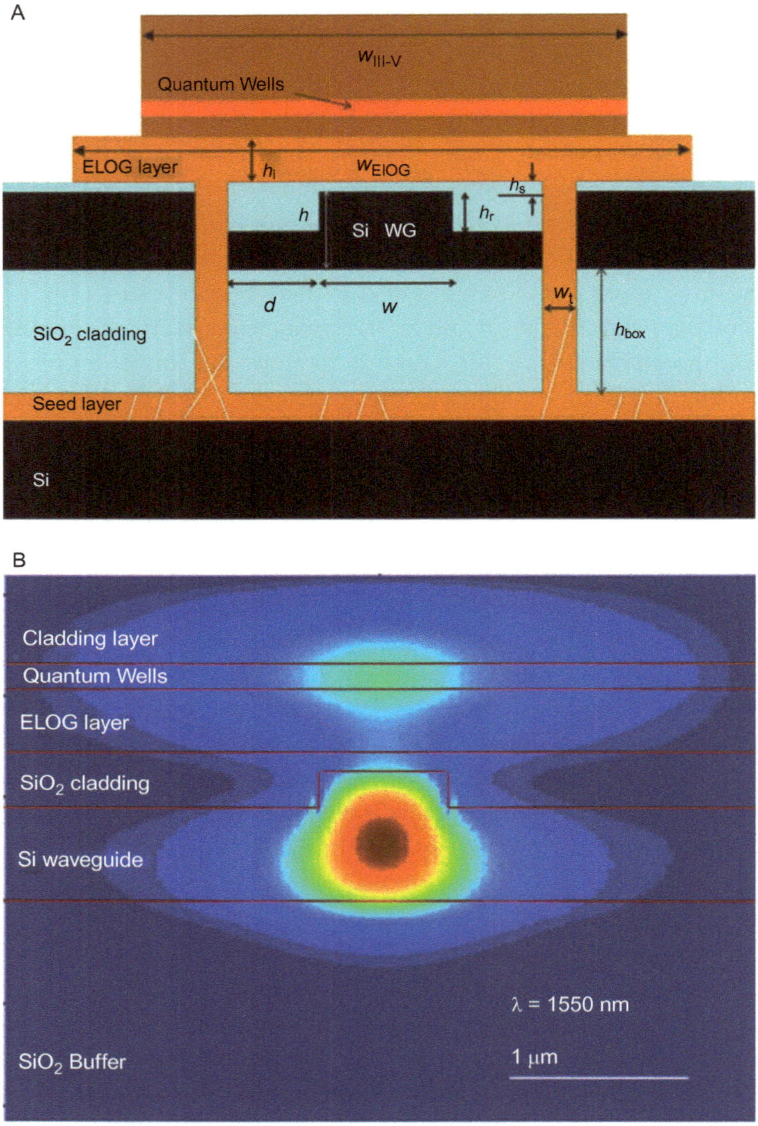

Fig. 2 (A) Schematic of the MECSL structure based on the ELOG integration platform. (B) A typical optical mode profile of 1.55 μm MECSL based on proposed monolithically integrated structure design. *Wang, Z., Junesand, C., Metaferia, W., Hu, C., Wosinski, L., Lourdudoss, S., 2012. III–vs on Si for photonic applications—a monolithic approach. Mater. Sci. Eng., B 177, 1551–1557; Copyright (2012), with permission from Elsevier.*

electrical losses. By exploiting CELOG approach, high crystalline quality atomically abrupt III-V/Si heterojunction can be achieved for high speed Si based photodiode applications in Si photonics.

Selective area growth (SAG) is the key for realizing ELOG and CELOG on masked III-V/Si templates. Hydride vapor phase epitaxy (HVPE) was investigated extensively for SAG of III-V semiconductors. HVPE exhibits very high growth rate for a small change of chemical potential of the precursors with respect to the product. By changing the chemical potentials, even etching can be achieved. Hence it was identified as a near-equilibrium process along with liquid phase epitaxy (LPE) (Stringfellow, 1991). Its supreme selectivity is understood to be due to the higher volatility of the physisorbed group III chloride molecules that are on the mask surface than those that are chemisorbed on the semiconductor surface (Lourdudoss and Kjebon, 1997). Together with very high growth rate and inherent growth selectivity over the dielectric masks, kinetically controlled anisotropic growth profiles are desired for dislocation filtration in ELOG and CELOG III-V/Si growth.

2. ELOG InP/Si

2.1 Optimization of ELOG

A high aspect ratio between the lateral overgrowth and the vertical growth of ELOG is essentially important for filtering out threading dislocations and form reasonably coalesced thin III-V semiconductors layer on Si. The optimal structure design of ELOG template and growth conditions were investigated for high aspect ratio (lateral/vertical growth) by conducting temporally resolved InP growth on InP substrates in HVPE reactor as described below.

The ELOG growth rates of III-V semiconductors as a function of orientation of the openings have been studied (Sun et al., 2001). The openings were formed on the Si_3N_4 mask layer on (001) InP substrate, which were oriented at various angles from [110] direction. Depending on the orientation of the openings on the substrate to initiate ELOG, different facet planes emerged leading to anisotropic lateral overgrowth rate on the patterned substrate. The aspect ratio is a key factor for ELOG growth. A high aspect ratio is desired for dislocation density reduction and device fabrication. The dependence of the lateral and vertical growth rates on the opening orientation is shown in Fig. 3. Whereas the vertical growth rate is almost constant for ELOG from the openings at all orientations, lateral overgrowth rate is

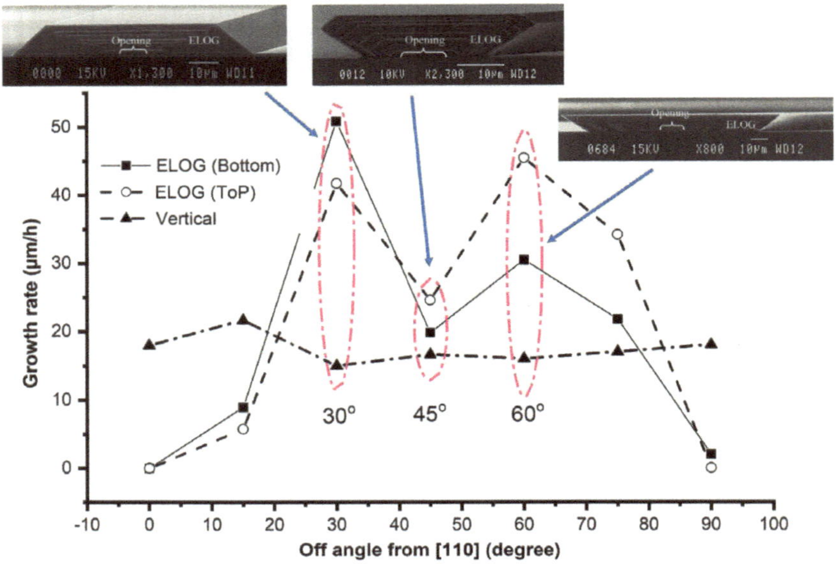

Fig. 3 Dependence of lateral and vertical growth rates on the ELOG opening orientation on InP substrate. *Sun, Y.-T., Messmer, E. R., Söderström, D., Jahan, D., Lourdudoss, S., 2001. Temporally resolved selective area growth of InP in the openings off-oriented from [110] direction. J. Cryst. Growth 225, 9–15; Copyright (2001), with permission from Elsevier.*

orientation dependent as seen in Fig. 3. The lateral overgrowth rates calculated at the top of ELOG and the bottom over Si_3N_4 mask are denoted as ELOG (Top) and ELOG (Bottom), respectively, in Fig. 3. As shown in the inset of SEM images, the growth along the 30° off [110] direction has a trapezoid profile with a higher lateral overgrowth rate over the Si_3N_4 mask, which is desired for coalesced layer formation. Higher lateral growth rate at the top of ELOG is obtained in the openings along the 45° and 60° off [110] directions. Comparing all the cases, high lateral overgrowth rates of InP are obtained in ELOG layers grown in openings oriented along the 30° and 60° off [110] directions, which will be used for achieving ELOG InP/Si with high aspect ratio.

2.2 ELOG of InP/Si in micro-opening

By exploiting the identified optimum opening orientations along 30° and 60° off [110] direction for high lateral overgrowth rate, the effect of opening dimension on defect filtering and the strain status in ELOG InP/Si were studied by various techniques. The mechanisms of dislocation and stacking fault filtration by ELOG approach have been devised based on experimental data.

The seed InP layer has a high density of threading dislocations up to $10^9\,cm^{-2}$. The threading dislocations exposed in openings will propagate to the surface of ELOG InP with growth. In zinc blende crystals, the dislocations can dissociate into partial dislocations bounding a stacking fault to facilitate the gliding of dislocations on $\{111\}$ planes. The threading dislocations and stacking faults intersecting the growth front of ELOG will affect the growth rate of the intersected planes and the ultimate morphology of ELOG InP on Si. In addition, the dynamic behavior of dislocations in III-V semiconductors are affected by the impurities in heteroepitaxial III-V semiconductor layers. It is reported that the sulfur doping at a level of $8 \times 10^{18}\,cm^{-3}$ can reduce the dislocation velocity in InP and achieve impurity hardening effect (Holt and Yacobi, 2007). It is essential to understand the relationships between growth morphology, defect structure and impurity incorporation during ELOG InP growth on Si for realizing device quality InP crystal on Si.

2.2.1 Morphology of ELOG InP/Si

Stripe openings of size 6 μm oriented at 30° and 60° off [110] directions were formed on the Si_3N_4 mask layer on the seed InP grown on (001) Si substrate with 4° off toward (111). ELOG InP growth was conducted in the HVPE reactor and broad lateral overgrowth can be obtained on the Si_3N_4 mask. As shown in Fig. 4A and B (Sun and Lourdudoss, 2003), facet planes are developed at the edge of the asymmetric ELOG layers, which could be due to the existence of high density threading dislocations in the MOVPE grown seed InP layer. The threading dislocations in opening will

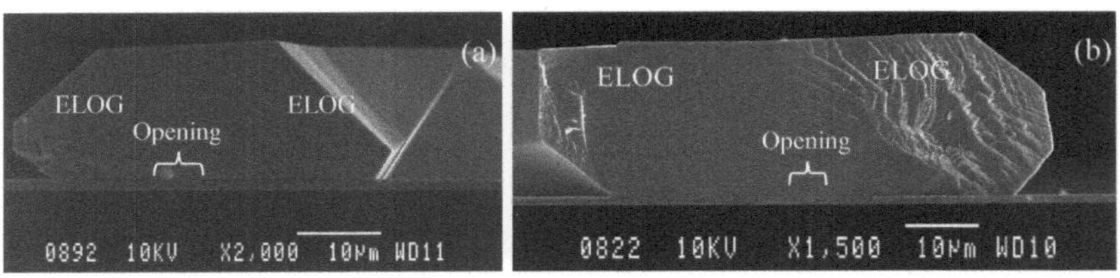

Fig. 4 SEM micrograph of ELOG InP/Si grown from the openings aligned at (A) 30° and (B) 60° off [110] directions. [110] view. *Sun, Y.-T., Lourdudoss, S., 2003. Effect of growth conditions on epitaxial lateral overgrowth of InP on InP/Si (001) substrate by hydride vapor phase epitaxy. Proc. SPIE 4997, 221–231. Copyright (2003) with permission from SPIE.*

propagate during the growth and intersect with the facet planes of ELOG to enhance the growth rate as predicted by Burton-Cabrera-Frank (BCF) model (Burton et al., 1951).

2.2.2 Influence of sulfur doping in ELOG InP/Si

The impact of impurity incorporation on ELOG growth morphology and dislocation dynamics was investigated by conducting ELOG of unintentionally-doped (UID) InP and sulfur-doped InP (InP:S) from ring shaped openings. As shown in Fig. 5A and B, fully coalesced octahedral shaped ELOG InP growths can be obtained from the ring shaped openings on Si and are bounded by {110}, {111}A and {111}B facet planes. The surface morphology of UID and sulfur-doped ELOG InP was characterized by

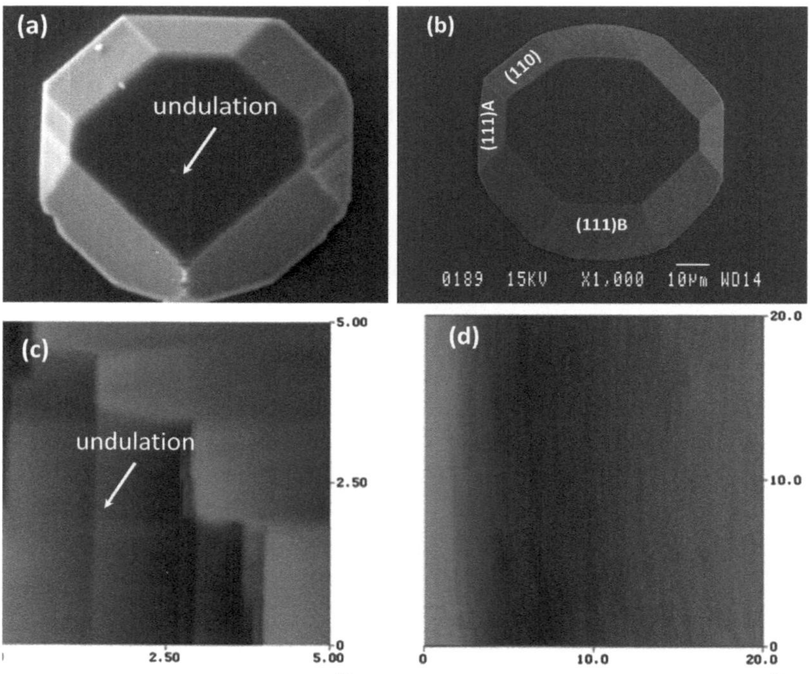

Fig. 5 SEM surface morphology images of and ELOG (A) UID InP and (B) InP:S from ring openings, respectively; (C) and (D) corresponding AFM images. *Olsson et al., 2007; Copyright (2007) with permission from Electrochemical Society; Sun, Y.-T., Baskar, K., Berggren, J., Lourdudoss, S., 2003a. InGaAsP multi-quantum wells at 1.5 μm wavelength grown on indium phosphide templates on silicon. International Conference on Indium Phosphide and Related Materials, 2003, Santa Barbara, CA, USA, 277–280; Copyright (2003) with permission from IEEE.*

scanning electron microscopy (SEM) and atomic force microscopy (AFM). Step like undulation along the $[\bar{1}10]$ direction can be observed on the surface of ELOG UID InP (Fig. 5A and C), which could be formed by the stacking faults that intersect the surface of the ELOG layer. By incorporating sulfur doping during ELOG InP growth, smoother morphology can be obtained (Fig. 5B and D). This is further discussed below.

The defect structure in heteroepitaxial InP/Si can be delineated by Huber's solution ($2H_3PO_4$:1HBr). As shown in Fig. 6A, high density etch pits are revealed on the surface of seed InP on Si grown by metal-organic vapor phase epitaxy (MOVPE). The etch pit density is reduced substantially on the surface of ELOG UID InP on Si grown from the openings at 30° off $[\bar{1}10]$ (Fig. 6B), where mainly etched grooves are visible. The etched grooves are along the $[\bar{1}10]$ direction, which are originated from stacking faults in ELOG InP. High density threading dislocations in seed InP that are exposed in the openings will propagate and glide during ELOG InP

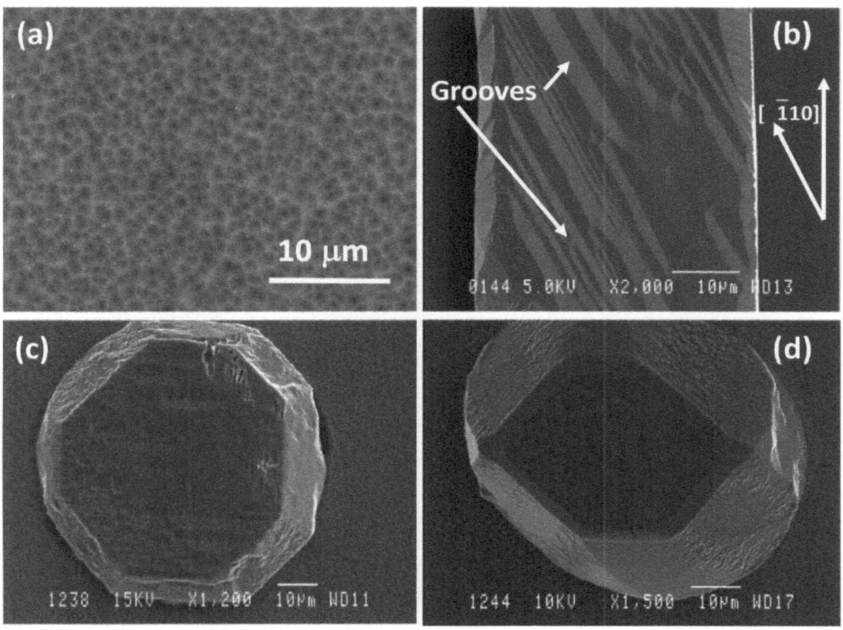

Fig. 6 SEM images of etch pit morphology after Huber's etching on (A) seed InP/Si grown by MOVPE, (B) ELOG UID InP/Si grown in stripe openings along 30° off $[\bar{1}10]$ direction with etched grooves, and (C) ELOG UID InP/Si and (D) ELOG InP:S/Si from ring shaped openings. The doped sulfur atoms reduce the gliding of dislocations in ELOG InP/Si and the formation of stack fault related etched grooves.

growth to generate stacking faults. The density of stacking faults in ELOG UID InP/Si is much lower than the original threading dislocation density in seed InP, which could be due to the interaction and annihilation of threading dislocations and stacking faults during growth. However, the sulfur dopant atoms can reduce the dislocation movement velocity in InP, thereby decelerate the gliding of dislocations that lead to the formation of stacking faults and associated surface steps. The impact of sulfur doping on the dynamics of threading dislocations during ELOG growth is revealed by the comparison of etch pit morphology on ELOG UID InP/Si and ELOG InP:S/Si grown from the ring shaped openings (Fig. 6C and D). For ELOG UID InP/Si, high density etched grooves are observed close to the edge (i.e., above the ring openings) of the octahedral shaped ELOG template in Fig. 6C, which has the same mechanism as in Fig. 6B. In contrast, isolated etch pits from threading dislocations are observed on the surface of ELOG InP:S/Si in Fig. 6D, which indicates the absence of dislocation gliding during ELOG InP:S growth.

InGaAsP multi-quantum wells (MQWs) at 1.55 μm were deposited on ELOG UID InP and InP:S templates grown from ring shaped openings on Si as shown in Fig. 5. For MQWs grown on the ELOG UID InP/Si, line shaped surface defects oriented at $[\bar{1}10]$ direction can be observed in the top view SEM image (Fig. 7A), which could be due to the extension of surface steps on the surface of the ELOG UID InP/Si (Fig. 5A). The surface of MQWs grown on the sulfur-doped ELOG InP/Si is rather clean but for certain amount of surface undulation close to the edge of template (Fig. 7C). The defect distribution in MQWs grown on the ELOG InP/Si templates are characterized by cathodoluminescence (CL) at 80 K. Dark line defects (DLDs) are revealed in the panchromatic CL (PCL) image of MQWs grown on the ELOG UID InP (Fig. 7B), which are caused by the surface defects due to stacking faults. The density of such dark line defects is higher at the edge of the ELOG UID InP/Si (i.e., just above the openings) and decreases toward the center of the grown region (i.e., ELOG region). For MQWs grown on the ELOG InP:S/Si, few dark line defects are observed and dark spot defects (DSDs) are dominant in the PCL image (Fig. 7D), which are threading dislocations from ELOG InP:S/Si template intersecting with the MQW layers. The density of DSD is higher close to the edge of ELOG InP:S/Si (i.e., just above the openings) and decreases toward the center of growth (i.e., ELOG region). Therefore, the incorporation of sulfur doping can inhibit the gliding of threading dislocations due to impurity hardening effect and prevent the formation of $\langle 110 \rangle$ oriented surface steps arising from

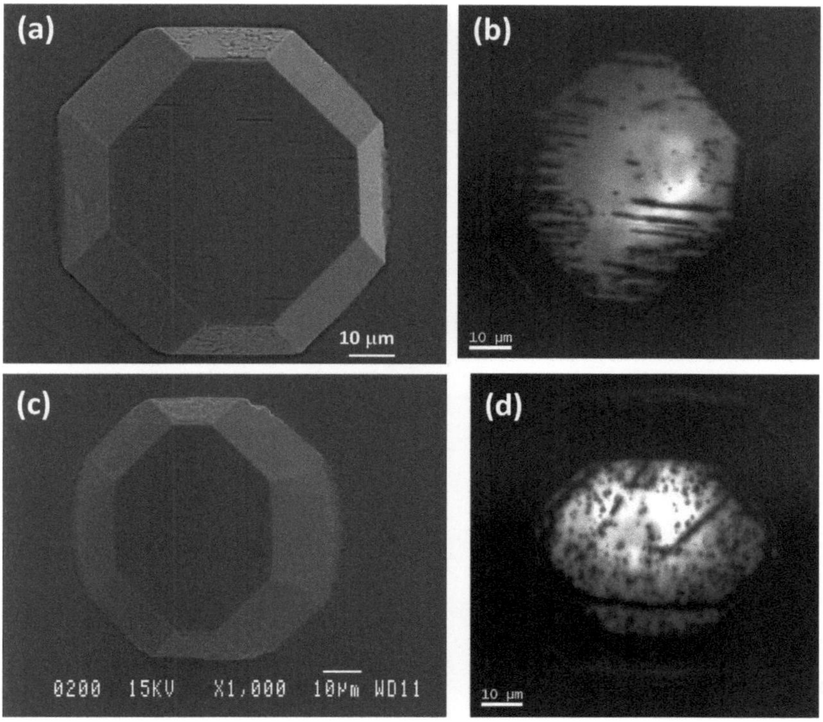

Fig. 7 (A) SEM and (B) 80K PCL micro-image of InGaAsP MQWs grown on ELOG UID InP/Si grown from ring shaped opening. (C) and (D) are, respectively, on ELOG InP:S/Si.

stacking faults. The observed impact of sulfur doping on growth morphology in ELOG from the ring shaped openings can also be utilized for achieving ELOG InP:S/Si from stripe openings with smooth morphology.

2.2.3 Strain in ELOG InP/Si

There are two major sources of strain in heteroepitaxial III-V semiconductors on Si: one is due to the lattice mismatch and the other due to the mismatch of the thermal expansion coefficients between the epitaxial layer and the substrate during cooling. For III-V semiconductors on Si, such as InP and GaAs, the lattice mismatch strain is compressive because the epitaxial layer has a larger lattice constant (InP: 5.8687 Å, GaAs: 5.65325 Å) than that of the Si substrate (Si: 5.431 Å), while the thermal strain is tensile because the epitaxial layer has a larger thermal expansion coefficient (InP: $4.60 \cdot 10^{-6}\,K^{-1}$, GaAs: $5.73 \cdot 10^{-6}\,K^{-1}$) than that of the substrate (Si: $2.6 \cdot 10^{-6}\,K^{-1}$). Almost entire lattice mismatch strain is released by the generation of misfit dislocations in the heteroepitaxial layers with lattice mismatch exceeding 4% (Nagai, 1974). In ELOG InP grown on Si, tensile strain has been constantly observed. In Sun et al. (2003b), strain status in ELOG InP/Si grown from the openings along $[\overline{1}10]$, 30° and 60° off [110] directions were investigated by high resolution X-ray diffraction (HRXRD) reciprocal lattice mapping (RLM) and low temperature photoluminescence (PL). Fig. 8A displays the HRXRDRLM of the InP seed and the ELOG UID InP. The strain in the ELOG InP/Si can be calculated by comparing

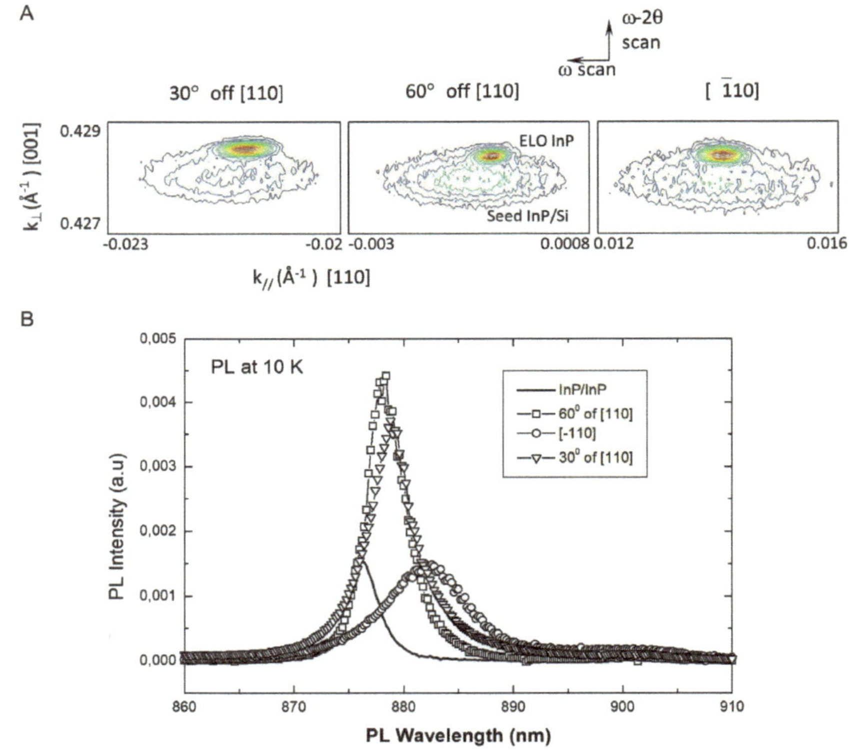

Fig. 8 (A) High resolution X-ray diffraction reciprocal lattice mapping of ELOG UID InP on Si obtained from the openings along various directions showing the (004) reflections of ELOG InP and seed InP on Si. (B) Low temperature photoluminescence of ELOG UID InP/Si grown from the openings along various orientations and the reference InP grown on plain InP:Fe substrate. ELO in (A) is same as ELOG. *Sun, Y.-T., Baskar, K., Lourdudoss, S., 2003b. Thermal strain indium phosphide on silicon obtained by epitaxial lateral overgrowth. J. Appl. Phys. 94, 2746–2748. Copyright (2003) American Institute of Physics (AIP) Publishing.*

the lattice constant of InP and that of the ELOG InP estimated from the separation between the ELOG InP peak and the Si substrate peak in HRXRDRLM around (004) reflection. The ELOG InP/Si along 30° and 60° [110] directions with higher aspect ratio between lateral and vertical growths have lower strain in the epitaxial layer. In the PL measurements, the emission wavelength of InP is affected by the strain in the grown material, since the hydrostatic component strain can change the volume of the crystal. A tensile strain in the epitaxial layer will narrow the bandgap and cause a red shift in the PL spectrum. As shown in Fig. 8B, for ELOG InP/Si along 60° [110] direction with a high aspect ratio, a red shift of 3.3 meV was obtained from the PL measurements at 10 K due to a 360 ppm tensile strain, which is lower than 9.7 meV observed for the ELOG InP grown along $[\overline{1}10]$ with low aspect ratio. The higher PL intensity observed for the ELOG InP/Si along 30° and 60° off [110] directions could be due to the orientation dependent preferential impurity incorporation resulting in high carrier concentration exhibiting Burstein-Moss effect. A good agreement is found in strain variation estimated from PL and HRXRD. These experiments clearly show that higher aspect ratio between lateral and vertical growths in ELOG InP/Si is desired to reduce the thermal strain in the epitaxial layer.

The strain status in the ELOG UID InP/Si along 60° [110] direction was analyzed by HRXRD and synchrotron X-ray topography techniques, both in large area and section geometries in Lankinen et al. (2006). HRXRD measurements reveal tensile strain in the ELOG InP layer and compressive strain in the seed InP grown by MOVPE, which are parallel to the surface and consistent with the results in Sun et al. (2003b). Synchrotron X-ray measurements reveal that the strain fields caused by the ELOG layer and the Si_3N_4 mask penetrate deeply into the Si substrate. A (713) transmission section topograph of an ELOG stripe is shown in Fig. 9. Fig. 9A is the optical microscope image of an ELOG stripe and Fig. 9B is its cross-sectional view. The influence of the ELOG structure on the distribution of the strain field in the Si substrate is revealed in Fig. 9C. In Fig. 9C the three discernible components of the strain field are also marked by three arrows, two arising from the mask regions at the side and one from the opening as indicated by the ELOG structure overlaid on Fig. 9C. It is evident that the strain is introduced during cooling after the growth because of the difference in the thermal expansion coefficients of InP and Si. Transmission topographs were also taken at (311) and (242) diffractions. The comparison of the topographs at different diffractions discerns an anisotropic distortion in the ELOG topographs with respect to the direction of ELOG stripes. This implies that

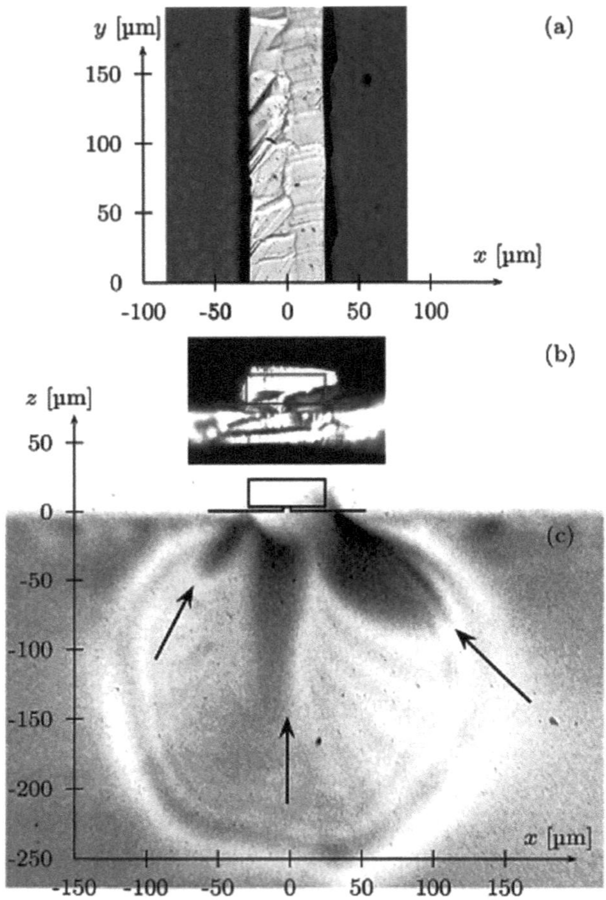

Fig. 9 (A) A top view optical microscope image of an ELOG stripe, (B) a cross-sectional optical micrograph of the stripe, and (C) a (713) transmission section topograph of the ELOG stripe showing three main components of strain field in the Si substrate marked with arrows. *Lankinen, A., Tuomi, T., Karilahti, M., Zytkiewicz, Z. R., Domagala, J. Z., McNally, P. J., Sun, Y.-T., Olsson, F., Lourdudoss, S., 2006. Crystal defects and strain of epitaxial InP layers laterally overgrown on Si. Cryst. Growth Des. 6, 1096–1100; Copyright (2006) American Chemical Society.*

the ELOG layer is strained at the direction perpendicular to stripe and not in the direction parallel to stripe. This has been explained by the small angle boundaries formed in the direction parallel to the stripe, which are absent in the direction perpendicular to the stripe.

The strain status in the coalesced ELOG InP:S/Si grown from 9 μm wide ring shaped openings was studied by photoluminescence at room temperature (Sun et al., 2003a). The derived bandgap energy of ELOG InP is 1.355 eV, which is comparable to that of reference undoped InP grown on planar substrate, 1.351 eV and indicates a material free of tensile type thermal strain. That the residual strain in the ELOG InP/Si can be reduced by coalesced growth is also observed from the ELOG InP grown from the nano-sized openings on Si as discussed below.

2.3 Nano-ELOG of InP/Si

By using micrometers wide seed window openings in ELOG InP/Si process, thick and nonuniform ELOG layers are often obtained because of the formation of stopping boundary planes, which may cause coalesced defects when two adjacent ELOG stripes merge together. In addition, wider openings allow large number of threading dislocations in the seed layer to propagate to the epitaxial layer. By making the openings size sufficiently small to nanometer range, the number of original threading dislocations propagating from the seed layer is limited. By aligning the seed window openings in the proper directions, v-shaped merging process can be realized, which will facilitate the reduction of coalesce defects by the so-called zippereffect (Yan et al., 2000).

Time resolved morphological evolution of ELOG InP grown from the mesh and the line openings were studied in Metaferia et al. (2011). The mesh and line opening patterns are shown in Fig. 10, where s is the width of the SiO_2 mask, a is the opening width of 200 nm, α and β are the angles that define the orientation of the openings. For the line openings, α angles of 15° and 30° off [110] direction were used. For the mesh openings, $\alpha = 15°$, $\beta = 105°$ and $\alpha = 15°$, $\beta = 120°$ were used. ELOG InP:S growth was conducted in HVPE reactor for 120 min. The samples were taken out for inspecting the surface morphology and its evolution at certain growth intervals. The results are shown in Fig. 11 for (A) $\alpha = 15°$ and $\beta = 105°$ (mesh), (B) $\alpha = 15°$ and $\beta = 125°$ (mesh), (C) $\alpha = 15°$ (line), and (D) $\alpha = 30°$ (line) at the growth intervals shown at the left. Full coalescence of ELOG takes place in 5 min for the mesh patterns (A and B) but 15 min for the line opening patterns (C and D). The ultimate surface roughness of the fully coalesced ELOG InP layers after 120 min growth is almost the same for both the cases (RMS = 16–25 nm).

2.3.1 *Crystalline quality of nano-ELOG InP/Si*

The quality of ELOG InP/Si grown from 200 nm and 1000 nm wide openings were analyzed by cross-sectional TEM characterization (Fig. 12).

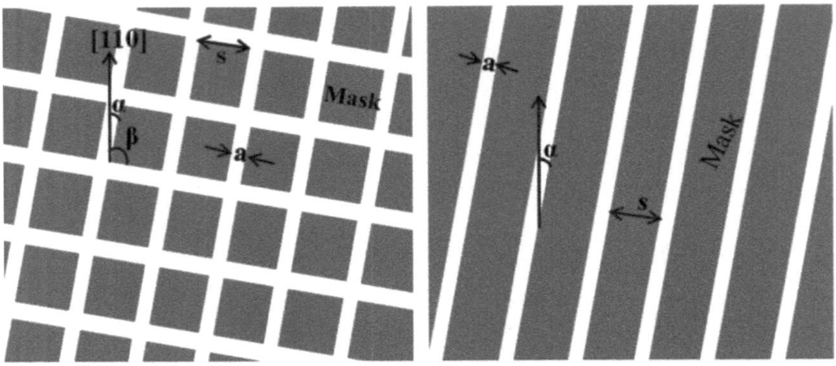

Fig. 10 Schematics of mesh (left) and line (right) openings used for ELOG; a = opening width and s = mask width; α and β are the angles that define the orientation of the openings. *Metaferia, W., Junesand, C., Gau, M.-H., Lo, I., Pozina, G., Hultman, L., Lourdudoss, S., 2011. Morphological evolution during epitaxial lateral overgrowth of indium phosphide on silicon. J. Cryst. Growth 332, 27–33; Copyright (2011), with permission from Elsevier.*

High density threading dislocations can be observed at the interface of seed InP and Si (2 in Fig. 12A), which are blocked by the mask layer but can escape through the 1000 nm wide opening (3 in Fig. 12A). According to the theoretical study by Olsson et al. (2008), if the opening size is smaller than the thickness of mask layer, filtering of dislocation can take place even above the opening. Significant dislocation filtering was observed in 200 nm wide opening even though the mask layer is only 40 nm thick. Stacking faults are found in the grown layers (1 in Fig. 12B), which can be due to imperfect coalescence caused by the surface roughness of the seed layer and/or of the mask.

The defect density in the ELOG InP:S/Si layer was studied by PCL mapping in which, dark spots and straight lines are observed (Junesand et al., 2014). The threading dislocation (TD) terminating at the surface of the grown InP layer acts as non-radiative recombination center and gives rise to dark spot in the PCL image, which has one-to-one correspondence to TD (Titchmarsh et al., 1977). According to the treatment in Junesand et al. (2014), the TD density in the ELOG InP/Si layer decreases with the grown layer thickness and the trend line of ELOG samples is displaced downward by one to two orders of magnitude as compared to the seed InP layer, which is attributed to the dislocation filtering mechanism of ELOG. However, it is possible that new TDs are formed by coalescence of ELOG from the adjacent openings, which could be due to the surface roughness of

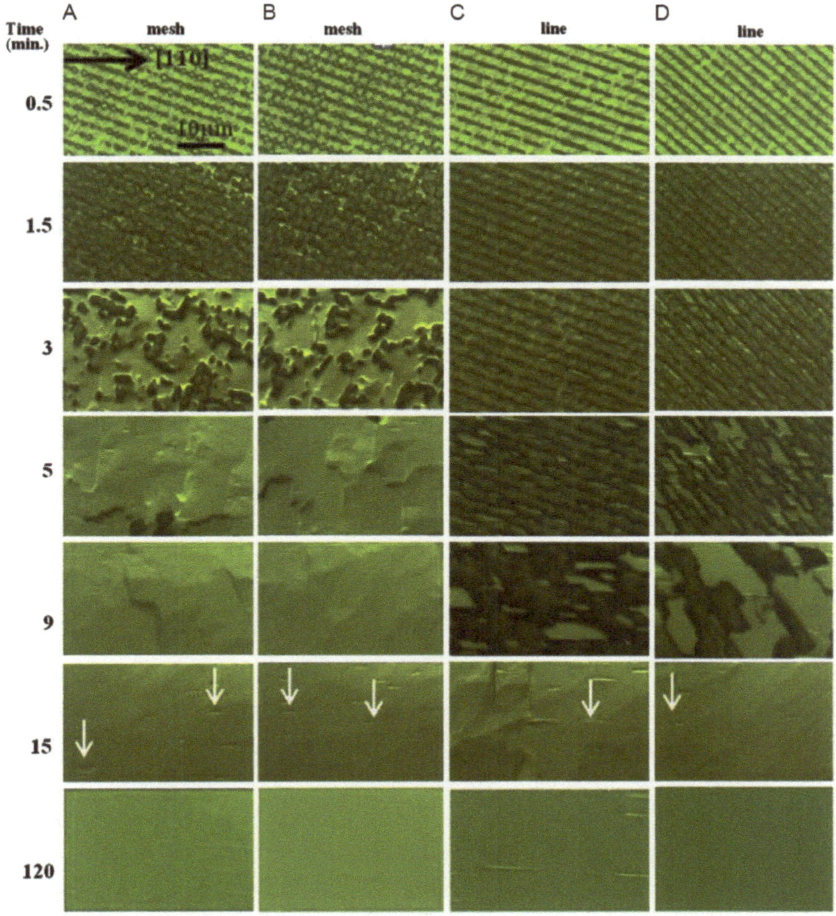

Fig. 11 Morphology evolution of ELOG InP from mesh and line patterns. (A) $\alpha = 15°$ and $\beta = 105°$ (mesh), (B) $\alpha = 15°$ and $\beta = 125°$ (mesh), (C) $\alpha = 15°$ (line), and (D) $\alpha = 30°$ (line). Time at the left is the total growth time. The arrows indicate surface steps arising from the intersection of planar defect with InP ELOG surface. *Metaferia, W., Junesand, C., Gau, M.-H., Lo, I., Pozina, G., Hultman, L., Lourdudoss, S., 2011. Morphological evolution during epitaxial lateral overgrowth of indium phosphide on silicon. J. Cryst. Growth 332, 27–33; Copyright (2011), with permission from Elsevier.*

the seed layer in the opening or the mask layer. By introducing chemical mechanical polishing of seed layer in the process flow, such coalescence defects can be reduced (Junesand et al., 2012). In addition to dark spots, dark line defects (DLDs) oriented along the [110] and $[\bar{1}10]$ directions were observed in PCL image obtained at 80 K. Such DLDs correspond to the intersection of stacking faults (SFs) with the (001) surface of the grown layer

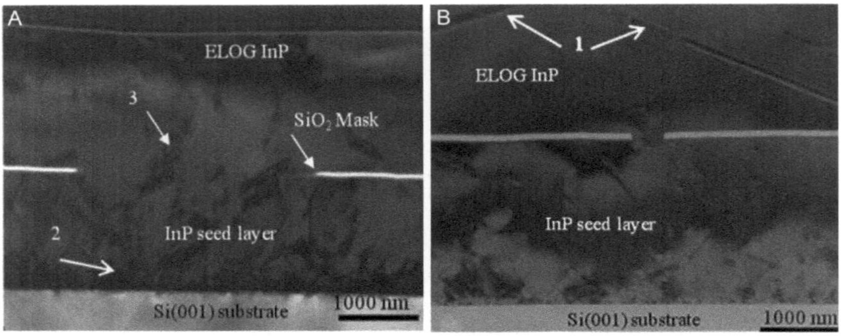

Fig. 12 TEM images of (A) sample with opening width of 1000 nm and (B) sample with opening width of 200 nm. Stacking faults (1), threading dislocations in the seed layer (2), and threading dislocations propagated through openings (3) are marked in the pictures. *Metaferia, W., Junesand, C., Gau, M.-H., Lo, I., Pozina, G., Hultman, L., Lourdudoss, S., 2011. Morphological evolution during epitaxial lateral overgrowth of indium phosphide on silicon. J. Cryst. Growth 332, 27–33; Copyright (2011), with permission from Elsevier.*

that are lying on $\{1\overline{1}1\}$ and $\{\overline{1}11\}$ planes. The DLD density is affected by V/III ratio and growth temperature. (Junesand et al., 2014). Both high and low V/III ratios will reduce the possibility of deposition error of P or In atoms. At a given V/III ratio, DLD density is lower at a higher growth temperature due to the annealing like effect on the planar defects (Cederberg et al., 2010). The length of the DLDs was measured from the room temperature PCL image of 10 μm and 100 μm thick ELOG InP layers. It was found that the SF width increases with layer thickness, which indicates that no new SFs are formed throughout the growth of ELOG but primarily formed at the early stage of growth. There are multiple SF formation mechanisms during ELOG InP/Si growth, which can be the SF propagation from seed InP and SF formation due to strain in the layers during early ELOG. SFs formed during ELOG might be possible to be removed by thermal annealing since the removal of SFs will reduce the total energy in the crystal, which has been observed in the early studies on heteroepitaxy of III-V/Si (Lee et al., 1987).

Pre-existing stacking faults in InP seed are more difficult to filter. They can propagate through openings and extend over the mask layer (Fig. 13) (Sun et al., 2015b). By analyzing the contrast of the stacking fault fringes in the cross-sectional TEM (XTEM) images with $g{\cdot}R$ criterion (Hirsch et al., 1965), it is identified that SF1, SF2 and SF3 in Fig. 13A and SF1, SF2, and SF3 in Fig. 13B are coplanar and have same inclination angle, although they are separated by the mask layers. The stacking fault in the InP seed lying on the {111} planes can intersect the openings oriented at

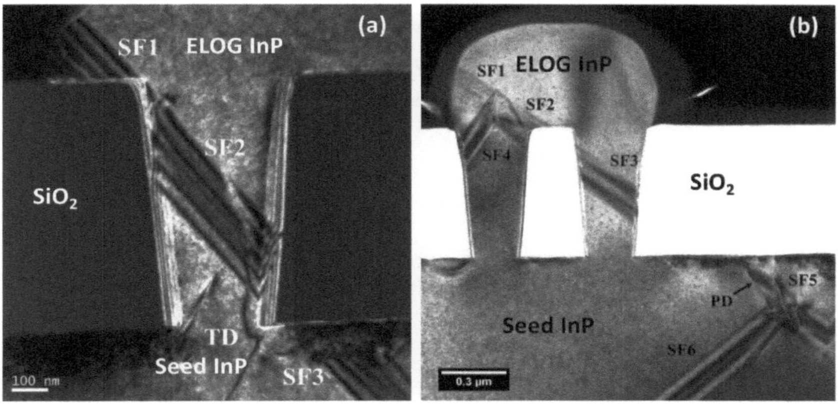

Fig. 13 (A) Dark field XTEM image of ELOG InP:S grown in parallel openings with spacing of 500 nm aligned at 30° off [110], g = [440]. (B) Bright field XTEM image of ELOG InP grown in parallel openings aligned at 60° off [110] with spacing of 300 nm, g = [220]. *Sun, Y.-T., Junesand, C., Metaferia, W., Kataria, H., Julian, N., Bowers, J., Pozina, G., Hultman, L., Lourdudoss, S., 2015b. Optical and structural properties of sulfur-doped ELOG InP on Si. J. Appl. Phys. 117, 215303-1-11; Copyright (2015) American Institute of Physics (AIP) Publishing.*

an angle off [110] direction, which is exposed in opening. Because of this, the stacking fault plane will extend in the opening with the InP vertical growth and extend laterally over mask as soon as the vertical growth reaches the surface of mask.

To account for the propagation of stacking fault through the mask opening, a model was proposed to establish the critical mask thickness with respect to the opening width necessary to hinder the SF propagation (Junesand et al., 2013). According to this model, the ratio of the mask height to the opening width should be >3.9 to ensure complete blocking of stacking faults for ELOG InP grown from openings oriented at 30° off [110] direction. However, this model has not been experimentally verified.

2.3.2 Coalesced vs uncoalesced ELOG

The crystalline quality of the coalesced and the uncoalesced ELOG InP/Si conducted in the openings 30° off [110] direction with mask thickness/opening width aspect ratio >2 was studied by μ-photoluminescence (μ-PL) at room temperature (Kataria et al., 2014). Fully coalesced ELOG InP layer was obtained from the patterns consisting of 300 nm wide openings with 1 μm spacing after a 2 min growth at 610 °C with V/III ratio of 10. As shown in Fig. 14A, crystalline defects were observed over the SiO_2 mask in the cross-sectional TEM image of coalesced ELOG InP/Si.

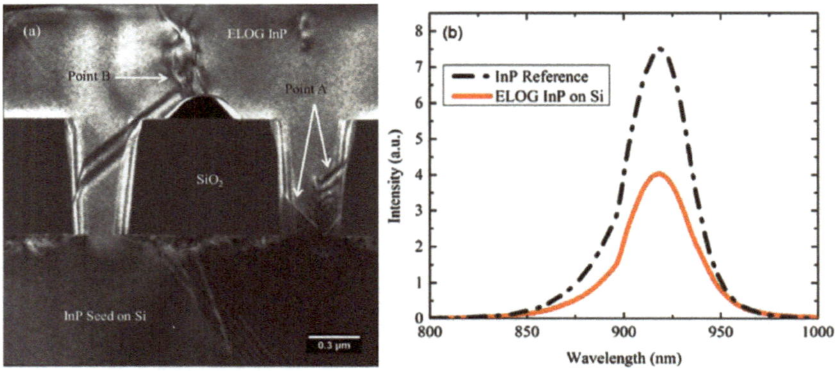

Fig. 14 (A) TEM image of cross-section of coalesced ELOG InP:S/Si. Point A indicates the defects blocked within the openings and point B indicates the coalescence defects. (B) Room temperature μ-PL measured from multiple openings on the coalesced ELOG InP:S/Si. *Kataria, H., Metaferia, W., Junesand, C., Zhang, C., Julian, N., Bowers, J. E., Lourdudoss, S., 2014. Simple epitaxial lateral overgrowth process as a strategy for photonic integration on silicon. IEEE J. Sel. Top. Quantum Electron. 20, 8201407-1-7; Copyright (2014) IEEE.*

Such coalescence defects were not observed in ELOG InP grown on InP substrate (Julian et al., 2012; Junesand et al., 2013). Therefore, the coalescence defects are not inherent to ELOG approach but determined by the seed layer quality. As shown in Fig. 14B, the PL intensity of the coalesced ELOG InP/Si is 50% lower than that of InP grown on InP substrate, which could be due to the coalescence defects in the grown layer. However, no wavelength shift of PL emission was observed indicating little strain in the grown layer, which is consistent with the thermal strain free status observed in the coalesced ELOG InP/Si grown from the ring shaped openings (Sun et al., 2003a).

Uncoalesced 20 μm wide ELOG InP layer was obtained from the pattern consisting of 1 μm openings on a 2 μm thick mask with 20 μm spacing after 17 min growth at 605 °C with V/III ratio of 10. The grown ELOG sample is characterized by cross-sectional TEM. As shown in Fig. 15A, threading dislocations are not only blocked by the SiO_2 mask layer but also inside the opening. This shows that the threading dislocations can be blocked by the sidewall of the mask when the aspect ratio of the mask thickness/opening width is larger than 1. The PL intensity and full width at half maximum (FWHM) of uncoalesced ELOG InP/Si are comparable to that of InP grown on InP substrate (Fig. 15B). This could be due to the absence of coalescence defects in this case. The emission wavelength of the uncoalesced ELOG InP/Si shows red shift which is arising from the tensile strain in the grown layer due to the larger thermal expansion coefficient of InP with respect to that of Si (which are $4.8 \times 10^{-6}\ K^{-1}$ and $2.59 \times 10^{-6}\ K^{-1}$, respectively, at ~300 K).

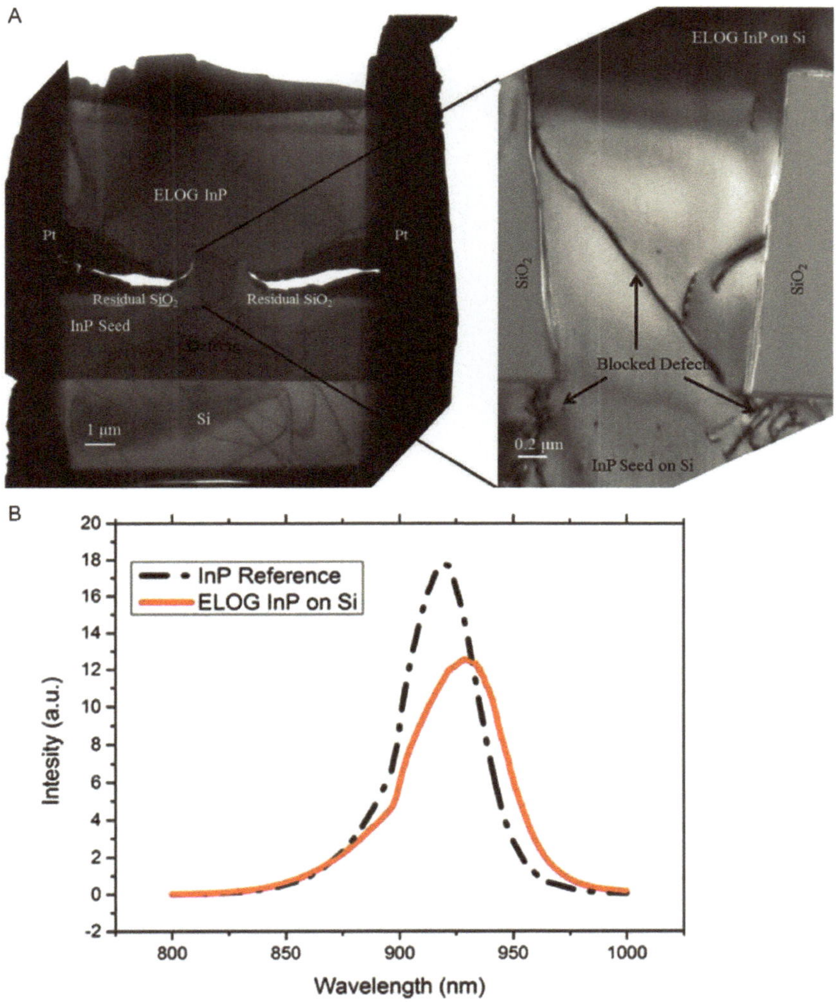

Fig. 15 (A) TEM image of the lamella taken from ELOG InP:S/Si with 1 µm wide opening in 2 µm thick SiO_2 mask with magnified view of the opening to show the defects blocked in such high aspect ratio openings defined by using optical lithography. (B) µ-PL spectra with comparable PL intensities and FWHM for ELOG InP on Si and InP reference. *Kataria, H., Metaferia, W., Junesand, C., Zhang, C., Julian, N., Bowers, J. E., Lourdudoss, S., 2014. Simple epitaxial lateral overgrowth process as a strategy for photonic integration on silicon. IEEE J. Sel. Top. Quantum Electron. 20, 8201407-1-7; Copyright (2014) IEEE.*

Multi-quantum wells (MQWs) with target wavelength of 1.55 µm were deposited on the above uncoalesced ELOG InP:S/Si and characterized by µ-PL. As shown in the cross-section TEM (Fig. 16A), highly uniform MQWs with smooth interfaces were achieved on ELOG InP:S/Si.

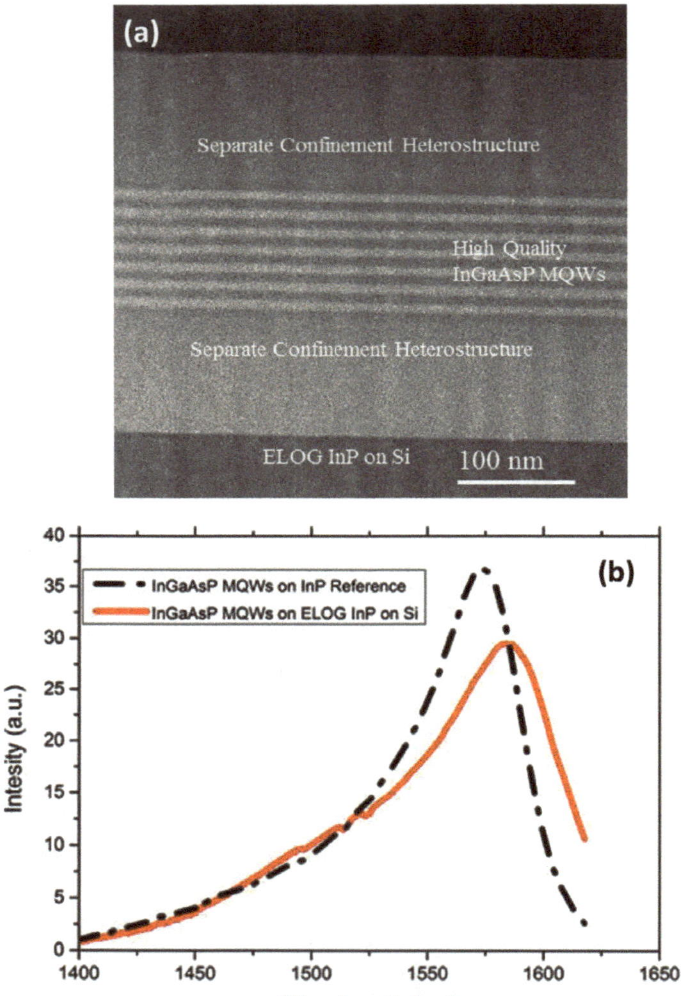

Fig. 16 (A) TEM image of InGaAsP MQWs on uncoalesced ELOG InP:S/Si. (B) μ-PL spectra of the MQWs grown on ELOG InP:S on Si and on reference InP substrate. *Kataria, H., Metaferia, W., Junesand, C., Zhang, C., Julian, N., Bowers, J. E., Lourdudoss, S., 2014. Simple epitaxial lateral overgrowth process as a strategy for photonic integration on silicon. IEEE J. Sel. Top. Quantum Electron. 20, 8201407-1-7; Copyright (2014) IEEE.*

PL intensity of the MQWs on the uncoalesced ELOG InP:S/Si was comparable to that of the reference MQWs InP substrate. However, the residual tensile strain in the uncoalesced ELOG InP:S/Si caused a red shift in the MQW emission wavelength (Fig. 16B) as was the case with InP (Fig. 15B).

As shown above, the coalescence of ELOG InP with extended growth time can give rise to a uniform large area InP layer on Si with little residual thermal tensile strain, which will facilitate the subsequent fabrication of photonic devices, such as ridge waveguide laser diodes, modulators, semiconductor optical amplifiers (SOAs) and arrayed waveguide photonic integrated circuits. However, coalescence of ELOG growths from neighboring mesas can nucleate defects at the coalescence boundaries, which has been investigated by Yan et al. (2000) and innovative opening geometeries to avoid the formation of coalescence defects in ELOG have been devised. In addition, the formation of coalescence defect is not necessarily inherent to ELOG approach but rather related to the crystalline quality of the III-V seed layer on Si and the growth conditions optimized for the formation of facets favorable for defect free coalescence, since coalescence defect free ELOG InP has been demonstrated on InP substrate (Julian et al., 2012; Junesand et al., 2013). Therefore, with careful optimization of growth conditions and opening geometry design, coalescence defect free ELOG InP on Si can be expected. The absence of thermal tensile strain in the coalesced ELOG InP/Si is beneficial for avoiding cracks in the layer for relatively large thickness (Oh et al., 2016; Yang et al., 2003).

3. CELOG III-V/Si

3.1 CELOG process

In the ELOG growth, the high quality regrown III-V semiconductor layer is separated from the Si substrate by the dielectric mask and the defective seed layer, which prevents the realization of direct III-V/Si heterojunctions with unique potential for photonic integration on silicon. As mentioned in the beginning of this chapter, CELOG offers this opportunity. The clue to obtain direct heterojunction of III-V with Si was obtained from scanning capacitance microscopy (SCM) analysis of an earlier study on temporally resolved orientation dependent dopant incorporation in ELOG (Sun et al., 2002). To understand the growth sequence of CELOG we undertook temporally resolved CELOG of InP on patterned InP substrate by growing alternating layers of UID InP and InP:S (Metaferia et al., 2015). The patterns consisted of etched InP stripe mesas of height 2 μm and width 20 μm along 30° off [110] direction on InP substrate. The mesas were covered with SiO_2 mask except at the top with an opening of 1 μm. The growth results shown in Fig. 17A indicate that InP growth gets initiated in the openings at the top and proceeds vertically and laterally. When the lateral growth reaches the

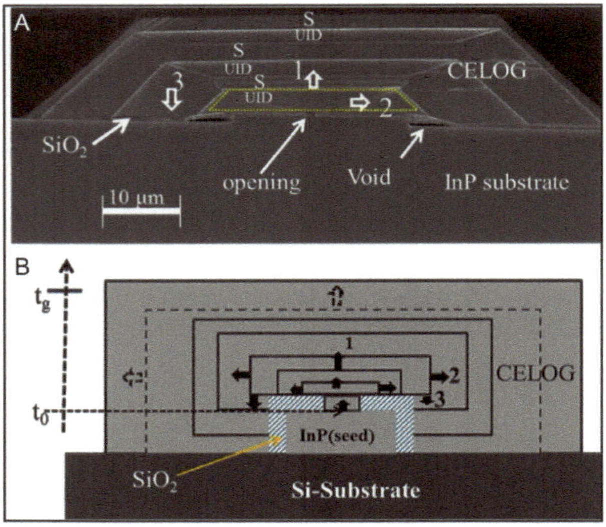

Fig. 17 (A) Cross-sectional SEM image of alternating CELOG InP:S and UID InP layers on InP substrate and (B) schematic of CELOG growth mechanism on silicon derived from the growth on InP substrate in (A). *Metaferia, W., Kataria, H., Sun, Y-T. and Lourdudoss, S., 2015. Growth of InP directly on Si by corrugated epitaxial lateral overgrowth. J. Phys. D Appl. Phys. 48, 045102-1-9; Copyright (2015) Institute of Physics.*

edge of InP mesa, the growth proceeds downward to the surface of SiO_2 coating on InP according to the clue obtained from the investigation in Sun et al. (2002). Fig. 17B describes schematically a similar situation with InP seed on Si substrate leading to InP growth on Si arising out of CELOG.

CELOG technology was exploited to fabricate III-V/Si direct heterojunction without introducing threading dislocations in the grown layers (Sun et al., 2014). The concept of CELOG for dislocation filtering was already shown in Fig. 1B. The process flow of a refined CELOG process is shown in Fig. 18. Growth is initiated with semi-insulating InP:Fe (SI-InP) on the seed to block any current through the seed in a subsequently processed device with direct III-V/Si heterojunction.

3.2 Crystalline quality of CELOG III-V/Si

TEM studies on UID InP/Si grown by CELOG show no threading dislocations but stacking faults are present (Metaferia et al., 2015). The absence of threading dislocations indicates the potential for creating a direct III-V/Si interface. Nevertheless the presence of stacking faults is disturbing. It appears that their origin is due to the deposition errors caused by the surface

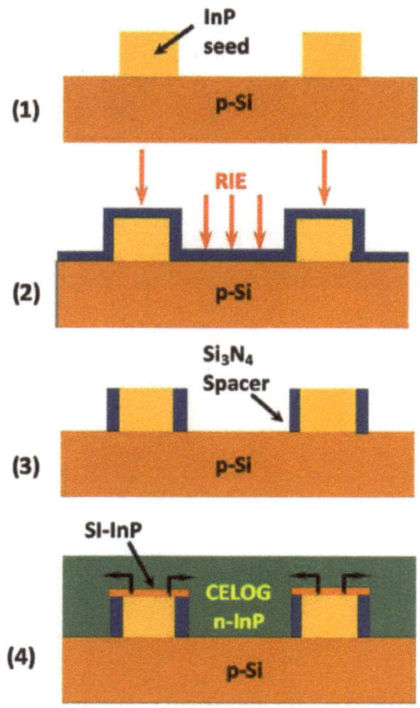

Fig. 18 Schematic description of a refined CELOG process for III-V/Si heterojunction formation exemplified with InP seed on Si. Growth is initiated with semi-insulating InP:Fe (SI-InP) on the seed to block current through the seed in a subsequently processed device with direct III-V/Si heterojunction.

roughness of Si surface formed during InP mesa etching (Ernst and Pirouz, 1988). Hence to avoid the formation of stacking faults, a more careful processing to maintain the original Si surface has to be optimized.

In another study (Sun et al., 2015a), self-aligned CELOG was investigated in which circular openings were formed in InP seed layer to expose Si surface for InP/Si direct heterojunction formation by initiating the CELOG InP growth on the surface of InP seed mesa with sidewalls buried by Si_3N_4 spacers (Fig. 19A). A direct InP/Si heterojunction is obtained with the fully coalesced InP layer on Si (Fig. 19B). The crystalline quality of CELOG InP on Si was studied by photoluminescence on the cross-sectional area InP/Si shown in Fig. 19B. Spectra taken on InP close to the CELOG InP/Si interface (S2 in Fig. 19B) and at the surface of CELOG InP (S3 in Fig. 19B) are shown in Fig. 19C. Spectrum S2 has higher PL intensity than the spectrum S3, which is even higher than that of spectrum S1 measured on InP grown on InP substrate (reference). A red shift of 4 meV is observed in the spectrum S2 with respect to spectrum S1, which could be due to band

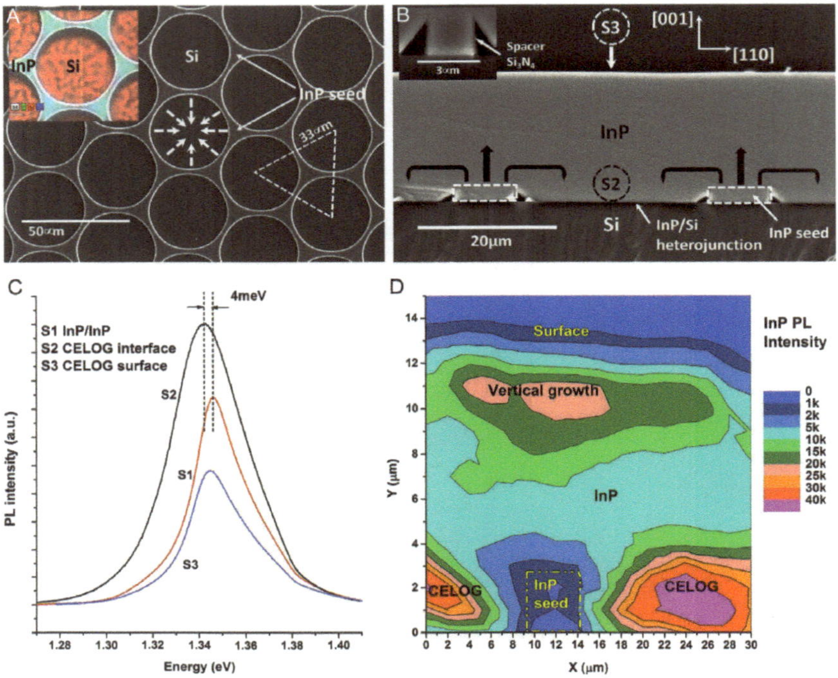

Fig. 19 (A) SEM top view of CELOG patterns with the inset of EDS elemental mapping. (B) Cross-sectional SEM image of CELOG UID InP/Si with direct heterojunction. (C) PL spectra of InP taken at positions of S2 and S3 along with reference sample S1. (D) PL mapping at the cross-section of CELOG UID InP/Si with peak wavelength of 920 nm. *Sun, Y.-T., Omanakuttan, G., Lourdudoss, S., 2015a. An InP/Si heterojunction photodiode fabricated by self-aligned corrugated epitaxial lateral overgrowth. Appl. Phys. Lett. 106, 213504-1-5; Copyright (2015) American Institute of Physics (AIP) Publishing.*

gap narrowing effect caused by the orientation dependent enhanced impurity incorporation (Bugajski and Lewandowski, 1985) as obsereved in Sun et al. (2002). The InP crystalline quality in CELOG was also studied by PL mapping of the cross-section of the CELOG InP/Si at the peak wavelength of 920 nm. As shown in Fig. 19D, the InP seed has low PL intensity due to high defect density as expected. PL intensity gradually increases along the vertical growth direction on the top of the InP seed due to the reduction of dislocation density with grown layer thickness. The most striking feature in the PL mapping is that the highest PL intensity is obtained in the CELOG region adjacent to the InP seed and above the InP/Si interface indicating efficient dislocation filtering mechanism in the CELOG approach.

The InP/Si heterojunction fabricated by self-aligned CELOG was characterized by TEM (Omanakuttan et al., 2019). Coherent InP/Si interface is

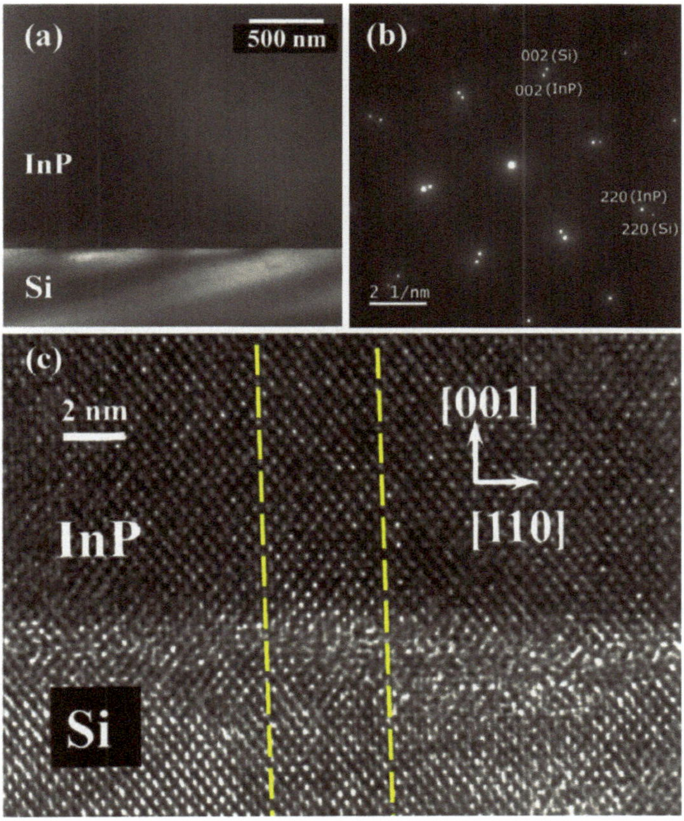

Fig. 20 (A) Bright field TEM image showing the smooth CELOG InP/Si interface. (B) Selected area electron diffraction patterns of InP and Si acquired at the InP-Si substrate interfacial region. (C) HRTEM cross section of CELOG InP/Si interface. *Omanakuttan, G., Sacristán, O. M., Marcinkevičius, S., Uždavinys, T. K., Jiménez, J., Ali, H., Leifer, K., Lourdudoss, S., Sun, Y.-T., 2019. Optical and interface properties of direct InP/Si heterojunction formed by corrugated epitaxial lateral overgrowth. Opt. Mater. Express 9, 1488–1500; Copyright (2019) OSA Publishing.*

observed in the cross-sectional TEM image (Fig. 20A). No defects are visible in the InP section above Si. Selective area electron diffraction pattern taken at the InP/Si interface shows characteristic reflections from Si substrate and InP layer (Fig. 20B) indicating high crystallinity of InP on Si. High resolution TEM image of the CELOG InP/Si heterojunction reveals the coherent InP/Si interface without intermediate amorphous layer (Fig. 20C). The (110) atomic planes were counted for Si and InP across the interface. Every 14 (110) InP planes correspond to 15 (110) Si planes in accordance with the lattice mismatch between Si and InP which creates a misfit dislocation confined to the InP/Si interface without introducing threading sections in grown layer. This situation resembles that of the III-V/Si interface formed by wafer-bonding and hence CELOG is in a way an epitaxial fusion or epitaxial bonding process.

In addition to PL and TEM characterization, we also undertook time resolved photoluminescence (TRPL) on the cross-section of the CELOG InP/Si heterostructure (i.e., on (110) plane) to quantify the minority carrier lifetime. The minority carrier lifetime of semiconductors in heterojunction is affected by the density of threading dislocations, which can act as non-radiative recombination centers to trap the photo-generated carriers and reduce their lifetime and diffusion length. It was found that the carrier life time measured at within 1–2 μm above the InP/Si interface is 710 ps, which is comparable to that measured on a 20 μm thick (001) surface of the same, 745 ps. The seed layer exhibited only 157 ps. These results are summarized in Omanakuttan et al. (2019).

The CELOG process is generic and can be applied to various III-V semiconductors for integration on Si. GaInP/Si heterojunction fabricated by CELOG approach has been investigated as a promising material system for high efficiency GaInP/Si tandem solar cell application in Omanakuttan et al. (2017). The cross-sectional SEM image of GaInP/Si achieved by CELOG is shown in Fig. 21(A). The crystalline quality of CELOG GaInP on Si was characterized by PL mapping of the cross-section of CELOG GaInP/Si heterostructure as done for InP/Si in Fig. 19(D). The distribution of PL intensity at peak wavelength of 765 nm corresponding to the bandgap of grown $Ga_{0.3}In_{0.7}P$ material is shown in Fig. 21B. As in Fig. 19(D), the seed layer has low PL intensity because of high density dislocations. The highest PL intensity is obtained in the region III of CELOG GaInP above the Si surface as marked in Fig. 21A, which indicates a reduced dislocation density in this region by CELOG. This demonstrates that GaInP initiated from InP seed layer can form interface with Si in a CELOG process.

4. Photonic devices on ELOG/CELOG III-V/Si

The ELOG and CELOG InP layers on Si with reduced dislocation density have been explored for photonic device fabrication. We present here proof-of-concept demonstration of strong photoluminescence from microdisks fabricated from multi-quantum well structure on ELOG InP/Si

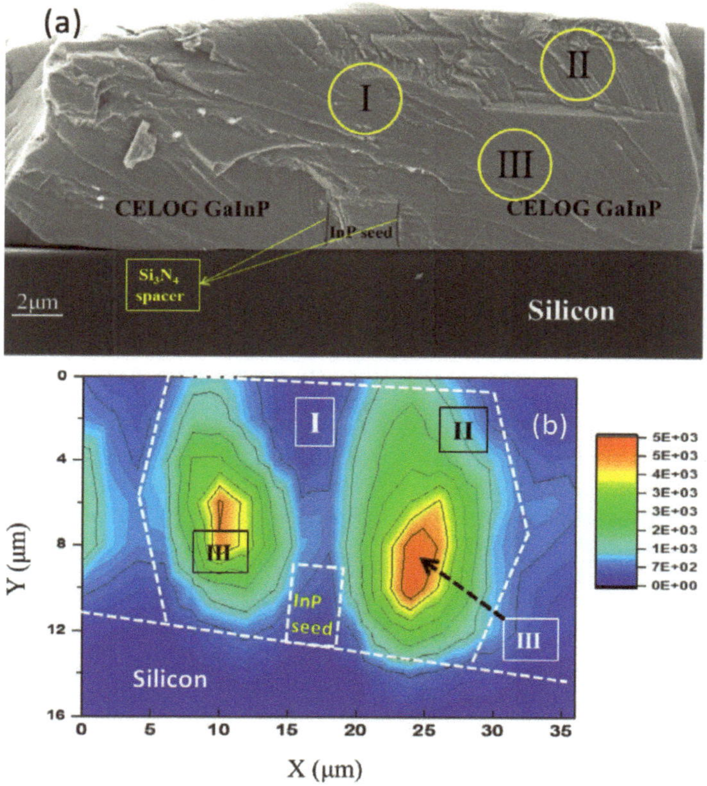

Fig. 21 (A) SEM cross-section image of CELOG GaInP/Si. (B) Cross-sectional PL intensity contour of CELOG GaInP/Si at peak wavelength of 765 nm. The regions I, II, III correspond to those marked in (A). *Omanakuttan, G., Stergiakis, S., Sahgal, A., Sychugov, I., Lourdudoss, S., Sun, Y.-T., 2017. Epitaxial lateral overgrowth of GaxIn1-xP toward direct GaxIn1-xP/Si heterojunction. Phys. Status Solidi A 214, 1600631; Copyright (2017) John Wiley and Sons.*

and ohmic behavior of heterojunction n-InP/n-Si and a heterojunction n-InP/p-Si photodiode fabricated via CELOG. All are exploitable for silicon photonics.

4.1 Multi-quantum well microdisks on nano-ELOG InP/Si

InGaAsP multi-quantum wells consisting of three 8 nm $In_{0.76}Ga_{0.24}As_{0.83}P_{0.17}$ wells surrounded by 7 nm $In_{0.48}Ga_{0.52}As_{0.83}P_{0.17}$ barriers and separate confinement heterostructure (SCH) targeting at 1.55 μm emission wavelength were grown on coalesced nano-ELOG (NELOG) patterns, which were further fabricated to microdisk (MD) structures with 8 and 12 μm diameters by

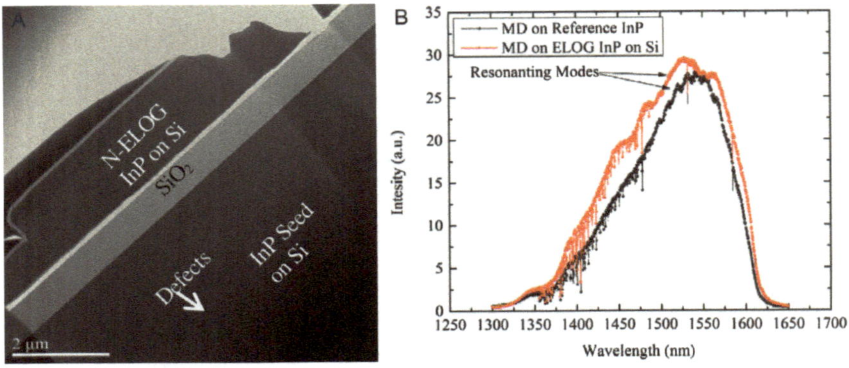

Fig. 22 (A) TEM images of etched MD along with the openings, showing that the SiO_2 mask blocks the defects. (B) PL spectra of optical pumped MQWs MDs and the reference sample with clearly visible resonant modes. *Kataria, H., Junesand, C., Wang, Z., Metaferia, W., Sun, Y. T., Lourdudoss, S., Patriarche, G., Bazin, A., Raineri, F., Mages, P., Julian, N., Bowers, J. E., 2013. Towards a monolithically integrated III–V laser on silicon: optimization of multi-quantum well growth on InP on Si. Semicond. Sci. Technol. 28, 094008-1-7; Copyright (2013) Institute of Physics.*

using e-beam lithography and inductively coupled plasma (ICP) etching (Kataria et al., 2013). As shown in Fig. 22A, the defects in the seed InP layer were blocked by SiO_2 mask layer and no sign of defects in the ELOG layer even though it was coalesced which demonstrates the feasibility of growing uniform MQW structure on coalesced NELOG InP. The etched MD with MQW was pumped by a 1.18 μm laser diode and the PL spectra of MD on NELOG and on reference InP substrate are shown in Fig. 22B, which have similar PL intensity and resonant modes. The lasing action is suppressed presumably due to sidewall roughness induced by ICP etching of MD structure. Enhanced scattering losses will cause insufficient feedback mechanism in the MD cavities required for lasing. In addition, the leakage loss through unetched ELOG InP is identified by mode solver based on finite difference method (Kataria et al., 2013). An increased etch depth into ELOG InP template will result in a low leakage loss in MQWs based microdisk laser. By further improvement of ICP process for deep mesa etching and smooth sidewall formation, monolithic integration of 1550 nm microdisk laser on Si can be realized by nano-ELOG InP/Si approach.

4.2 Heterojunction photodiode on CELOG III-V/Si

The direct III-V/Si heterojunctions with low optical and electrical losses are the essential elements in high performance heterojunction photodiodes.

Besides, in a silicon heterojunction solar cell, a wider bandgap III-V semiconductor material on Si subcell can serve as a window layer to reduce the recombination loss and enhance the open circuit voltage (Wagner et al., 2014). The CELOG technology has demonstrated InP/Si direct heterojunction by epitaxial fusion with low threading dislocations and enhanced carrier lifetime, which is promising for silicon based heterojunction photodiode applications. The electrical properties of the InP/Si heterojunction fabricated by the CELOG approach are critical for electronic and photonic device applications, which can be tuned by changing the doping polarity in the constituent materials (Sun et al., 2014). For CELOG UID (n^-) InP/n-Si heterojunction, an I-V curve with rectifying characteristic was obtained (Fig. 23A) and a highly sulfur doped CELOG n-type InP/n-Si heterojunction shows an ohmic behavior (Fig. 23B). The advantage of the absence of an amorphous layer in the InP/Si heterojunction is evident.

An InP/Si heterojunction has a type-II band discontinuity alignment, as shown in Fig. 24A, which can be explored for high efficiency Si based III-V heterojunction photodiode applications (Sun et al., 2015a). For n^-InP/p^-Si heterojunction, depletion region is formed at the InP/Si interface. By exposing the n-InP/p-Si heterojunction to light, photogenerated carriers in the depletion region will be separated by the built-in electric field and

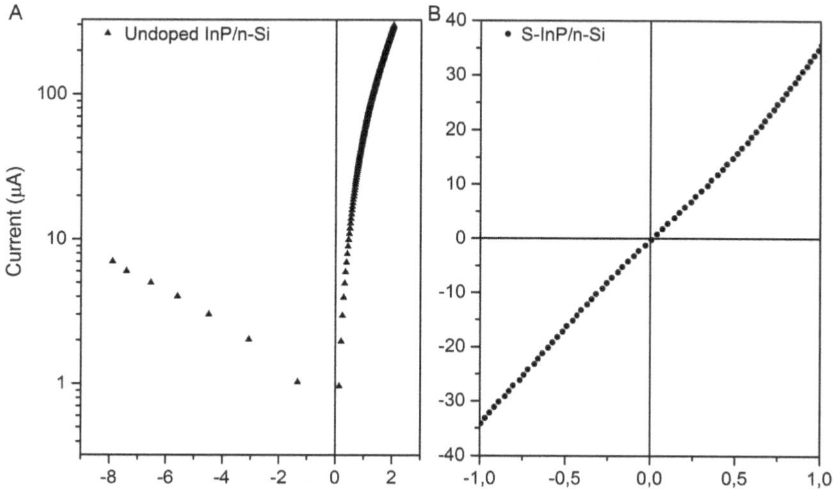

Fig. 23 (A) I-V curve of UID (n^-) InP/n-Si heterojunction. (B) I-V curve of n-InP/n-Si ohmic heterojunction. *Sun, Y.-T., Kataria, H., Metaferia W., Lourdudoss, S., 2014. Realization of an atomically abrupt InP/Si heterojunction via corrugated epitaxial lateral overgrowth. CrstEngComm 16, 7889–7893; Copyright (2014) Royal Society of Chemistry.*

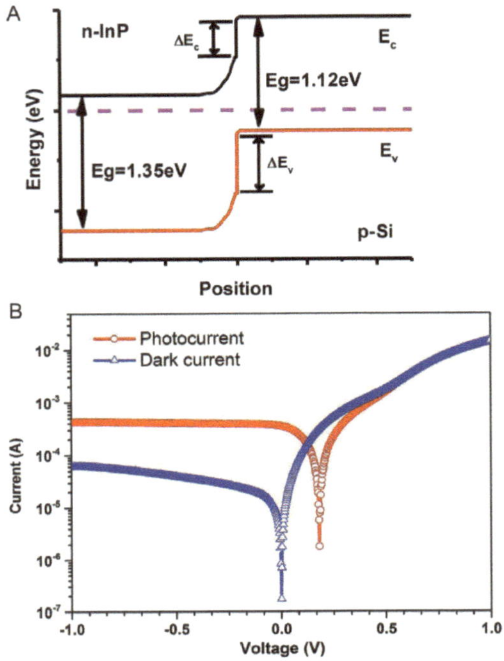

Fig. 24 (A) Type-II band alignment of n-InP/p-Si heterojunction. (B) I-V curves of n-InP/p-Si heterojunction under dark and AM1.5 illumination condition. *Sun, Y.-T., Omanakuttan, G., Lourdudoss, S., 2015a. An InP/Si heterojunction photodiode fabricated by self-aligned corrugated epitaxial lateral overgrowth. Appl. Phys. Lett. 106, 213504-1-5; Copyright (2015) American Institute of Physics (AIP) Publishing.*

the effective force fields arising from the favorable conduction and valence band discontinuities of ΔEc and ΔEv, respectively. Electron will accumulate on the n-InP side and hole will accumulate on the p-Si side. Thus, an open circuit voltage can be developed at n-InP/p-Si heterojunction as shown in the photo current-voltage (I-V) curve in Fig. 24B experimentally demonstrated in Sun et al. (2015a).

5. Summary and conclusions

Monolithic III-V/Si integration by heteroepitaxial approach is promising to realize photonic functionalities such as III-V based laser sources and photodetectors, and silicon based waveguides on Si platform cost effectively. The synergetic optoelectronic properties arising from the direct III-V/Si integration beyond add-on fashion is attractive for the applications of high efficiency Si based heterojunction photodiode. Among several approaches existing for III-V integration on Si, in this chapter we have indicated that ELOG is one among them. ELOG has been enabled by HVPE technology to reduce the threading dislocation density in III-V layers on Si. By aligning the openings in dielectric mask layer along the optimum orientation on (001) Si substrate precoated with InP seed layer, ELOG InP on Si with high aspect ratio between lateral overgrowth and vertical growth can be obtained. The wide lateral overgrowth of InP on the dielectric mask has a reduced dislocation density compared to that in the seed InP layer and the residual thermal strain in the ELOG InP grown layer arising during cooling after growth is relaxed in the coalesced layer. Although a very thick layer (10–20 μm) is forcibly formed during ELOG using microsized openings, by exploiting the nano-size openings in linear and mesh patterns, fully coalesced ELOG InP on Si can be formed with lower thickness (~5 μm). The PL spectrum of the coalesced ELOG InP/Si does not show any wavelength shift caused by thermal strain in the grown layer. Multi-quantum wells were grown on coalesced ELOG from multiple nanosized openings and subsequently processed to microdisk laser structure. Strong luminescence with resonance at cavity mode is observed in the optically pumped microdisk MQW structures on Si, which is promising for realizing lasers on Si with further development. The aspect ratio of mask structure (i.e., mask thickness with respect to the opening width) is critical for dislocation filtering in ELOG approach: aspect ratio larger than 2 is efficient for dislocation filtering. In contrast, planar defects such as stacking faults grown in openings off-oriented from [110] direction are more difficult to be completely blocked by mask wall and aspect ratio up to ~4 is required for ELOG along 30° off [110] direction. Hindering the stacking faults and their impact on the device performance is an ongoing study in this community. In order to achieve direct III-V/Si heterojunction without defective seed layer and dielectric mask layer between high quality InP and Si, CELOG approach is employed to grow InP from the surface of seed mesa with sidewalls protected by spacers. In the CELOG approach, the dislocations in the seed layer can propagate to the surface of the grown layer or escape to the lateral overgrowth region, but cannot bend toward the surface of Si, which assures the effective filtration of dislocations in the CELOG region above Si surface. Near dislocation-free CELOG InP was revealed by TEM on Si. The CELOG InP/Si template has been processed to heterojunction photodiode and photo-response arising from the type-II InP/Si heterojunction was observed under illumination. The CELOG approach is generic and can be applied to other III-V semiconductors. As an example, direct GaInP/Si heterojunction has been demonstrated with this method which is promising for high efficiency GaInP/Si dual junction solar cell application. We have indicated that ELOG offers the potential to integrate III-V gain medium on silicon waveguide and CELOG to obtain direct III-V/Si heterojunction interface, both by epitaxial means. Thereby these techniques should be further explored and optimized for large scale integration as an alternative to bonding methods.

Acknowledgments

We would like to thank Fredrik Olsson, Carl Junesand, Wondwosen Metaferia, Himanshu Kataria, Giriprasanth Omanakuttan, Zhechao Wang, Lech Wosinski, Srinivasan Anand, and Krishnan Baskar for individually contributing several important results found in this chapter. We also thank Swedish Research Council (VR), Swedish Foundation for Strategic Research (SSF), Swedish Energy Agency, KIC Innoenergy, and EU-SOLAR-ERA.NET for partially financing our research.

References

Bugajski, M., Lewandowski, W., 1985. Concentration-dependent absorption and photoluminescence of n-type InP. J. Appl. Phys. 57, 521–530.

Burton, W.K., Cabrera, N., Frank, F.C., 1951. The growth of crystals and the equilibrium structure of their surfaces. Phil. Trans. R. Soc. A 243, 299–358.

Campbell, J.C., 2007. Recent advances in telecommunications avalanche photodiodes. J. Lightwave Technol. 25, 109–121.

Cederberg, J.G., Leonhardt, D., Sheng, J.J., Li, Q., Carroll, M.S., Han, S.M., 2010. GaAs/Si epitaxial integration utilizing a two-step, selectively grown Ge intermediate layer. J. Cryst. Growth 312, 1291–1296.

Chang, Y.S., Naritsuka, S., Nishinaga, T., 1998. Optimization of growth condition for wide dislocation-free GaAs on Si substrate by microchannel epitaxy. J. Cryst. Growth 192, 18–22.

Ernst, F., Pirouz, P., 1988. Formation of planar defects in the epitaxial growth of GaP on Si substrate by metal organic chemical-vapor deposition. J. Appl. Phys. 64, 4526–4530.

Hawkins, A.R., Wu, W., Abraham, P., Streubel, K., Bowers, J.E., 1996. High gain-bandwidth-product silicon heterointerface photodetector. Appl. Phys. Lett. 70, 303–305.

Hirsch, P.B., Howie, A., Nicholson, R.B., Pashley, D.W., Whelan, M.J., 1965. Electron Microscopy of Thin Crystals. William Clowes & Sons Ltd.

Holt, D.B., Yacobi, B.G., 2007. Extended Defects in Semiconductors. Cambridge University Press.

Julian, N., Mages, P., Zhang, C., Zhang, J., Kraemer, S., Stemmer, S., Denbaars, S., Coldren, L., Petroff, P., Bowers, J., 2012. Coalescence of InP epitaxial lateral overgrowth by MOVPE with V/III ratio variation. J. Electron. Mater. 41, 845–852.

Junesand, C., Hu, C., Wang, Z., Metaferia, W., Dagur, P., PozinaLars Hultman, G., Lourdudoss, S., 2012. Effect of the surface morphology of seed and mask layers on InP grown on Si by epitaxial lateral overgrowth. J. Electron. Mater. 41, 2345–2349.

Junesand, C., Kataria, H., Metaferia, W., Julian, N., Wang, Z., Sun, Y.-T., Bowers, J., Pozina, G., Hultman, L., Lourdudoss, S., 2013. Study of planar defect filtering in InP grown on Si by epitaxial lateral overgrowth. Opt. Mater. Express 3, 1960–1973.

Junesand, C., Gau, M.-H., Sun, Y.-T., Lourdudoss, S., Lo, I., Jimenez, J., Postigo, P.A., Morales, F.M., Hernandez, J., Molina, S., Abdessamad, A., Pozina, G., Hultman, L., Pirouz, P., 2014. Defect reduction in heteroepitaxial InP on Si by epitaxial lateral overgrowth. Mater. Express 4, 41–53.

Kataria, H., Junesand, C., Wang, Z., Metaferia, W., Sun, Y.T., Lourdudoss, S., Patriarche, G., Bazin, A., Raineri, F., Mages, P., Julian, N., Bowers, J.E., 2013. Towards a monolithically integrated III–V laser on silicon: optimization of multi-quantum well growth on InP on Si. Semicond. Sci. Technol. 28. 094008-1-7.

Kataria, H., Metaferia, W., Junesand, C., Zhang, C., Julian, N., Bowers, J.E., Lourdudoss, S., 2014. Simple epitaxial lateral overgrowth process as a strategy for photonic integration on silicon. IEEE J. Sel. Top. Quantum Electron. 20. 8201407-1-7.

Langdo, T.A., Leitz, C.W., Currie, M.T., Fitzgerald, E.A., Lochtefeld, A., Antoniadis, D.A., 2000. High quality Ge on Si by epitaxial necking. Appl. Phys. Lett. 76, 3700–3702.

Lankinen, A., Tuomi, T., Karilahti, M., Zytkiewicz, Z.R., Domagala, J.Z., McNally, P.J., Sun, Y.-T., Olsson, F., Lourdudoss, S., 2006. Crystal defects and strain of epitaxial InP layers laterally overgrown on Si. Cryst. Growth Des. 6, 1096–1100.

Lee, J.W., Shichijo, H., Tsai, H.L., Matyi, R.J., 1987. Defect reduction by thermal annealing of GaAs layers grown by molecular beam epitaxy on Si substrates. Appl. Phys. Lett. 50, 31–33.

Lourdudoss, S., Kjebon, O., 1997. Hydride vapor phase epitaxy revisited. IEEE J. Sel. Top. Quantum Electron. 3, 749–767.

Matsunaga, Y., Naritsuka, S., Nishinaga, T., 2002. A new way to achieve dislocation-free heteroepitaxial growth by molecular beam epitaxy: vertical microchannel epitaxy. J. Cryst. Growth 237–239, 1460–1465.

Metaferia, W., Junesand, C., Gau, M.-H., Lo, I., Pozina, G., Hultman, L., Lourdudoss, S., 2011. Morphological evolution during epitaxial lateral overgrowth of indium phosphide on silicon. J. Cryst. Growth 332, 27–33.

Metaferia, W., Kataria, H., Sun, Y.-T., Lourdudoss, S., 2015. Growth of InP directly on Si by corrugated epitaxial lateral overgrowth. J. Phys. D Appl. Phys. 48. 045102-1-9.

Michel, J., Liu, J., Kimerling, L.C., 2010. High-performance Ge-on-Si photodetectors. Nat. Photonics 4, 527–534.

Nagai, H., 1974. Structure of vapour-deposited $Ga_xIn_{1-x}As$ crystal. J. Appl. Phys. 45, 3789–3794.

Oh, S., Jun, D.H., Shin, K.W., Choi, I., Jung, S.H., Choi, J., Park, W., Park, Y., Yoon, E., 2016. Control of crack formation for the fabrication of crack-free and self-isolated high-efficiency gallium arsenide photovoltaic cells on silicon substrate. IEEE J. Photovoltaics 6, 1031–1035.

Olsson, F.K., Aubert, A., Avella, M., Jiménez, J., Barrios, C.A., Berggren, J., Lourdudoss, S., 2007. Heteroepitaxy of InP on silicon-on-insulator for optoelectronic integration. ECS Trans. 3, 23–29.

Olsson, F., Xie, M., Lourdudoss, S., Prieto, I., Postigo, P.A., 2008. Epitaxial lateral overgrowth of InP on Si from nano-openings: theoretical and experimental indication for defect filtering throughout the grown layer. J. Appl. Phys. 104. 093112-1-6.

Omanakuttan, G., Stergiakis, S., Sahgal, A., Sychugov, I., Lourdudoss, S., Sun, Y.-T., 2017. Epitaxial lateral overgrowth of $GaxIn_{1-x}P$ toward direct $GaxIn_{1-x}P$/Si heterojunction. Phys. Status Solidi A 214. 1600631.

Omanakuttan, G., Sacristán, O.M., Marcinkevičius, S., Uždavinys, T.K., Jiménez, J., Ali, H., Leifer, K., Lourdudoss, S., Sun, Y.-T., 2019. Optical and interface properties of direct InP/Si heterojunction formed by corrugated epitaxial lateral overgrowth. Opt. Mater. Express 9, 1488–1500.

Stringfellow, G.B., 1991. Fundamental aspects of vapor growth and epitaxy. J. Cryst. Growth 115, 1–11.

Sun, Y.-T., Lourdudoss, S., 2003. Effect of growth conditions on epitaxial lateral overgrowth of InP on InP/Si (001) substrate by hydride vapor phase epitaxy. Proc. SPIE 4997, 221–331.

Sun, Y.-T., Messmer, E.R., Söderström, D., Jahan, D., Lourdudoss, S., 2001. Temporally resolved selective area growth of InP in the openings off-oriented from [110] direction. J. Cryst. Growth 225, 9–15.

Sun, Y.-T., Anand, S., Lourdudoss, S., 2002. Crystallographic orientation dependence of impurity incorporation during epitaxial lateral overgrowth of InP. J. Cryst. Growth 237 (239), 1418–1422.

Sun, Y.-T., Baskar, K., Berggren, J., Lourdudoss, S., 2003a. InGaAsP multi-quantum wells at 1.5 μm wavelength grown on indium phosphide templates on silicon. In: International Conference on Indium Phosphide and Related Materials, 2003, Santa Barbara, CA, USA, pp. 277–280.

Sun, Y.-T., Baskar, K., Lourdudoss, S., 2003b. Thermal strain indium phosphide on silicon obtained by epitaxial lateral overgrowth. J. Appl. Phys. 94, 2746–2748.

Sun, Y.-T., Kataria, H., Metaferia, W., Lourdudoss, S., 2014. Realization of an atomically abrupt InP/Si heterojunction via corrugated epitaxial lateral overgrowth. CrstEngComm 16, 7889–7893.

Sun, Y.-T., Omanakuttan, G., Lourdudoss, S., 2015a. An InP/Si heterojunction photodiode fabricated by self-aligned corrugated epitaxial lateral overgrowth. Appl. Phys. Lett. 106. 213504-1-5.

Sun, Y.-T., Junesand, C., Metaferia, W., Kataria, H., Julian, N., Bowers, J., Pozina, G., Hultman, L., Lourdudoss, S., 2015b. Optical and structural properties of sulfur-doped ELOG InP on Si. J. Appl. Phys. 117. 215303-1-11.

Titchmarsh, J.M., Booker, G.R., Harding, W., Wight, D.R., 1977. Carrier recombination at dislocations in epitaxial gallium phosphide layers. J. Mater. Sci. 12, 341–346.

Wagner, H., Ohrdes, T., Dastgheib-Shirazi, A., Puthen-Veettil, B., König, D., Altermatt, P.P., 2014. A numerical simulation study of gallium phosphide/silicon heterojunction passivated emitter and rear solar cells. J. Appl. Phys. 115. 044501-6.

Wang, Z., Junesand, C., Metaferia, W., Hu, C., Wosinski, L., Lourdudoss, S., 2012. III–Vs on Si for photonic applications—a monolithic approach. Mater. Sci. Eng., B 177, 1551–1557.

Yan, Z., Hamaoka, Y., Naritsuka, S., Nishinaga, T., 2000. Coalescence in microchannel epitaxy of InP. J. Cryst. Growth 212, 1–10.

Yang, V.K., Groenert, M., Leitz, C.W., Pitera, A.J., Currie, M.T., Fitzgerald, E.A., 2003. Crack formation in GaAs heteroepitaxial films on Si and SiGe virtual substrates. J. Appl. Phys. 93, 3859.

CHAPTER SIX

Monolithic integration of lattice-matched Ga(NAsP)-based laser structures on CMOS-compatible Si (001) wafers for Si-photonics applications

Kerstin Volz[a], Peter Ludewig[b], Wolfgang Stolz[a,b,*]
[a]Structure and Technology Research Laboratory, Material Sciences Center and Department of Physics, Philipps-University, Marburg, Germany
[b]NAsP III/V GmbH, Marburg, Germany
*Corresponding author: e-mail address: Wolfgang.Stolz@Physik.Uni-Marburg.de

Contents

1. Introduction and motivation

Si-photonics is based on the mature process technology developed over decades for the realization of ever more complex complementary metal oxide semiconductor (CMOS) integrated circuits on Si (001) exact wafer orientations. One of the key components for full scale functionality of Si-photonics circuitry is the integrated laser emitter, as also exemplified in this volume of "Semiconductors and Semimetals," where the present status of development of various concepts for both hybrid as well as monolithic

Semiconductors and Semimetals, Volume 101
ISSN 0080-8784
https://doi.org/10.1016/bs.semsem.2019.07.003

integrated laser sources on CMOS-compatible Si (001)-substrate are presented and discussed. These concepts for laser integration can be categorized into three main areas: (i) hybrid integration concepts by laser die or wafer bonding (see also volume 99 of this series, chapters "Quantum dot lasers for silicon photonics" by Arakawa et al.; "Epitaxial lateral overgrowth of III-V semiconductors on Si for photonic integration" by Sun and Lourdudoss; "Monolithic integration of lattice-matched Ga(NAsP)-based laser structures on CMOS-compatible Si (001) wafers for Si-photonics applications" by Volz et al.; "Growth of III-V semiconductors and lasers on silicon substrates by MOCVD" by Shi and Lau), (ii) heteroepitaxial deposition of lattice-mismatched, standard III/V-semiconductor laser stacks (see also volume 99 of this series chapters "Building blocks of silicon photonics" by Vivien et al.; "Heterogeneously integrated III–V photonic devices on Si" by Matsuo), and (iii) lattice-matched epitaxial growth of Ga(NAsP)-based laser on CMOS-compatible Si (001) in this chapter.

All heteroepitaxial approaches have to create at least a misfit dislocation array at the interface of the respective III/V-material and Si in order to accommodate the significant differences in the lattice constants. This strain relaxation mechanism also leads to the formation of a significant amount of threading dislocations, which have hindered the realization of long lifetime laser devices in the past (Chen et al., 1987; Choi et al., 1991; Deppe et al., 1990; Egawa et al., 1990, 1994; Sugo et al., 1992). Remaining threading dislocation densities typically in the range of mid 10^6 cm^{-2} have been reported (Chen et al., 2016; Jung et al., 2017) in recent years. For reliable laser devices also for integrated lasers on Si continuous wave (cw) operation at least up to 85 °C with lifetimes at these high temperatures in excess of 10^5 h are required. These characteristics have not been reported for any heteroepitaxial laser material to date. In contrast, in the lattice-matched Ga(NAsP)-based approach no dislocations are formed from the very beginning, thus, this concept offers the potential to realize monolithic integrated laser emitters on Si (001) with a potentially high performance and long lifetime for Si-photonics applications.

As a proof-of-principle this novel material class has been realized and investigated on GaP (001)-substrates, which possess almost the same lattice constant as Si. Following this study on GaP-substrate, the most important prerequisites for a monolithically integrated efficient laser material have been verified. The direct band gap character of this material class has been proven by photoluminescence (PL) and photoluminescence excitation spectroscopy (PLE) for As-concentrations in excess of about 70% (Kunert et al., 2006).

After optimization of the metal organic vapor phase epitaxy (MOVPE) growth conditions (Kunert et al., 2007) high internal quantum efficiencies of Ga(NAsP) quantum well heterostructures (QWH) approaching 100% at low temperature and 50% at room temperature could be obtained (Rosemann et al., 2013). For these Ga(NAsP)-QWH electrical injection lasers at room temperature with reasonable threshold current densities of 4.3 kA/cm^2 could be demonstrated (Hossain et al., 2011a,b). In a detailed evaluation of these laser devices, the performance limiting recombination path could be identified as carrier leakage in this novel Ga(NAsP)/GaP-QWH material class (Hossain et al., 2012). All of these properties obtained at an early stage of development for this lattice-matched Ga(NAsP)/GaP-based laser material clearly point to great potential for the subsequent monolithic integration onto CMOS-compatible Si (001) substrate wafers.

In this chapter, the present status of understanding of material properties, when integrated on CMOS-compatible Si (001) substrates, with respect to structural, electrical and optoelectronic properties as basis for efficient laser sources in the important data com wavelength range around 850 nm, the extended data com range around 950 nm as well as toward the first telecom window around 1.3 μm for Si-photonics applications are summarized. This chapter is organized as follows. In Section 2 the MOVPE growth conditions of Ga(NAsP)-based heterostructures on (001) Si are described briefly. Section 3 details the atomic scale structural analysis and material stability of Ga(NAsP)-heterostructures, when grown lattice-matched on Si (001) substrate. The properties of the (BAlGa)(AsP)-material class, which is applied as barrier, electrical contact and waveguide as well as separate confinement heterostructure (SCH) material, are given in Section 4. In Section 5 the optoelectronic properties of Ga (NAsPSb)-based heterostructures and lasers on (001) Si-substrate are summarized. Finally, the monolithic integration concept for Ga(NAsPSb)-based laser structures into the standard Si-CMOS-process flow are exemplified and discussed in Section 6. Section 7 ends this chapter with conclusions.

2. Epitaxial growth of (BGa)((AsP)/Ga(NAsP)-hetero- and laser structures on (001) Si by metal organic vapor phase epitaxy

This section motivates and summarizes the epitaxial growth conditions by MOVPE of the applied GaP-on-Si (001) templates as well as the subsequent realization of Ga(NAsP)/GaP/(BGa)(AsP)-heterostructures on

CMOS-compatible Si (001) wafer orientation. In addition, the detailed structural characteristics as obtained by HR-XRD, AFM as well as (S)TEM are presented and discussed with respect to the challenges of the monolithic integration of III/V-compounds on Si.

The MOVPE growth studies for Ga(NAsP)/(BGa)(AsP)-heterostructures on Si (001) have been performed in research MOVPE equipment (Aix 200-gas foil rotation reactor from Aixtron) for deposition on 2″ substrate wafers as well as a production-type MOVPE-twin reactor system for deposition on 300 mm Si (001) wafers. This twin MOVPE cluster tool (Crius R system from Aixtron) is equipped with one 300 mm Si-reactor as well as one separate 300 mm III/V-reactor, which are connected via an automated 300 mm wafer transfer with a FOUP-cassette system for wafer loading. The specific MOVPE growth conditions for the various materials and heterostructures are detailed in the subsequently mentioned publications.

From the very beginning we have concentrated our efforts for the Si-integration on standard CMOS-compatible wafer specifications. In particular, the Si (001)-substrate wafer orientation is exact within $\pm 0.5°$. When integrating III/V-semiconductors on Si (001), one faces several challenges, which have to be solved leading to a defect-free III/V-nucleation as a prerequisite for a defect-free subsequent deposition of the active lattice-matched III/V-laser stack. In the design of the epitaxial deposition process one has to establish a charge-neutral interface configuration due to the polar character of all III/V-semiconductors on the nonpolar Si, one has to eliminate the antiphase disorder, originating from possible monoatomic steps on the Si-substrate, at a very early state of III/V-deposition. In addition it is important to establish 2D-growth conditions for the III/V-material, avoiding the formation of stacking faults and twins during the coalescence of 3D-islands formed under not optimized growth conditions. In order to solve these mentioned challenges we have chosen GaP as the nearest lattice-matched binary III/V—material in the starting phase of III/V-deposition on CMOS-compatible Si (001) wafers. Details of this complex process have been reported in Volz et al. (2011). The microscopic local arrangement at the GaP/Si-interface leading to a charge-neutral interface configuration has been elaborated both experimentally by aberration-corrected scanning transmission electron microscopy (STEM) and theoretically by density functional theory (DFT) modeling in Beyer et al. (2016).

The understanding of this complex nucleation and overgrowth process (Volz et al., 2011) has led to a defect-free GaP-on-Si (001) template also on 300 mm substrate wafers as exemplified in the subsequent structural analysis

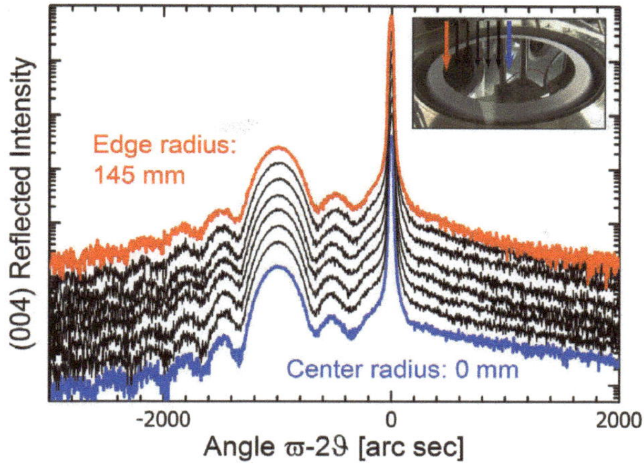

Fig. 1 High-resolution X-ray diffraction (HR-XRD) pattern of GaP-template layers on Si around the (400) reflection. The inset shows the position of the HR-XRD-measurements on the 300 mm wafer surface.

by high-resolution X-ray diffraction (HR-XRD), dark-field (DF)-TEM as well as atomic force microscopy (AFM). The HR-XRD pattern around the (004) reflex of Si is shown in Fig. 1 as a function of the position on the 300 mm Si-wafer as indicated in the inset from the wafer center to the edge of the wafer. The slightly compressively strained GaP-nucleation and buffer layer is clearly separated from the Si-substrate pattern. The distinct observation of Pendellösungs-fringes indicates that a high quality, plane-parallel GaP-layer with smooth surface morphology has been realized on Si.

The DF-TEM analysis of the deposited GaP-layer stack is exemplified in Fig. 2, where a [1-10]-cross section is depicted applying (002)-DF-TEM imaging conditions. The self-annihilation of antiphase domains (APD), visible as black contrast at the GaP-Si-Interface in the first 10's of nanometers as well as the smooth surface morphology of the GaP-layer are clearly detected in this TEM micrograph. The excellent surface morphology of this GaP-on-Si (001) template has also been verified by detailed AFM-analysis as a function of MOVPE growth conditions. Identical surface morphologies have been achieved in the 300 mm wafer center, the middle periphery as well as the wafer edge, as verified by 5 μm by 5 μm range AFM pictures in Fig. 3A–C. The actual position of AFM images on the wafer is indicated in the inset of Fig. 1. The excellent surface smoothness of these GaP-on-Si (001) templates is verified by obtaining surface roughnesses of 0.25 nm root-mean-square (rms).

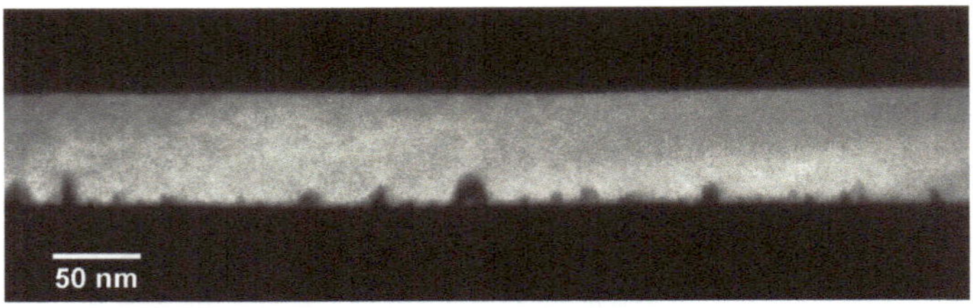

Fig. 2 Cross-sectional (viewing direction [1–10]) dark field (DF) transmission electron micrograph (TEM) of GaP-template layers on Si-substrate ((002)-reflex).

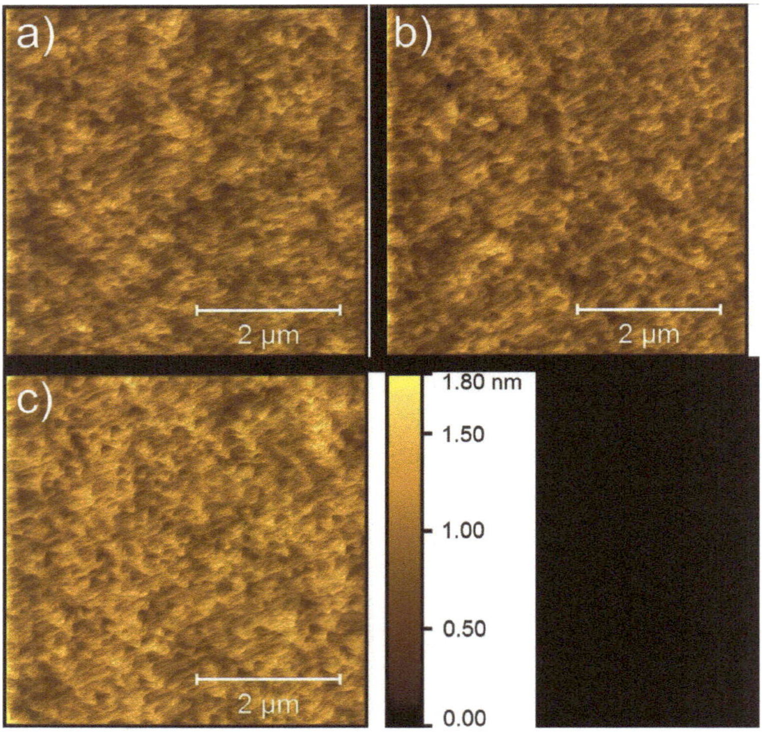

Fig. 3 Surface morphology of GaP-template layers on Si-substrate as determined by atomic force microscopy (AFM) at the wafer edge (A), at half radius (B) as well as in the wafer center (C), respectively.

These high-quality GaP-on-Si (001) templates are a perfect platform for subsequent deposition of III/V-layer stacks both for the heteroepitaxial, lattice-mismatched as well as for lattice-matched concepts. As an example for a heteroepitaxial deposition, Jung et al. (2017) have reported a significant reduction in remaining threading dislocation densities for their quantum dot

(QD) laser structures grown on Si (001)-substrate from $10^8\ cm^{-2}$ for direct growth of GaAs on Si to $7 \times 10^6\ cm^{-2}$ for the growth of the heteroepitaxial GaAs-based QD layer stack on a GaP-on-Si (001) template.

For the optimal design of the epitaxial growth process in particular for the lattice-matched approach one has to take the difference in thermal expansion coefficient between III/V-materials and Si into account. This is exemplified for GaP as well as the ternary (BGa)P-mixed crystalline system in Fig. 4, where the change in lattice constants of the respective materials is depicted for the temperature range from 250 K to 1000 K. Linear thermal expansions coefficients of $3.3 \times 10^{-6}\ K^{-1}$ for GaP and $2.1 \times 10^{-6}\ K^{-1}$ have been assumed from Kudman and Paff (1972) and Okada and Tokumaru (1984). As we would like to integrate also strained layer stacks in the active laser part of our integrated laser stack, as it is usually applied in the state of the art standard III/V-laser designs (i.e., Singh, 1995), careful strain-compensation schemes have to be developed also for the present Ga(NAsP)-based laser material on Si-substrate. These schemes require to integrate B-containing layer stacks in order to realize also tensile-strained (BGa)((As)P)-materials for precise strain compensation of potentially compressively strained active Ga(NAsP)-QWH layers. It is important to note

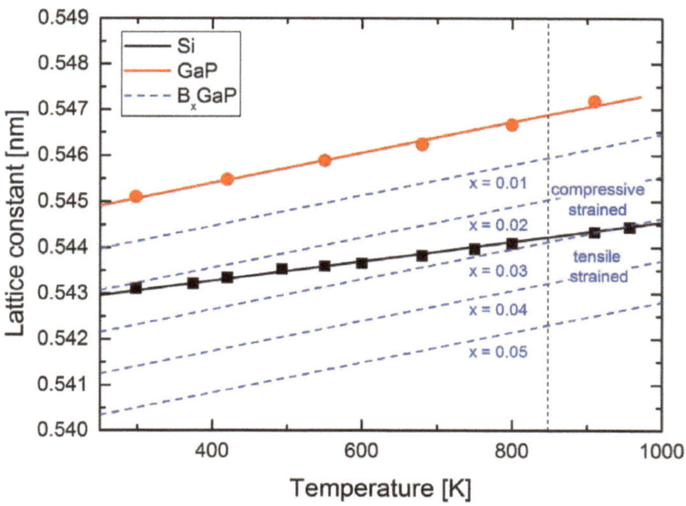

Fig. 4 Lattice constant of GaP (red circles) and Si as a function of temperature. The thermal expansion coefficients of $3.3 \times 10^{-6}\ K^{-1}$ for GaP and $2.1 \times 10^{-6}\ K^{-1}$ for Si has been taken from the literature (Kudman and Paff, 1972; Okada and Tokumaru, 1984). Four additional dashed lines represent the temperature dependence of (BGa)P with varying B-contents, assuming the same thermal expansion coefficient as GaP.

that one has to keep both the actual strain level at growth temperature as well as at room temperature or operation temperature of the entire laser stack into account (Kunert et al., 2008) to guarantee high-crystalline perfection layer stacks free from any dislocation defects and cracks.

In order to evaluate the incorporation efficiency of the constituent elements in the Ga(NAsP)-material system, detailed MOVPE growth investigations as a function of growth temperature in the wide temperature range of 525 °C and 700 °C have been performed (Ludewig et al., 2016). As test layer stack a triple Ga(NAsP)-QWH with intermediate (BGa)(AsP)-barriers has been deposited for various growth temperatures on the above discussed GaP-on-Si (001) template. For these growth series all other MOVPE growth parameters except the growth temperature have been kept constant. The subsequent analysis is based on a combination of detailed HR-XRD, PL and high-angle annular dark field (HAADF) scanning transmission electron microscopy (STEM) studies (Ludewig et al., 2016). For the entire, comparatively wide growth temperature range high-quality MQWH could be realized, as indicated by the HR-XRD pattern around the (004) reflex of Si for one test structure exemplarily shown in Fig. 5. The obtained high crystalline perfection allows for a detailed analysis of the MQWH layer by full dynamical XRD-modeling applying the model described in Tapfer and Ploog (1989)) and De Caro et al. (1996). From this analysis the precise strain-state as well as the individual layer thicknesses are obtained. The layer thicknesses from the XRD-analysis are also verified by

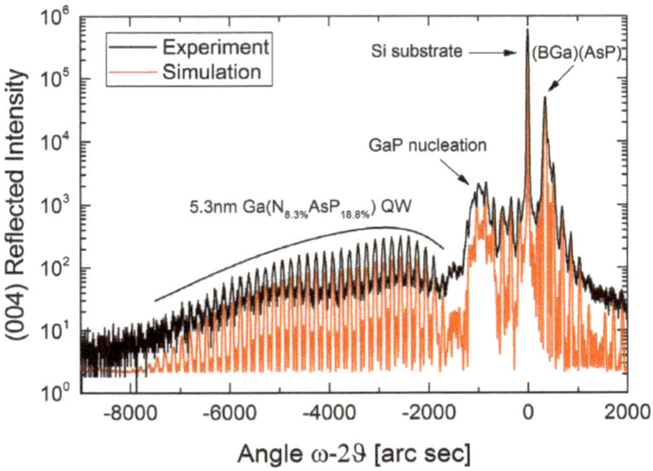

Fig. 5 High-resolution X-ray diffraction of a Ga(NAsP)/(BGa)(AsP)-layer structure on Si-substrate.

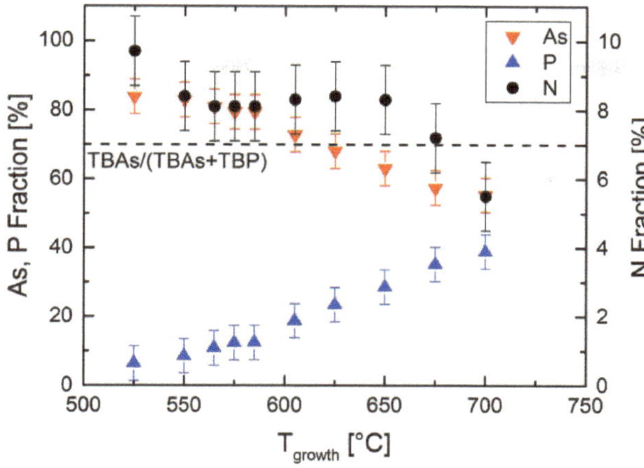

Fig. 6 Evaluated chemical composition of the Ga(NAsP) material as a function of MOVPE growth temperature.

TEM-studies (see also Section 3 for details). These informations are combined with room temperature PL-investigations for the characteristic emission wavelength of the Ga(NAsP)-MQWH also taking quantization effects into account to determine the individual chemical composition of the quaternary Ga(NAsP)-QW layers as detailed in Jandieri et al. (2015).

From these combined studies the evaluated chemical compositions of the Ga(NAsP)-QW layers as a function of MOVPE growth temperature are summarized in Fig. 6. The incorporation efficiencies in particular for N are very well controlled. For a wide range of growth temperatures in between 550 °C and 650 °C the N- up-take is constant (Fig. 6). For the As-content we observe a steady decrease with increasing temperature, while the P-concentration shows the opposite behavior (Fig. 6). With increasing growth temperature an increase in growth rates of the constituent Ga(NAsP)- as well as (BGa)(AsP)-layers is clearly detected and quantified (Ludewig et al., 2016).

3. Atomic-scale structural analysis and stability of Ga(NAsP) on (001) Si

In this section, the atomic-scale structural characteristics of Ga(NAsP) on (001) Si are elaborated in particular with respect to chemical stability during epitaxial growth and subsequent annealing steps, which might be applied during subsequent CMOS-process steps for the realization of Si-photonics circuits.

It is interesting to note that the Ga(NAsP)-material class is thought to be a metastable system like for example (GaIn)(NAs) grown on GaAs-substrate. However, the high crystalline perfection of Ga(NAsP)/(BGa)(AsP)-MQW grown on Si (001) substrate at high growth temperature of 700 °C without any sign of phase separation from HR-XRD analysis seems to be in contradiction to the assumed metastable character of this quaternary material class. Therefore, detailed atomic scale structural investigations of Ga(NAsP)/(BGa)(AsP)-MQWH have been performed by HAADF-STEM methodology. The atomic scale chemical composition has been studied for samples grown at very different growth temperatures in the range of 525–700 °C with quantum well thicknesses varying in between 1.4 nm and 12 nm. Series of samples have subsequently been annealed under equilibrium conditions in the MOVPE reactor or applying rapid thermal annealing (RTA) procedures in the temperature range of 700–1000 °C.

As a first step the changes in atomic scale chemical analysis under RTA procedures are examined (Wegele et al., 2016b). Exemplarily these characteristics are elaborated using Ga(NAsP)-MQW samples with a chemical composition of 7% N and 11% P and a well width of 5.5 nm grown at a temperature of 575 °C. The STEM-micrographs of the as-grown sample as well as samples annealed for 10 s applying a RTA-process at 850 °C, 925 °C and 975 °C are summarized in Fig. 7A–D together with the corresponding compositional fluctuation maps (Fig. 7E–H), respectively.

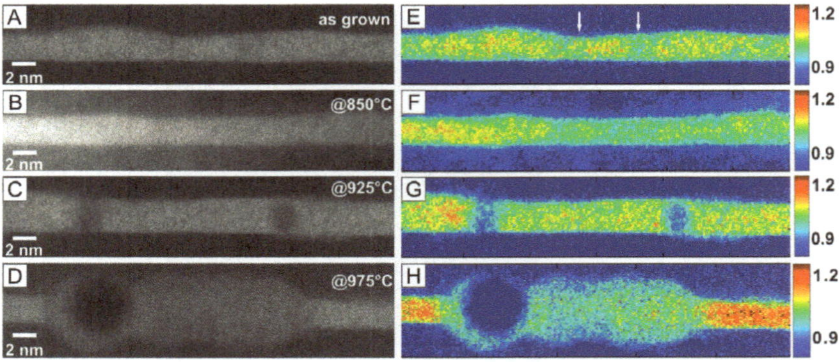

Fig. 7 STEM-micrographs of Ga(NAsP)-layers as grown (A), annealed at 850 °C (B), 925 °C (C) and 975 °C (D) and the corresponding compositional fluctuation maps (intensity at each point normalized to the average intensity in the Ga(NAsP)-layer) (E–H). The growth direction is from bottom to top. The two white arrows in (E) mark regions with lower dark field intensity. As for this image a large inner detector angle has been used, these regions might contain more N than the environment, presumably in form of N-chains.

The composition fluctuation maps are obtained by normalizing the intensity at each point by the average intensity in the QW. Other details of the evaluation procedure are given in Wegele et al. (2016b). While the lower interface of the Ga(NAsP)-QW is very abrupt also on the atomic scale, the upper interface of Ga(NAsP) appears slightly corrugated. This corrugation of the upper interface increases for increased quantum well width as well as for increased growth temperature, as quantified in Wegele et al. (2016a,b).

The as-grown sample as well as the sample RTA-annealed at 850 °C exhibit a rather homogeneous compositional contrast with no sign of any phase separation effects. For RTA-process temperatures of 925 °C and above nanostructural changes are detected. Interdiffusion processes set in at these high temperatures. In addition, circular shaped, dark regions are detected, which increase in size with increasing RTA-temperature. Detailed high-resolution STEM micrographs with varying defocus conditions point to the formation of circular voids in Ga(NAsP)-QW layers at these high RTA-temperatures (Wegele et al., 2016b). It is important to note that the chemical homogeneity in the Ga(NAsP) QW layer in between the voids is continuing to improve with increasing RTA-temperature.

The same detailed STEM-analysis has also been performed on the Ga(NAsP)/(BGa)(AsP)-MQW samples deposited in the MOVPE growth temperature range of 525 °C and 700 °C (Wegele et al., 2016a), which have been discussed also in the preceding paragraph of this chapter. Details of the chosen experimental HAADF-STEM conditions as well as the evaluation procedure for these samples are summarized in Wegele et al. (2016a). The compositional fluctuation maps of samples grown at 525 °C and 700 °C are shown in Fig. 8A and B, respectively. The increased corrugation of the upper interface of the Ga(NAsP)-QW layer grown at higher temperature is clearly detected. In addition, the standard deviation from the mean intensity is summarized in Fig. 8C as a function of the inner detector angle for GaP as well as the Ga(NAsP) QW material grown at 525 °C and 700 °C. It is interesting to note that the sample grown at higher temperature of 700 °C has a significant smaller standard deviation of the chemical composition than the one grown at lower temperature of 525 °C for all chosen inner detector angles. The larger fluctuations observed in both cases for smaller inner detector angles indicates that the N-composition is fluctuating stronger than the respective As-concentration in the Ga(NAsP)-layer (Wegele et al., 2016a). This trend in improving the chemical homogeneity of the quaternary Ga(NAsP) QW material with increasing growth temperature is verified for the entire wide growth temperature window as

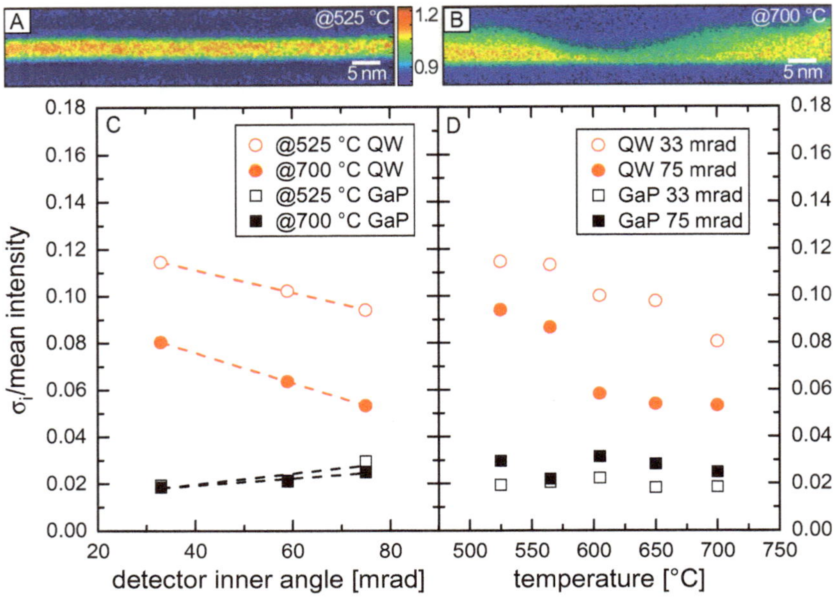

Fig. 8 Compositional fluctuation in dependence of growth temperature. Compositional fluctuation maps of samples grown at 525 °C (A), and 700 °C (B) for STEM-images measured at 59 mrad inner detector angle. (C) There is a decreasing trend of the compositional fluctuation for increasing inner detector angle for 525 °C (red open circles) as well as 700 °C (red filled circles) meaning that As-atoms are distributed more homogeneously than N-atoms in Ga(NAsP)-QWs. (D) Compositional fluctuation in Ga(NAsP)-QWs becomes lower with increasing growth temperatures for 33 mrad (red open circles) as well as 75 mrad (red filled circles inner detector angles). The open and filled squares are the local composition fluctuations in GaP for 33 mrad and 75 mrad inner detector angles, respectively.

exemplified in Fig. 8D, where the standard deviation of the intensity, i.e., the composition clearly reduces with increasing growth temperature. Thus, increasing the growth temperature definitely improves the chemical homogeneity of the quaternary Ga(NAsP) QW material. This experimental finding in addition to the above mentioned behavior under RTA-annealing are clear experimental evidences that Ga(NAsP) is not metastable but rather a stable III/V-compound, if grown lattice-matched on Si (001) substrate.

In addition, we have, thus, performed detailed theoretical studies applying density functional theory (DFT) modeling. The computational details of the used DFT are specified in Wegele et al. (2016a). The DFT-simulations for the strain relaxation energies (SRE) and phase separation energies (PSE) for $Ga(N_xAs_{0.85-x}P_{0.15})$ on Si- and on hypothetical $Ga(As_{0.85}P_{0.15})$-substrates are summarized in Fig. 9 as a function of the N-content.

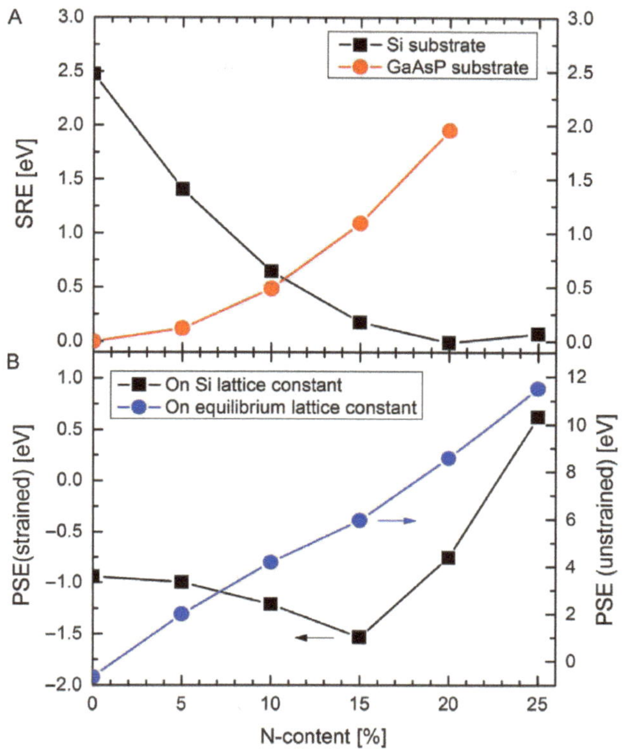

Fig. 9 Computational results for (A) strain relaxation energy (SRE) of $Ga(N_xAs_{0.85-x}P_{0.15})$ on Si (filled black symbols) and virtual $Ga(As_{0.85}P_{0.15})$ (filled red symbols). In (B) the values of the phase separation energy (PSE) of strained to Si lattice constant (filled black symbols, left hand side axis) and on to equilibrium lattice constant (filled blue symbols, right hand side axis) are plotted. Energy values refer to on $(2 \times 4 \times 5)$ supercell. The lines are guides to the eye.

It is important to note that the constraints for the epitaxial layer due to the respective substrate material has explicitly been taken into account. The SRE is defined as the energy difference between the fully relaxed Ga(NAsP) bulk cell and the strained system of the same composition on the respective substrate material. The PSE is given by the energy difference between the strained Ga(NAsP) and the likewise strained components of GaAs, GaP, GaN (all in the zinc blende crystalline structure) with the lattice parameters of the respective substrates.

While the SRE is continuously increasing with increasing N-content for growth on Ga(AsP)-substrates, the SRE is first decreasing significantly for the growth on the Si-lattice constant for N-contents up to about 20% (Fig. 9A). For higher N-concentrations the SRE increases again due to the change from compressive strain to tensile strain in the Ga(NAsP)-material. The PSE into the binary components for the growth on Si-substrate first decreases for N-concentrations up to about 15% (Fig. 9B). This implies that the compressive strained Ga(NAsP)-material system becomes increasingly stabilized with respect to phase separation for the growth on the lattice constants of Si.

Altogether both from the experimental point of view as well as from the theoretical DFT-modeling the Ga(NAsP)-material class is stabilized for the lattice-matched growth on Si (001) substrates. These findings are extremely valuable also from application point of view to emphasize the future potential of this material system as active laser source material on Si (001) substrates.

4. Properties of the (BGa)(PAs)-material class

The necessity to introduce the (BGa)(AsP)-material system for strain compensation purposes has been elaborated already in Section 2. This material class is also applied to realize (BGa)P/(BGa)(AsP)-waveguiding and separate confinement heterostructure (SCH) layer stacks. Thus, the refractive index properties as well as the electrical transport characteristics of the (BGa)(AsP)-material class as a function of composition are summarized in this section.

The optical properties of (BGa)(AsP) with respect to the compositional dependence of the indirect and direct energy gaps as well as the related refractive indices have been clarified by means of spectroscopic ellipsometry (SE) and surface photovoltage (SPV) spectroscopy (Hossain et al., 2011a,b; Rogowsky et al., 2011). With increasing B-content in ternary (BGa)P-layers a significant reduction in the indirect band energy from that of GaP is observed. A significant bowing parameter of the indirect band gap of $6.2 \pm 0.2\,\text{eV}$ is found for B-concentrations up 6%, while the change in the direct band gap is smaller and reduces by about 58 meV per percent of B in this composition range from that of GaP (Hossain et al., 2011a,b). The higher lying energy gaps and their respective dependences on chemical composition in (BGa)(AsP)-material class have been evaluated by SE (Rogowsky et al., 2011).

For the design of electrical injection lasers the refractive index is of key importance in the intended application wavelength ranges for the lattice-matched Ga(NAsPSb)-laser material around 850, 950 and 1300 nm. Therefore, in situ reflectance measurements during MOVPE-growth of various

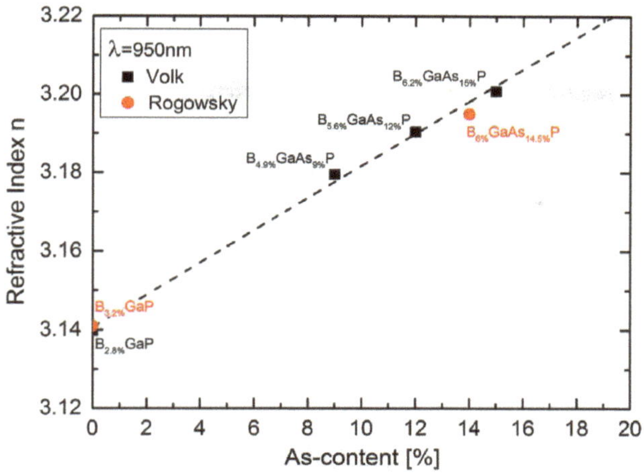

Fig. 10 Calculation of the refractive index of lattice-matched (BGa)(AsP) on Si based on a shift of 3.35% due to the temperature change from growth temperature at 600 °C to room temperature.

(BGa)(AsP)-layer structures have been performed at a wavelength of 950 nm at typical growth temperatures around 600 °C (Volk and Stolz, 2017). The measurement system was calibrated with the well-known Si- and GaP-materials. The determined refractive indices were extrapolated to room temperature for (BGa)(AsP) and are summarized in Fig. 10 as a function of the As-concentration. The ternary $B_{0.028}Ga_{0.972}P$, which is lattice-matched to Si (001) at the chosen growth temperature, has a refractive index of 3.14 at room temperature. This value is slightly larger than that of GaP of 3.13. With increasing As-content, which requires also an increase of the B-concentration due to the required lattice-matching condition on (001) Si-substrate at growth temperature, the refractive index increases linearly to a value of 3.20 for an As-content of 15% in the quaternary (BGa)(AsP)-material. These determined values of the refractive index allow for the design of the optical waveguide for electrical injection laser structures, applying n- and p-doped (BGa)P-contact and waveguide layers as well as the quaternary (BGa)(AsP)-separate confinement heterostructure (SCH) layers, respectively. Detailed wavelength dependent SE studies are underway to quantify the refractive index in the entire range of emission wavelengths from 850 nm to 1.3 μm of the active Ga(NAsP)-QW laser material on Si (001) substrate.

For the effective p- and n-type doping of the (BGa)(AsP)-material class Zn as acceptor and Si or Te as donor impurities have successfully been applied. The interplay and effect of the B-related localized states on the conduction band transport have been clarified for both the ternary (BGa)P- (Petznick et al., 2014) as well as the quaternary (BGa)(AsP)-material (Ostheim et al., 2016) by temperature-dependent magneto-transport studies also applying hydrostatic pressure. With increasing B-concentration an increase in the density of B-localized states in the vicinity of the conduction band edge of (BGa)((As)P) is observed, which leads to a reduction of the free carrier concentration and, thus, the n-type-doping efficiency. For the optimization of electrical injection structures with n- and p-doped (BGa)P-waveguide and contact as well as (BGa)(AsP)-SCH materials, respectively, the correlation of the B-content and the actual doping level, achievable in the different functional layers, has to be taken into account explicitly.

After careful optimization the electrical characteristics of the full pn-junction emitter structures have significantly been improved also by considering the discussed annealing procedures in Section 3 into account. This is exemplarily detailed in Fig. 11, where the dependence of the onset voltage U_0 as well as the differential resistance R_{diff} for lateral current injection as a function of the temperature in the range of 0–100 °C is summarized. The electrical I/V-characteristics of the broad area emitter structure at room temperature are indicated in the inset of this Fig. 11. The ideality factor is found to be around 2, the onset voltage U_0 is slightly below 2 V with values

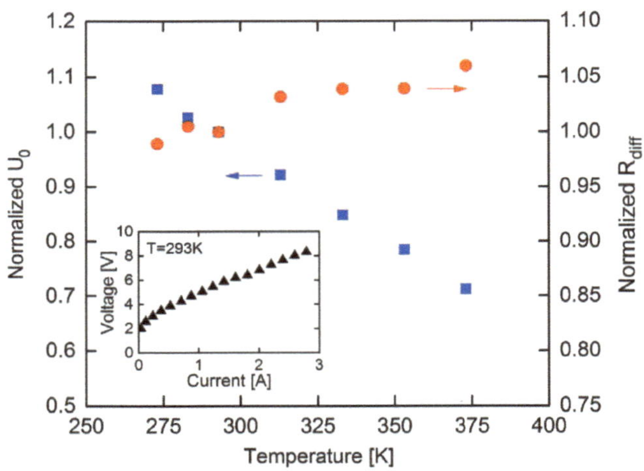

Fig. 11 Temperature dependence of the onset voltage U_0 and the differential resistance R_{diff} (normalized to 2.3 V at room temperature and its corresponding differential resistance, respectively). The inset shows a current-voltage characteristics at room temperature.

of R_{diff} of 1.5 Ω at room temperature. This value of R_{diff} is similar to certain standard III/V-laser stacks for comparable geometries on GaAs-substrate. With increasing temperature up to 100 °C the value of R_{diff} is only slightly increasing, while U_0 is significantly reduced in this temperature range to values below 1.5 V for temperatures above 85 °C. All of these results underline the very good electrical characteristics obtained for these novel lattice-matched material systems on Si (001)-substrate.

5. Optoelectronic properties of Ga(NAsP)-based heterostructures

In this section, the optoelectronic properties of Ga(NAsP)-based QWH will be presented and discussed. First the composition dependence of the band gap of Ga(NAsP) is described. The low-temperature, low excitation density excitonic characteristics and disorder parameters are only briefly summarized for this material system. Here, we concentrate, however, on the modal gain characteristics and lateral electrical injection lasers under high-excitation conditions. The hetero-offset in this novel system is studied and discussed with respect to possible loss mechanisms, in particular carrier leakage. Finally, the electroluminescence properties of emitter structures in the wavelength range from 850 nm to 1.2 μm are presented and discussed with respect to a possible extension to the 1.3 μm range.

The band gap dependence on the composition of Ga(NAsP)-MQWH grown lattice-matched on either GaP- or Si-substrates has been studied experimentally by a combination of HR-XRD, STEM as well as by PL-spectroscopy. The N-content has been varied in between about 4% and 10%, while the P-concentrations have been chosen in the range of 0% and 23% in the quaternary Ga(NAsP)-compound. Part of the samples have also been evaluated by STEM to obtain an independent determination of the Ga(NAsP)-quantum well width and its fluctuations. These experimental findings have been compared to a theoretical modeling based on a 8-band k.p-analysis which has been extended by a band anticrossing (BAC) model (Jandieri et al., 2015). The respective strain value as determined by HR-XRD and the related modifications in band gap have explicitly been taken into account in this modeling approach. It is important to note that for the wide composition range mentioned above, which extends by far outside the dilute concentration limits of N, the experimental results can be modeled by assuming a constant BAC-coupling parameter

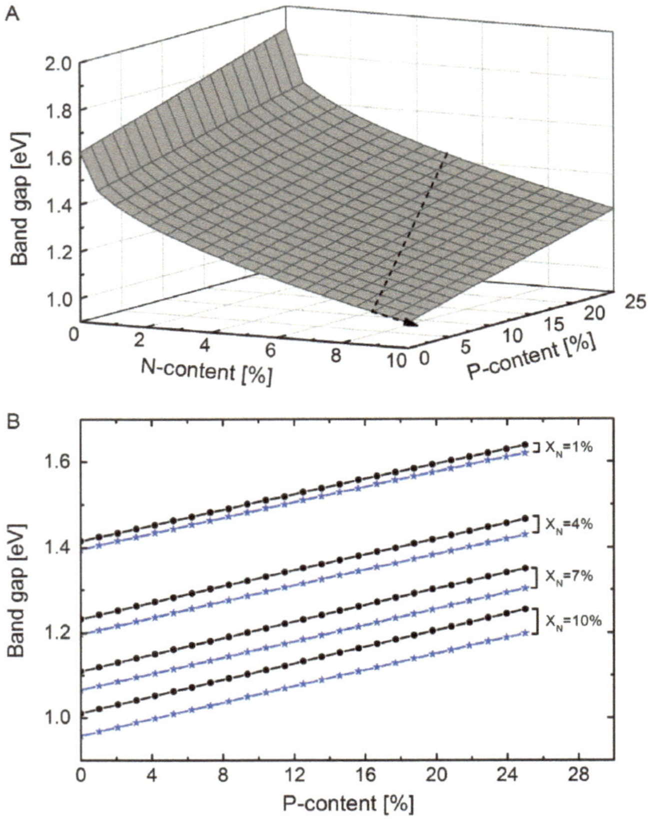

Fig. 12 Band gap in Ga(NAsP) quaternary compound pseudo-morphically grown on Si substrate as a function of N and P contents (A). Several cross-sections of this dependence calculated at various N concentrations are shown in (B) by solid circles (Si-substrate). The band gap calculated for Ga(NAsP) samples grown on GaP substrates are shown by stars. The lines are guides to the eye. The arrow line in (A) exhibits the chosen chemical compositions for the different Ga(NAsP)-samples in Fig. 13.

of 1.9 and 1.75 eV for the pseudomorphic growth of Ga(NAsP) on GaP- and Si-substrates, respectively (Jandieri et al., 2015).

The energy gap of the quaternary Ga(NAsP) grown lattice-matched on Si (001) substrate, related to the near-infrared emission wavelength range, are summarized in Fig. 12A as a function of the N- as well as P-concentrations in the above mentioned concentration range. Various cross-sections of the band gap dependence for specific N-contents were calculated and are depicted in Fig. 12B by solid circles as a function of the incorporated P-concentration. The corresponding results, calculated for the growth on GaP-substrates, are included additionally by stars in this Fig. 12B. This band gap diagram, spanning in the entire range from 850 nm (1,46 eV) to 950 nm (1.29 eV) and out to 1. 3 μm (0.95 eV), serves as a basis for the selection of the appropriate chemical composition for the intended application wavelength by also taking quantization effects in the MQWH into account. Exemplarily, we have summarized the electroluminescence (EL) of Ga(NAsPSb)-MQWH on Si (001) substrate at room temperature with varying chemical composition in Fig. 13, as indicated by the arrow line in Fig. 12A. In these samples a quantum well width of 5 nm and a compressive strain value around 1.5%–2% have been kept constant. The emission wavelength range can be tuned all the way from the data com range around 850 nm down to 1.2 μm, which exhibits an emission wavelength below the band gap of Si already. This characteristic is particularly important for emitters in combination with Si/SiO_2-optical waveguide structures for Si-photonics based applications. Studies are underway to also expand the emission wavelength range to the first telecom window around 1.3 μm at room temperature.

The optical properties of quaternary Ga(NAsP)-MQWH at low excitation conditions have been investigated experimentally by PL as well as PL excitation (PLE) including time-resolved PL spectroscopy (Gies et al., 2014; Karcher et al., 2010; Shakfa et al., 2016; Woscholski et al., 2016a,b).

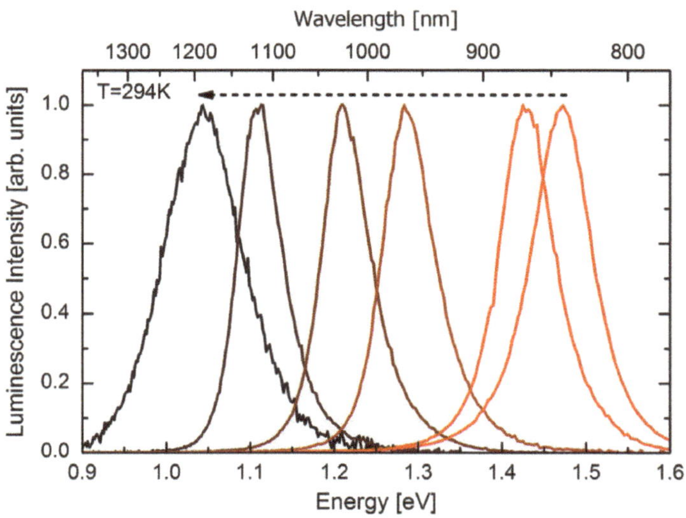

Fig. 13 Room temperature electroluminescence of Ga(NAsP)-QWH with varying chemical composition, as indicated by the arrow line in Fig. 12A, in the wavelength range of 850 nm–1.2 μm.

These experimental studies have been modeled by extending the standard excitonic disorder method (Rubel et al., 2005), which has been applied successfully to a wide range of ternary and quaternary compound semiconductor material systems, by introducing an additional long-range disorder scale complementing the small range disorder due to intrinsic statistical alloy fluctuations. This additional long-range disorder scale has been related to well width fluctuations in this material system (Karcher et al., 2010). Very recently the modeling of such multinary material systems has been revisited assuming also different energy dependences of the density of localized states in the model (Valkovskii et al., 2018). It is important to note that these low excitation density excitonic characteristics are important in particular as optimization criterion for MOVPE growth conditions, however, cannot directly be applied to high-density electron-hole plasma type characteristics as present in lasers.

In the following, we will, thus, concentrate on the optical properties of these Ga(NAsP)-MQWH on Si (001) substrate under high-density optical as well as electrical excitation conditions. Applying the variable stripe-length method under optical excitation conditions, high values of the modal gain of up to around $80\,cm^{-1}$ have been verified both experimentally as well as by first principle theoretical modeling at room temperature in lattice-matched Ga(NAsP)-MQWH with 4 individual QWs on Si (001) substrate (Koukourakis et al., 2012). Assuming a typical optical confinement factor of 2% for these MQWH with 4 individual QWs, material gain values of up to $4.000\,cm^{-1}$ have been obtained experimentally. These material gain values are typical for high-efficiency direct band gap III/V-laser materials and clearly underline the great potential of this novel lattice-matched Ga(NAsP)-laser material on Si (001) for possible Si-photonics applications.

Also, electrical injection lasers in this material system have successfully been realized with threshold current densities of $500\,A/cm^2$ at low temperatures around 80 K (Liebich et al., 2011). While increasing the temperature the threshold current density of the lasers is increasing and at the same time the differential efficiency is decreasing. In order to understand the performance limiting processes, these Ga(NAsP)-lattice-matched lasers on Si (001) have been analyzed by means of temperature-dependent as well as pressure-dependent studies of the laser characteristics (Read et al., 2015). Based on this detailed analysis, the primary cause for the increase in threshold current density is assumed to be related to carrier leakage out of the Ga(NAsP)-QW states at elevated temperatures.

The precise hetero-offset in the multinary Ga(NAsP)/GaP/(BGa)(AsP)-MQWH has been unknown at the time of the first laser studies. Thus, the first modeling activities have relied on the linear interpolation of the theoretical valence band offset of the binary GaAs- and GaP-semiconductors. Based on the theory as elaborated by Van de Walle and Neugebauer (2003), the valence band offset of the binaries GaAs and GaP is given by a relatively large value of 0.56 eV. As the chemical concentration of As is typically larger than 75% in the Ga(NAsP)-material system, this implies that the valence band offset should be relatively large in the range of 300–500 meV for the quaternary Ga(NAsP)-material system depending on the exact chemical composition with respect to GaP. Based on these assumptions, yielding a deep hole QW structure, the leakage process, which resulted from the above mentioned laser study, should be associated to a shallow electron QW structure. Various attempts in structure design have been pursued to reduce the determined leakage process, however, no significant improvement under the assumption of a large valence band offset could be achieved.

Thus, we have studied the dependence of the valence band offset on composition, strain and quantization experimentally. Various Ga(NAsP)-MQWH with intermediate GaP-barriers as well as (BGa)(AsP)-strain-compensating barrier layers have been realized by MOVPE with varying chemical compositions and well widths. These MQWH have been analyzed structurally by HR-XRD and STEM as well as optically by PL- and PLE-spectroscopy. As sensitive measure for the valence band offset the energy splitting of the fundamental electron to heavy hole (e-hh_1) and the electron to light hole (e-lh_1) resonances in the PLE-spectra have been studied as a function of the chemical composition, strain and the well width of the Ga(NAsP)-laser material experimentally. In the complementary modeling activities of these fundamental optical transitions both the hh_1–lh_1 splitting as well as the absolute values of the e-hh_1- and e-lh_1-transitions have explicitly been taken into account. The hh_1–lh_1 splitting for Ga(NAsP)-MQWH with varying P-contents have been summarized in Fig. 14 as a function of the quantum well width. Different growth series of samples with P-concentrations of 15% (N-content of 7.5%) (full squares) and 0% (N-concentration of 10%) (full circles) are shown. The best description of the experimental values leads to strain-free valence band offsets of 120 meV for P-content of 15% and 200 meV for P-content of 0%, respectively. The error bars in the determination of the valence band offset is estimated to be about ± 25 meV, as various other parameters entering the

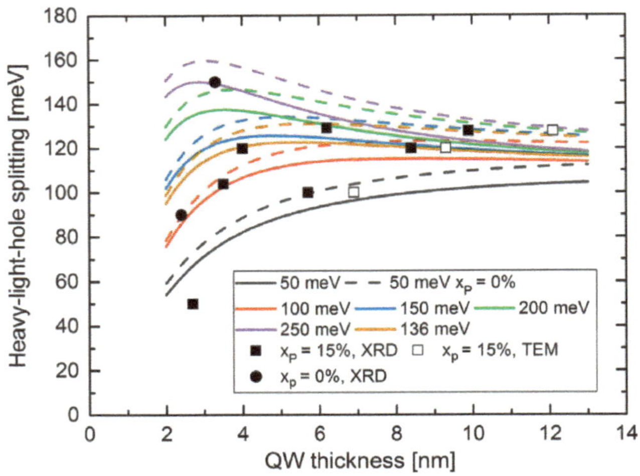

Fig. 14 The energy difference of electron-heavy hole and electron-light hole transitions in Ga(NAsP)-QWH for different valence band offsets (50–200 meV) as a function of well width L_z for chemical compositions of 15% P (7.5% N) and 0% P (10% N), respectively.

theoretical modeling (deformation potentials, effective carrier masses) are not known with high precision. However, it is important to note that the determined valence band offset is drastically different to the above assumptions from the theoretical calculation based on the binaries GaAs and GaP, as elaborated to 0.56 eV by Van de Walle and Neugebauer (2003). Therefore, it can be concluded that the carrier leakage process is presumably caused by holes (not electrons), which are excited out of their respective QW states at elevated temperatures. At present optimization procedures in the design of Ga(NAsP)-based MQWH lasers, which take these different valence band offset values into account, are underway for the different emission wavelengths in order to improve the optical efficiency of Ga(NAsP)-lasers also at elevated temperatures and, thus, verify the determined significantly smaller values of the valence band offset in this material system.

6. Monolithic integration of Ga(NAsP)-based emitter structures into the Si-CMOS process flow

In this part of the present chapter, the manufacturability as well as the ease of integration of the monolithic integrated, lattice-matched Ga(NAsP)-laser material into the standard Si-based CMOS-process flow is introduced and discussed.

For Si-photonics applications it is of key importance, that all concepts used can be integrated as easily as possible into a standard CMOS-process flow in order to use the available highly sophisticated CMOS-process technology, i.e., in Si-foundry production schemes. The present Ga(NAsP)-based laser technology offers some key advantages as it is a monolithic integration concept on the lattice constant of Si applying the (001) wafer orientation within the specifications of the CMOS-process. Due to the unique lattice-matching concept to the Si-substrate wafer any formation of defects as dislocations, twins or stacking faults can clearly be avoided in this concept. The principles of this monolithic integration concept are schematically shown in Fig. 15.

In this concept, n-doped and p-doped Si-epitaxial layers are used as buffer and contact layers, respectively. In order to guarantee the planarity of the processed wafer at intermediate processing steps the epitaxial Ga(NAsP)-based laser stack is deposited under selective area epitaxial growth (SAG) conditions in recesses, which are either V-shaped or rectangularly shaped depending on the application. The detailed description of the complex SAG parameters with respect to lateral chemical composition and lateral thickness gradients is beyond the scope of this review article. In addition the height of the emitted laser light can also be adjusted to subsequently monolithically processed waveguide, detector structures or other active or passive elements. In order to achieve single mode emission of the Ga(NAsP)-laser, the laser stack will be processed into ridge-waveguide (RWG) laser structures. Thus, the active Ga(NAsP)-laser ridge will be laterally passivated by SiO_x-passivation layers. In addition, the end facets of the laser ridge will be coated by CMOS-compatible high reflection (HR) and anti-reflection (AR) coatings. These process steps lead to a complete embedding of the Ga(NAsP)-based laser ridge with CMOS-compatible materials.

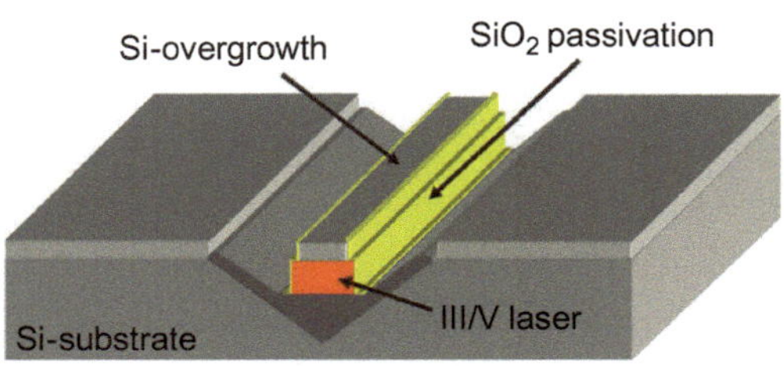

Fig. 15 Schematical integration concept of Ga(NAsP)-based laser structures on Si substrate in recess opening to guarantee flat wafer surfaces for subsequent lithography steps in the CMOS-process flow.

After cleaning of the Si-wafer after deposition and coating of the Ga(NAsP)-based laser ridge, any cross-contamination in the subsequent CMOS-process steps can clearly be avoided due to this complete embedding concept.

One additional advantage of the suggested monolithic integration concepts is the application of doped Si-buffer and cap layers as electrical contact layers for the Ga(NAsP)-based laser ridge. Thus, no III/V-semiconductor based metallizations are required as the established CMOS-compatible metallization schemes for Si can be applied directly to the Si-epitaxial buffer and cap layers. It is important to note that due to the epitaxial overgrowth of the Ga(NAsP)-laser stack by Si-epitaxial layers, this laser integration will be a front-end process. All of these distinct advantages of the presented Ga(NAsP)-based laser integration concept facilitate the monolithic realization of Si-photonics based integrated circuitry applying the standard CMOS-process flow with specific front-end "detour" process steps to realize the lattice-matched Ga(NAsP)-based laser ridge.

7. Conclusions

The present status of understanding for the novel lattice-matched Ga(NAsP)-based active laser material with (BGa)(AsP)-based barrier, waveguide as well as separate confinement heterostructure layers are discussed as a possible solution for monolithically integrated optoelectronic and photonic functionalities on CMOS-compatible Si (001) substrates. This concept avoids the formation of any misfit and threading dislocations from the very beginning, which are hindering the long-term stability of standard heteroepitaxial III/V-laser materials based on GaAs, InP, GaSb or GaN, respectively. The epitaxial growth by metal organic vapor phase epitaxy (MOVPE) is studied in detail. A wide range of structural, electrical as well as electronic and optoelectronic investigation techniques are applied to evaluate and verify the respective properties of this novel lattice-matched Ga(NAsP)/(BGa)(AsP)-material combination on (001) Si substrate also in comparison to first principle theoretical modeling. It is important to note that the novel active Ga(NAsP) is becoming a stable material system when grown lattice-matched to Si-substrates. The proof-of-concept of this novel laser material approach is verified by the successful realization of high model gain as well as electrically injected lasers (001) Si-substrate. In addition, the unique integration scenario into the standard CMOS-production process for the realization of Si-photonics integrated circuits is outlined and discussed.

Acknowledgments

The authors are indebted to a large number of coworkers and cooperation partners as well as funding agencies. The active support of A. Beyer, M. Diederich, S. Gupta, K. Jandieri, M. Kröner, S. Liebich, J.O. Oelerich, S. Reinhard, T. Wegele, M. Zimprich at the Structure and Technology Research Laboratory (STRL) (Material Sciences Center and Department of Physics, Philipps-University Marburg), of B. Kunert, M. Volk, S. Ziegler, J. Zimmet at NAsP III/V (Marburg), of A. Bäumner, S.D. Baranovskii, C. Bückers, S. Gies, W. Heimbrodt, C. Karcher, M. Koch, S.W. Koch, M.K. Shakfa, R. Woscholski at the Department of Physics (Philipps-University Marburg), of P. Rosenow, A. Stegmüller, R. Tonner at the Department of Chemistry (Philipps-University Marburg), of S. Chatterjee, F. Dobener, P.J. Klar, L. Ostheim, N. Rosemann at the Department of Physics (Justus-Liebig University Giessen), of N.C. Gerhard, M.R. Hofmann, N. Koukourakis at Department of Electrical Engineering (Ruhr-University Bochum), of R. Ostendorf, S. Rogowski, J. Wagner at the Fraunhofer-Institute Applied Solid State Physics (Freiburg), and of J. Chamings, T. Eales, N. Hossain, C. Fitch, S.J. Sweeney at the Department of Physics (University of Surrey, United Kingdom) is highly appreciated.

Various parts of this work have been supported by the German Science Foundation (DFG) within the Collaborative Research Group 483 "Metastable compound semiconductors and heterostructures" and the Research Training Group 1782 "Functionalization of semiconductors" as well as by the Federal Ministry of Education and Research (BMBF) within the Collaborative Research Project "Monolithic integration of Ga(NAsP)-based laser structures on Si (MonoLaSi)."

References

Beyer, A., Stegmüller, A., Oelerich, J.O., Jandieri, K., Werner, K., Mette, G., Stolz, W., Baranovskii, S.D., Tonner, R., Volz, K., 2016. Pyramidal structure formation at the interface between III/V semiconductors and silicon. Chem. Mater. 28, 3265–3275.

Chen, H.Z., Ghaffari, A., Wang, H., Morkoc, H., Yariv, A., 1987. Operation of continuous-wave GaAs/AlGaAs broad-area injection-lasers on (100) Si-substrates at room temperature. Opt. Lett. 12, 812–814.

Chen, S.M., Li, W., Wu, J., Jiang, Q., Tang, M.C., Shutts, S., Elliott, S.N., Sobiesierski, A., Seeds, A.J., Ross, I., Smowton, P.M., Yiu, H.Y., 2016. Electrically pumped continuous wave III-V quantum dot lasers on silicon. Nat. Photonics 10, 307–312.

Choi, H.K., Wang, C.A., Karam, N.H., 1991. GaAs-based diode-lasers on Si with increased lifetime obtained by using strained InGaAs active layer. Appl. Phys. Lett. 59, 2634–2635.

De Caro, L., Giannini, C., Tapfer, L., et al., 1996. Determination of the lattice strain and chemical composition of semiconductor heterostructures by high-resolution X-ray diffraction. J. Appl. Phys. 79, 4101–4110.

Deppe, D.G., Chand, N., van der Ziel, J.P., Zydzik, G.J., 1990. $Al_xGa_{1-x}As$-GaAs vertical cavity surface-emitting laser grown on Si-substrate. Appl. Phys. Lett. 56, 740–742.

Egawa, T., Tada, H., Kobayashi, Y., Soga, T., Jimbo, T., Umeno, M., 1990. Low-threshold continuous wave room temperature operation of $Al_xGa_{1-x}As$/GaAs single quantum well lasers grown by metalorganic chemical vapour deposition on Si-substrates with SiO_2 back coating. Appl. Phys. Lett. 57, 1179–1181.

Egawa, T., Jimbo, T., Hasegawa, Y., Umeno, M., 1994. Optical and electrical degradation of GaAs-based laser diodes grown on Si-substrates. Appl. Phys. Lett. 64, 1401–1403.

Gies, S., Zimprich, M., Wegele, T., Kruska, C., Beyer, A., Stolz, W., Volz, K., Heimbrodt, W., 2014. Annealing effects on the composition and disorder of Ga(NAsP) quantum ells on silicon substrates for laser applications. J. Cryst. Growth 402, 169–174.

Hossain, N., Sweeney, S.J., Rogowski, S., Ostendorf, R., Wagner, J., Liebich, S., Zimprich, M., Volz, K., Kunert, B., Stolz, W., 2011a. Reduced threshold current dilute-nitride Ga(NAsP)/GaP quantum well lasers grown by MOVPE. Electron. Lett. 47, 931–932.

Hossain, N., Hosea, T.J.C., Sweeney, S.J., Liebich, S., Zimprich, M., Volz, K., Kunert, B., Stolz, W., 2011b. Band structure properties of novel $B_xGa_{1-x}P$ alloys for silicon integration. J. Appl. Phys. 110, 063101. 6 pages.

Hossain, N., Jin, S.R., Liebich, S., Zimprich, M., Volz, K., Kunert, B., Stolz, W., Sweeney, S.J., 2012. Efficiency-limiting processes in Ga(NAsP)/GaP quntum well lasers. Appl. Phys. Lett. 101, 011107. 4 pages.

Jandieri, K., Ludewig, P., Wegele, T., Beyer, A., Kunert, B., Springer, P., Baranovskii, S.D., Koch, S.W., Volz, K., Stolz, W., 2015. Compositional dependence of the band gap in Ga(NAsP) quantum well heterostructures. J. Appl. Phys. 118, 065701. 7 pages.

Jung, D., Norman, J., Kennedy, M.J., Shang, C., Shin, B., Wan, Y., Gossard, A.C., Bowers, J.E., 2017. High efficiency low threshold current 1.3 μm InAs quantum dot lasers on on-axis (001) GaP/Si. Appl. Phys. Lett. 11 122107 (4pages).

Karcher, C., Jandieri, K., Kunert, B., Fritz, R., Zimprich, M., Volz, K., Stolz, W., Gebhard, F., Baranovskii, S.D., Heimbrodt, W., 2010. Pecularities of the photoluminescence of metastable Ga(NAsP)/GaP quntum well structures. Phys. Rev. B Condens. Matter 82, 245309. 13 pages.

Koukourakis, N., Bückers, C., Funke, D.A., Gerhardt, N.C., Liebich, S., Chatterjee, S., Lange, C., Zimprich, M., Volz, K., Stolz, W., Kunert, B., Koch, S.W., Hofmann, M.R., 2012. High room temperature optical gain in Ga(NAsP)/Si heterostructures. Appl. Phys. Lett. 100, 092107. 4 pages.

Kudman, I., Paff, R.J., 1972. Thermal expansion of $In_xGa_{1-x}P$ alloys. J. Appl. Phys. 43, 3760–3762.

Kunert, B., Volz, K., Koch, J., Stolz, W., 2006. Direct-band-gap Ga(NAsP)-material system pseudomorphically grown on GaP-substrates. Appl. Phys. Lett. 88, 182108 (3 pages).

Kunert, B., Volz, K., Koch, J., Stolz, W., 2007. MOVPE growth conditions oft he novel, direct band gap dilute nitride Ga(NAsP) material system pseudomorphically strained on GaP-substrates. J. Cryst. Growth 298, 121–125.

Kunert, B., Zinnkann, S., Volz, K., Stolz, W., 2008. Monolithic integration of Ga(NAsP)/(BGa)P multi-quantum well structures on (001) silicon substrate by MOVPE. J. Cryst. Growth 310, 4776–4779.

Liebich, S., Zimprich, M., Beyer, A., Volz, K., Kunert, B., Stolz, W., Lange, C., Chatterjee, S., Hossain, N., Sweeney, S.J., 2011. Laser operation of Ga(NAsP) lattice-matched to (001) Si-substrate. Appl. Phys. Lett. 99 071109 (3 pages).

Ludewig, P., Reinhard, S., Jandieri, K., Wegele, T., Beyer, A., Tapfer, L., Volz, K., Stolz, W., 2016. MOVPE growth studies of Ga(NAsP)/(BGa)(AsP) multi quantum well heterostructures (MQWH) for the monolithic integration of laser structures on (001) Si substrate. J. Cryst. Growth 438, 63–69.

Okada, Y., Tokumaru, Y., 1984. Precise determination of lattice parameter and thermal expansion coefficient of Si between 300 K and 1500 K. J. Appl. Phys. 56, 314–320.

Ostheim, L., Klar, P.J., Liebich, S., Ludewig, P., Volz, K., Stolz, W., 2016. Interplay of boron localized states and electron transport in $B_xGa_{1-x}As_{0.11}P_{0.89}$:Te. Semicond. Sci. Technol. 31, 07LT01. 5 pages.

Petznick, S., Ostheim, L., Klar, P.J., Liebich, S., Volz, K., Stolz, W., 2014. Effect of boron localized states on the conduction band transport in $B_xGa_{1-x}P$. Appl. Phys. Lett. 105, 222105 (4 pages).

Read, G.W., Marko, I.P., Hossain, N., Sweeney, S.J., 2015. Physical properties and characteristics of III/V lasers on silicon. IEEE J. Sel. Top. Quantum Electron. 21, 1502208. 8 pages.

Rogowsky, S., Baeumler, M., Wolfer, M., Kirste, L., Ostendorf, R., Wagner, J., Liebich, S., Stolz, W., Volz, K., Kunert, B., 2011. Vibrational mode and dielectric function spectra of BGaP probed by Raman scattering and spectroscopic ellipsometry. J. Appl. Phys. 109, 053504. 6 pages.

Rosemann, N., Metzger, B., Kunert, B., Volz, K., Stolz, W., Chatterjee, S., 2013. Temperature-dependent quantum efficiency of Ga(NAsP) quantum wells. Appl. Phys. Lett. 103, 252105. 4 pages.

Rubel, O., Galluppi, M., Baranovskii, S.D., Volz, K., Geelhaar, L., Riechert, H., Thomas, P., Stolz, W., 2005. Quantitative description of disorder parameters in (GaIn)(NAs) quantum wells from temperature-dependent photoluminescence spectroscopy. J. Appl. Phys. 98, 063518. 7 pages.

Shakfa, M.K., Woscholski, R., Gies, S., Wegele, T., Wiemer, M., Ludewig, P., Jandieri, K., Baranovskii, S.D., Stolz, W., Volz, K., Heimbrodt, W., Koch, M., 2016. Carrier dynamics in Ga(NAsP)/Si multi-quantum well heterostructures with varying well thickness. Superlattice. Microst. 93, 67–72.

Singh, J., 1995. Semiconductor Optoelectronics: Physics and Technology. McGraw-Hill, New York.

Sugo, M., Mori, H., Sakai, Y., Itoh, Y., 1992. Stable cw operation at room temperature of a 1.5 μm wavelength multiple quantum well laser on Si substrate. Appl. Phys. Lett. 60, 472–473.

Tapfer, L., Ploog, K., 1989. X-ray interference in ultrathin epitaxial layers—a versatile method for the structural analysis of single quantum wells and heterointerfaces. Phys. Rev. B Condens. Matter 40, 9802–9810.

Valkovskii, V., Jandieri, K., Gebhard, F., Baranovskii, S.D., 2018. Rethinking the theoretical description of photoluminescence in compound semiconductors. J. Appl. Phys. 123, 055703. 8 pages.

Van de Walle, C.G., Neugebauer, J., 2003. Universal alignment of hydrogen levels in semiconductors, insulators and solutions. Nature 423, 626–628.

Volk, M., Stolz, W., 2017. Determination of refractive index and direct band gap of lattice-matched (BGa)P and (BGa)(AsP) materials on exact oriented silicon. J. Appl. Phys. 122, 235702 (7 pages).

Volz, K., Beyer, A., Witte, W., Ohlmann, J., Nemeth, I., Kunert, B., Stolz, W., 2011. GaP-nucleation on exact Si (001) substrates for III/V device integration. J. Cryst. Growth 315, 37–47.

Wegele, T., Beyer, A., Ludewig, P., Rosenow, P., Duschek, L., Jandieri, K., Tonner, R., Stolz, W., Volz, K., 2016a. Interface morphology and composition of Ga(NAsP) quantum well structures for monolithically integrated LASERs on silicon substrates. J. Phys. D Appl. Phys. 49, 075108 (9 pages).

Wegele, T., Beyer, A., Gies, S., Zimprich, M., Heimbrodt, W., Stolz, W., Volz, K., 2016b. Correlation of the nanostructure with optoelectronic properties during rapid thermal annealing of Ga(NAsP) quantum wells grown on Si (001) substrates. J. Appl. Phys. 119, 025705 (6 pages).

Woscholski, R., Shakfa, M.K., Gies, S., Wiemer, M., Rahimi-Iman, A., Zimprich, M., Reinhard, S., Jandieri, K., Baranovskii, S.D., Heimbrodt, W., Volz, K., Stolz, W., Koch, M., 2016a. Time-resolved photoluminescence of Ga(NAsP) multiple quantum wells grown on Si substrate: effects of rapid thermal annealing. Thin Solid Films 613, 55–58.

Woscholski, R., Gies, S., Wiemer, M., Shakfa, M.K., Rahimi-Iman, A., Ludewig, P., Reinhard, S., Jandieri, K., Baranovskii, S.D., Heimbrodt, W., Volz, K., Stolz, W., Koch, M., 2016b. Influence of the growth temperature and disorder on spectral and temporal properties of Ga(NAsP) heterostructures. J. Appl. Phys. 119, 145707 (5 pages).

CHAPTER SEVEN

Growth of III–V semiconductors and lasers on silicon substrates by MOCVD

Bei Shi, Kei May Lau*
Department of Electronic and Computer Engineering, Hong Kong University of Science and Technology, Kowloon, Hong Kong
*Corresponding author: e-mail address: eekmlau@ust.hk

Contents

1. Introduction

Because of the direct bandgap nature and availability of various alloys for band structure engineering and quantum size effects, current existing active photonic devices have been implemented based on III–V compound semiconductors, covering an extremely wide spectral range. With the recent flourishment of silicon photonics, passive devices including waveguides and modulators have been well developed and integrated with the support of advanced complementary metal oxide semiconductor (CMOS) foundry

Semiconductors and Semimetals, Volume 101
ISSN 0080-8784
https://doi.org/10.1016/bs.semsem.2019.07.005

technologies (Zhou et al., 2015). III–V materials, with unique advantages including higher intrinsic electron mobility and smaller effective electron mass relative to Si and Ge, are on the other hand showing great promise for high frequency and high speed electron devices. However, volume scaling has been limited by challenges in device processing on native lattice-matched III–V substrates, with fragile mechanical properties and small sizes, high costs and low throughput in manufacturing. In this regard, silicon serves as a more promising integration platform to build large-scale and cost-effective photonic integrated circuits (PICs) for inter/intra-chip interconnects. Monolithic integration of high performance III–V lasers on the highly lattice-mismatched Si substrates has always been treated as a "holy grail" and the long-awaited missing piece of silicon photonics (Heck et al., 2013).

This chapter thus focuses on direct epitaxy of III–Vs (mainly GaAs, InP, and some ternary alloys) on Si by metalorganic chemical vapor deposition (MOCVD). MOCVD is well-established for its flexibility and versatility in growing different material systems for various optoelectronic devices. Moreover, the reactor design and flow dynamics enable scaling to industrial large volume production, similar to silicon foundries. Considering Si-based lasers are among the most significant building blocks in silicon photonics, we will update a comprehensive review of efforts in realizing semiconductor lasers epitaxially grown on Si, emitting at near-infrared wavelengths. Other photonic devices, e.g., photodetectors, semiconductor optical amplifiers (SOAs), can be extended with the same technique but will not be discussed here.

2. Heteroepitaxy of III–V compound semiconductors on silicon by MOCVD

In this section, the fundamental challenges associated with epitaxial growth of III–V on lattice-mismatched Si substrates will be addressed, prior to a review of GaAs and InP materials grown on Si by MOCVD. The review is separately categorized to GaAs-on-Si and InP-on-Si growth, based on the different lattice mismatches. A further advance toward micro- and nano-scale monolithic integration of III–Vs on Si is introduced in the final part of this section.

2.1 Fundamental challenges

Several fundamental challenges are limiting the metamorphic growth of high quality III–V layers on Si. Fig. 1 shows the bandgap and lattice constant

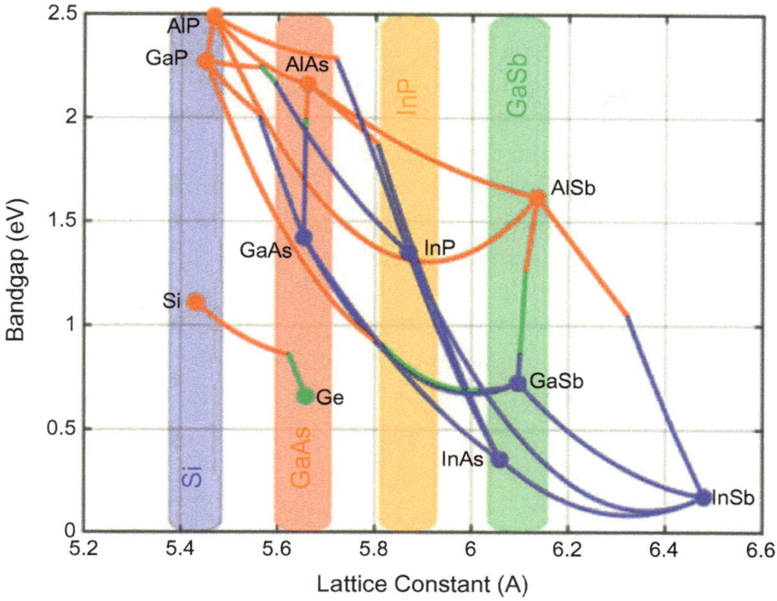

Fig. 1 Graph displaying the bandgaps and lattice constants of III–V semiconductors and Si-Ge alloys. Blue lines represent direct bandgap materials. Indirect bandgap materials with sixfold conduction-band symmetry are shown in red while the ones with eightfold conduction-band symmetry are displayed in green. Vertical color bars represent different commercial semiconductors based alloys with a lattice mismatch range of 1%. *Richardson, C. J., Lee, M. L., 2016. Metamorphic epitaxial materials. MRS Bull. 41, 193–198, reproduced with permission.*

relationship of various compound semiconductors (Richardson and Lee, 2016). The lattice constant mismatches between GaAs/Si, InP/Si and GaSb/Si are approximately 4%, 8%, and 12%, respectively. High density of threading dislocations (TDs) and planar defects such as stacking faults and twins form in the as-grown arsenide and phosphide materials to accommodate the large lattice constant mismatch between the III–V materials and Si. These TDs, with the density up to 10^9–10^{10}/cm^2, acting as non-radiative recombination and carrier scattering centers, are detrimental to the minority carrier devices such as lasers, solar cells, photodetectors, bipolar transistors, etc. (Li and Lau, 2017; Mahajan, 2000). As a result, III–V photonic devices on Si tend to exhibit lower internal efficiencies and much shorter device lifetimes than those on native substrates (Norman et al., 2018).

The thermal expansion coefficient mismatch is another challenge that needs to be tackled, especially for GaAs/Si with a high residual stress over10^9 dyn/cm^2 (Sugo et al., 1991). The residual stress not only leads to the rapid degradation in GaAs-based lasers on Si (Van der Ziel et al., 1987a), but also limits the maximum thickness of GaAs buffer up to approximately 7 μm for a ΔT of −575 °C, 5.1 μm for a ΔT of −675 °C, and 4.9 μm for a ΔT of −725 °C during the cooling stage from the growth temperature (Yang et al., 2003). In addition, the presence of micro-cracks on thin films is undesirable for devices because they can act as scattering centers for light propagation, resist in-plane electrical current flow, and introduce electrical shorting paths in vertical current spreading (Zamir et al., 2001).

Another difficulty in III–V/Si epitaxy is related to antiphase boundary (APB) stemming from the presence of polar/non-polar interfaces. Fig. 2 schematically shows the formation of APBs on a (100) non-polar surface with single-height atomic terraces. The boundaries between different domains consist of either Ga-Ga bonds or As-As bonds and therefore possess a net charge, which uncontrollably dope the GaAs and act as strong scattering centers (Adomi et al., 1990). Meanwhile, the impact of APBs on the optical properties of III–V materials has been studied by photoluminescence (PL), showing a lower emission intensity with a broader spectrum, compared to a single domain GaAs-on-Si reference (Akiyama et al., 1984; Alcotte et al., 2016). For this scenario, APBs serve as nonradiative recombination centers. The APBs can be typically avoided by growing III–V layers on misoriented Si (100) substrates (Kohama et al., 1989a; Seki et al., 1987). The mechanism is to generate a surface containing double-step terraces by a proper pre-growth annealing of the misoriented Si substrate (Kroemer, 1987). Some early research has also reported the realization of single domain

Fig. 2 Mechanism of APB formation in polar-on-non-polar growth due to the presence of single-height steps on the substrate surface, and the typical surface morphology. *Reprinted from Kroemer, H., 1987. Polar-on-nonpolar epitaxy. J. Cryst. Growth 81, 193–204; Li, Q., Lau, K. M., 2017. Epitaxial growth of highly mismatched III-V materials on (001) silicon for electronics and optoelectronics. Prog. Cryst. Growth Charact. Mater. 63, 105–120, with permission from Elsevier.*

GaAs films on on-axis (001) oriented Si substrates by both molecular beam epitaxy (MBE) and MOCVD growth (Akiyama et al., 1986; Kawanami et al., 1987). The successful accomplishment of single domain GaAs was attributed to the combination of surface cleaning by heat treatment (over 900 °C anneal in a mixed ambient of H_2 and AsH_3), a thin low-temperature (LT) GaAs layer (<20 nm) and the nucleation temperature (400–450 °C) (Akiyama et al., 1984, 1986).

2.2 High quality GaAs directly grown on Si by MOCVD

GaAs-on-Si technology has been intensively investigated and impressive progress has been made since the 1980s. This section will introduce various kinds of approaches to well address the fundamental challenges mentioned above, mainly focusing on the direct epitaxy by MOCVD. Generally, the methods can be divided into two approaches: direct epitaxy of GaAs on Si substrates, or growing with various buffers/transition layers between GaAs/Si. Fig. 3 classifies the epitaxial GaAs-on-Si growth methods (Bolkhovityanov and Pchelyakov, 2008).

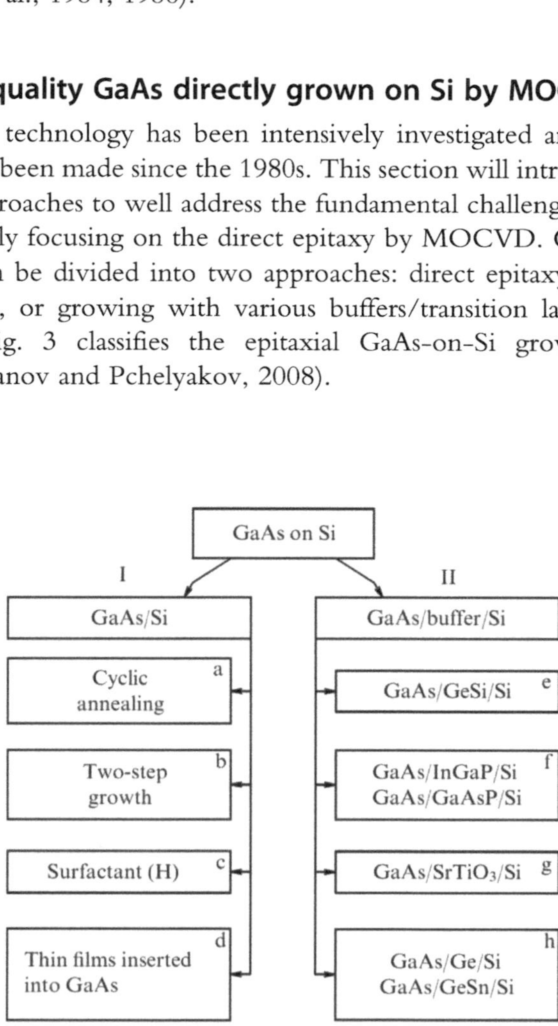

Fig. 3 Classification of epitaxial GaAs-on-Si growth methods. *Reproduced from Bolkhovityanov, Y. B., Pchelyakov, O. P., 2008. GaAs epitaxy on Si substrates: modern status of research and engineering. Physics-Uspekhi. 51, 437–456, © 2008 IOP Publishing Ltd.*

2.2.1 Two-step growth method

For GaAs/Si heterostructures, one method to obtain a smooth surface as well as a low TD density is by employing the so-called "two-step" growth mode (Akiyama et al., 1984). In such a growth mode, a very thin GaAs is first nucleated on the Si substrates at very low temperatures (normally 400 °C or even lower), before the thick GaAs film is deposited at elevated temperatures to significantly reduce the surface defect density. At the initial stage, the low-temperature GaAs usually nucleates on Si in a three-dimensional Volmer-Weber or Stranski-Krastanov type (Tachikawa et al., 1993). The lower the growth temperature is, the shorter diffusion length is acquired for the adatoms to self-assemble themselves into islands, resulting in a higher GaAs island density (Bolkhovityanov and Pchelyakov, 2008). Meanwhile, the thickness of the LT buffer also plays an important role in controlling the crystalline perfection of the GaAs films and a thin buffer on the order of 20 nm has been reported (Freundlich et al., 1988; Lum et al., 1987). Furthermore, the deposited LT-GaAs layer is desirable to be a continuous film since misfit dislocations (MDs) and planar defects tend to nucleate at the edge of the GaAs islands, which can be incorporated into the coalesced thin films. The surface roughness is also influenced by the growth condition of the initial GaAs nucleation layer (Fang et al., 1990).

A high-quality GaAs buffer can also be achieved by an in-situ two-step MOCVD growth method using amorphous-Si (a-Si) and amorphous-GaAs (a-GaAs) as the buffer layers (Uen et al., 2006). The structure presented in Fig. 4 includes a 20 Å a-Si and a 250 Å a-GaAs re-crystallized at a high temperature of 780 °C. Combining three periods of thermal cycle annealing and the two-step growth mode, the 4-μm-thick crack-free GaAs buffer on Si demonstrates a low etch-pit density (EPD) in the order of $10^6/cm^2$, together with a record-low full-width at half maximum (FWHM) of 102 arcsec by X-ray diffraction (XRD) rocking scan. However, mechanism on the effects and roles of the ultrathin amorphous buffers remains to be investigated.

Once the Si surface has been covered with a uniform GaAs layer, the growth temperature is subsequently ramped to a higher level as the two-step process. The high temperature (HT) growth (typically >550 °C) has an annealing effect on the quality of the initial buffer by healing out the point defects in the GaAs nuclei and creating a less defective GaAs buffer. Further increasing the thickness of HT-GaAs h would also decrease the TD density by a factor of $1/h$ due to an enhanced interaction of TDs, as shown in Fig. 5A.

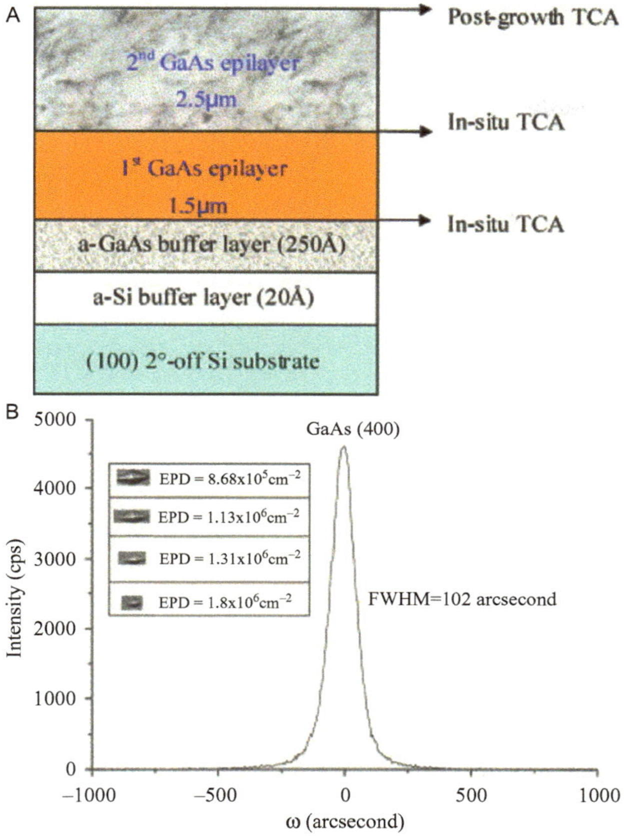

Fig. 4 (A) Schematic of the 4-μm-thick GaAs on Si structure. (B) XRD rocking curve of the GaAs epilayer grown on Si substrates, and inset shows the counts of different types of etch pits. *Reprinted from Uen, W.-Y., Li, Z.-Y., Huang, Y.-C., Chen, M.-C., Yang, T.-N., Lan, S.-M., Wu, C.-H., Hong, H.-F., Chi, G.-C., 2006. Heteroepitaxial growth of GaAs on Si by MOVPE using a-GaAs/a-Si double-buffer layers. J. Cryst. Growth 295, 103–107, with permission from Elsevier.*

2.2.2 Post-growth thermal cycle annealing

While the conventional two-step growth method generally yields $\sim 10^8/cm^2$ dislocations in the single-crystal epitaxial layer, thermal cycle annealing (TCA) has been proven as a very effective method to lower the TDD in GaAs/Si, in both MOCVD and MBE growths (Jung et al., 2017; Yamaguchi et al., 1990). To date, this approach has been widely applied to obtain high crystalline quality III–V buffers, with temperature cycling between a very low temperature (e.g., 200–300 °C) and a high temperature (over 750 °C) to enable the thermally driven dislocation migration,

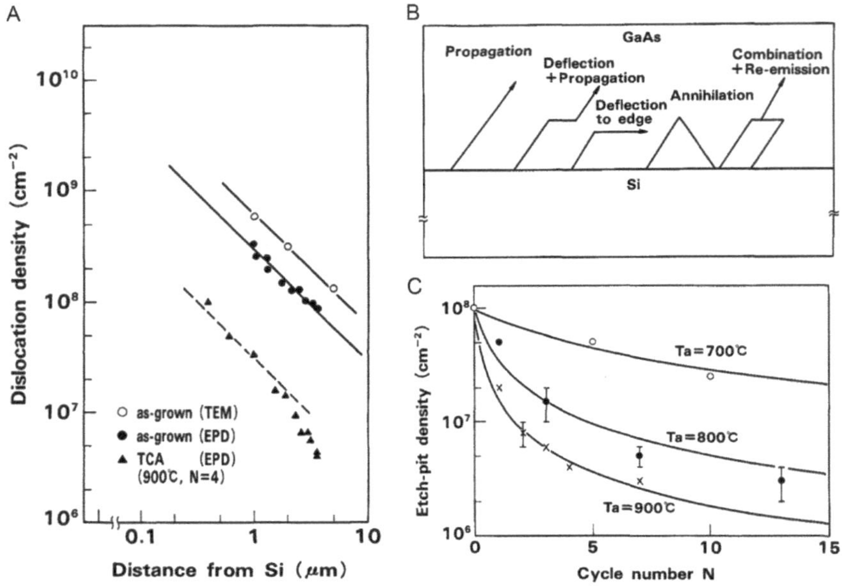

Fig. 5 (A) Dislocation density evolution for as-grown and thermally annealed GaAs/Si, determined by both TEM and EPD. (B) Mechanisms of dislocation motion in GaAs-on-Si, and (C) EPD in thermal cycle annealed GaAs/Si as a function of annealing temperature T_a and cycle number N. Panel A: *Reproduced from Yamaguchi, M., Tachikawa, M., Itoh, Y., Sugo, M., Kondo, S. 1990. Thermal annealing effects of defect reduction in GaAs on Si substrates. J. Appl. Phys. 68, 4518–4522, with the permission of AIP Publishing.* Panel C: *Yamaguchi, M., 1991. Dislocation density reduction in heteroepitaxial III-V compound films on Si substrates for optical devices. J. Mater. Res. 6, 376–384, reproduced with permission.*

leading to the annihilation of those dislocations. The mechanism of TCA on dislocation density reduction for GaAs-on-Si is attributed to the large thermal expansion coefficient between GaAs and Si (5.73×10^{-6}/K, and 2.59×10^{-6}/K). The periodic oscillations of the temperature apply compressive and tensile strains in the GaAs film, reversing the motion of TDs (Bolkhovityanov and Pchelyakov, 2008). As shown in Fig. 5B, the interactions of dislocations are enhanced in the thermally annealed samples, resulting in dislocation coalescence, annihilation and re-emission (Yamaguchi, 1991). Pure edge dislocations are mainly observed in thermal cycle annealed GaAs-on-Si samples.

The dislocation density after TCA, characterized by both plan-view transmission electron microscopy (TEM) observation and the EPD via molten KOH etching, is lowered by an order of magnitude (shown in Fig. 5A). It is thus concluded that the dislocation coalescence is the main driving force for dislocation annihilation in TCA (Yamaguchi, 1991). By further enlarging the annealing temperature range and increasing the cycle number, significant dislocation reduction was achieved, reaching a minimum EPD of 2×10^6 /cm^2, as presented in Fig. 5C. However, severe surface desorption has been observed for GaAs annealed at 800 °C (Shi et al., 2019), and one should also note that the large thermal mismatch introduced by TCA could consequently enhance the possibility of cracks formation on the GaAs surface.

The threading dislocations for samples with and without TCA are also characterized by cross-sectional TEM, as demonstrated in Fig. 6. It is noted that before TCA, large numbers of defects appear in the upper region of the epilayer, while the dislocation densities were considerably reduced for post-TCA samples (Kakinuma et al., 1999). Most of the dislocations are confined below the intermediate GaAs layer and near the (Al)GaAs/Si interface.

Low temperature PL was carried out to evaluate the efficiency of the TCA process on the GaAs intermediate buffer on Si by MOCVD (Akahori et al., 2001). Fig. 7A shows the InGaP layer grown on two-step GaAs/Si for solar cell applications. Five-period TCA was applied after 0.7 μm GaAs growth, with a static annealing at the high temperature (900 °C) for 5 min. The low-temperature PL intensity of the TCA sample, as shown in Fig. 7B, is increased by 100 times compared to the control sample without TCA. The PL intensity enhancement is attributed to the reduction of the dislocations, which act as nonradiative recombination centers in the forbidden band and quench the PL.

In addition to the TCA procedure, post-growth annealing (PGA) and in-situ annealing (ISA) were also attempted to achieve low TD density in GaAs crystal (Itoh et al., 1988). Using a sequence of 13 periods of in-situ annealing steps at 800 °C by MOCVD, the TD density was reduced to 2×10^6 to 5×10^6/cm^2.

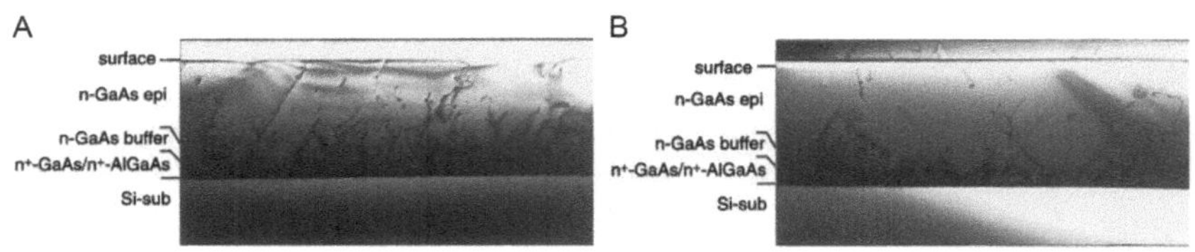

Fig. 6 Cross-sectional TEM images for (A) as-grown GaAs buffer on Si and (B) after TCA. *Reprinted from Kakinuma, H., Ueda, T., Gotoh, S., Yamagishi, C., 1999. Reduction of threading dislocations in GaAs on Si by the use of intermediate GaAs buffer layers prepared under high V–III ratios. J. Cryst. Growth 205, 25–30, with permission from Elsevier.*

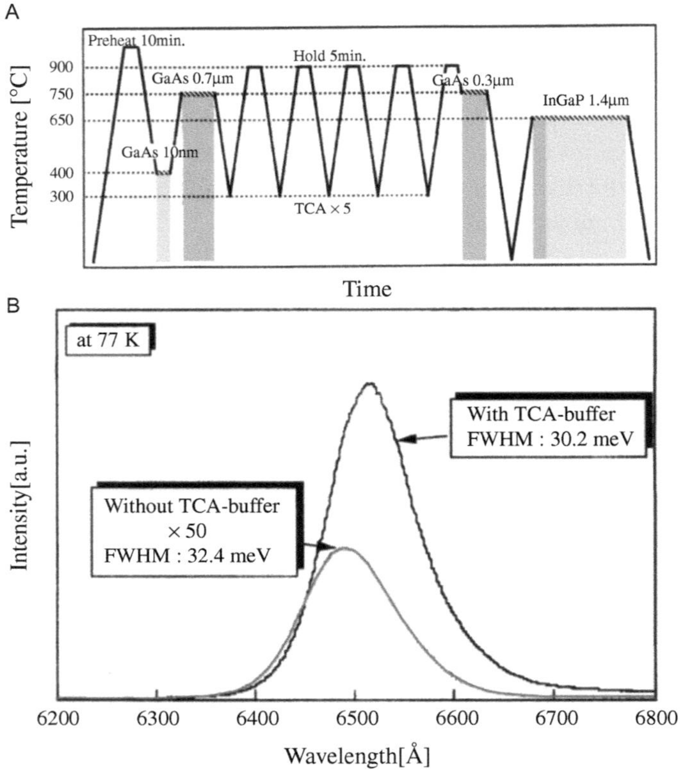

Fig. 7 (A) Growth sequence of InGaP on Si with GaAs intermediate buffer and TCA. (B) PL spectra of InGaP on Si with and without TCA. *Reprinted from Akahori, K., Wang, G., Okumura, K., Soga, T., Jimbo, T., Umeno, M., 2001. Improvement of the MOCVD-grown InGaP-on-Si towards high-efficiency solar cell application. Sol. Energy Mater. Sol. Cells 66, 593–598, with permission from Elsevier.*

However, experimental results have also revealed the creation of extra defects in the cooling stage after GaAs growth, due to a large thermal stress of $>10^9$ dyn/cm^2 at room temperature (Sugo et al., 1989). The EPD of GaAs at growth temperature of 650 °C was obtained using an in-situ HCl vapor phase etching (Tachikawa and Mori, 1990). It was uncovered that the GaAs buffer with TCA at the growth temperature exhibited an EPD density of $\sim10^4$/cm^2. Yet, after the crystal was cooled down to room temperature, the EPD formed by KOH etching was increased to 10^6–10^7/cm^2. This observation suggests that thermal stress induced dislocation propagation to the GaAs surface is the main factor of the dislocation generation in the cooling stage.

2.2.3 Dislocation filter layers

The effects of dislocation filter layers, including strained interlayers and strained-layer superlattices (SLSs), have been intensively investigated in the past decades (Choi et al., 1987; El-Masry et al., 1987; Takano et al., 1998). The general mechanism of how dislocation filter layer works can be explained by the interaction between the strain fields of the insertion layers with the dislocations. The Peach-Koehler force, determined by the strain (ε) and layer thickness (h), will either repel the dislocations laterally toward the edge of the sample and parallel to the growth plane, or bend the dislocations and enhance the dislocation coalescence and annihilation (Fang et al., 1990). Fig. 8 describes several situations of interaction between SLS and dislocations. Only when the strain field is strong enough can the dislocations be influenced, as shown in Fig. 8C–E. Excessive strain field would inversely introduce more mobile misfit dislocations (Shi et al., 2019).

Ternary-binary SLSs based on $In_xGa_{1-x}As$-GaAs, $GaAs_{1-y}P_y$-GaAs, $Al_x(Ga_{1-x})As$-GaAs and $Al_xGa_{1-x}P$-AlP have been attempted to improve the GaAs crystalline quality (Fischer et al., 1986a; Hayafuji et al., 1990, 1988; Soga et al., 1990, 1984) by MOCVD and MBE growths. Oftentimes, SLSs and the TCA process are combined to reduce TDs. Systematic study and optimization have been carried out by Hayafuji et al using MOCVD growth (Hayafuji et al., 1990). Their experiments have revealed that it is most efficient to conduct TCA at a GaAs buffer thickness of ~1.5 μm, and to insert SLS when the buffer reaches ~2 μm, as summarized in Fig. 9C. This is explained by the dependence of residual tensile thermal stress and the additional compressive strain generated by the SLS on the thickness of GaAs, respectively (Hayafuji et al., 1990). By optimizing the positions of TCA and SLS, a low EPD of $1.5 \times 10^6/cm^2$ was demonstrated. However, there is a debate on the effectiveness of Al(Ga)As/GaAs SLS insertion, which

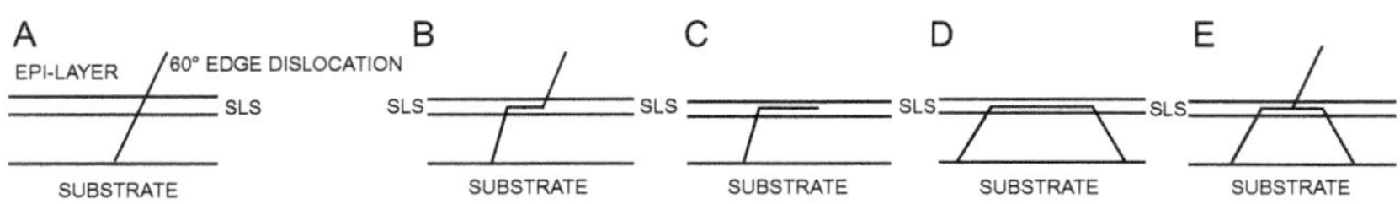

Fig. 8 The interaction between SLS and dislocations: (A) no interaction due to too weak strain field, (B) strain field not strong enough to bend the dislocations into lateral movement, (C) strong strain field to bend the dislocation toward edge of the sample, (D) mutual annihilation of dislocations, and (E) re-emission of a dislocation after dislocation coalescence. *Reproduced from Fang, S., Adomi, K., Iyer, S., Morkoc, H., Zabel, H., Choi, C., Otsuka, N., 1990. Gallium arsenide and other compound semiconductors on silicon. J. Appl. Phys. 68, R31–R58, with the permission of AIP Publishing.*

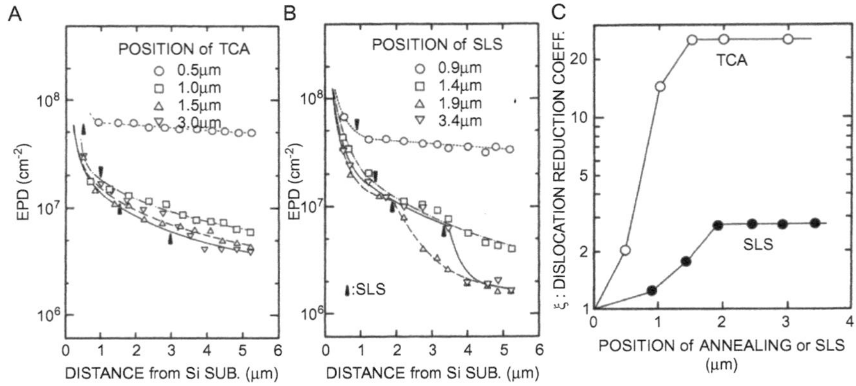

Fig. 9 EPD evolutions as a function of KOH etching depth for (A) TCA-treated and (B) SLS-treated samples. (C) Dependence of dislocation reduction coefficient for after employing TCA and SLS at different growth positions. *Reproduced from Hayafuji, N., Miyashita, M., Nishimura, T., Kadoiwa, K., Kumabe, H., Murotani, T. 1990. Effect of employing positions of thermal cyclic annealing and strained-layer superlattice on defect reduction in GaAs-on-Si. Jpn. J. Appl. Phys. 29, 2371, © 1990 IOP Publishing Ltd.*

barely introduces the strain field due to a negligible lattice mismatch between Al(Ga)As and GaAs. The dislocation filtering principle is either explained by the thermal mismatch between AlGaAs/GaAs and consequently crystal hardening of the AlGaAs layer (Hayafuji et al., 1988), or due to the built-in electric field of modulation doped AlAs (Shinohara, 1988).

There are some inherent shortcomings in the use of ternary-binary SLS system, such as the limited number of periods used to avoid the generation of new misfit dislocations. Therefore, the ternary-ternary strain-compensated $In_xGa_{1-x}As$-$GaAs_{1-y}P_y$ ($y \approx 2\times$) structure was proposed and attempted (El-Masry et al., 1987). Another effective approach was to insert a single strained interlayer (Takano et al., 1998). Coupled with the four-cycle TCA on a 2-μm GaAs epilayer, the InGaAs interlayer was optimized at a 200 nm $In_{0.09}Ga_{0.91}As$. Apparent dislocation blocking effect was revealed by the cross-sectional TEM in Fig. 10A. The insertion position was also optimized at a distance of 1.8 μm, below the GaAs top surface. The defect density of the 4-μm-thick GaAs buffer is characterized as low as $1.2 \times 10^6/cm^2$ by plan-view TEM, and the EPD is even two times lower. However, this approach is to some extent problematic with the film cracking issue.

2.2.4 GaAs growth on nanopatterned Si

The Aspect Ratio Trapping (ART) concept, based on selective-area epitaxy (SAE) technique, was proposed to improve the quality of III–V

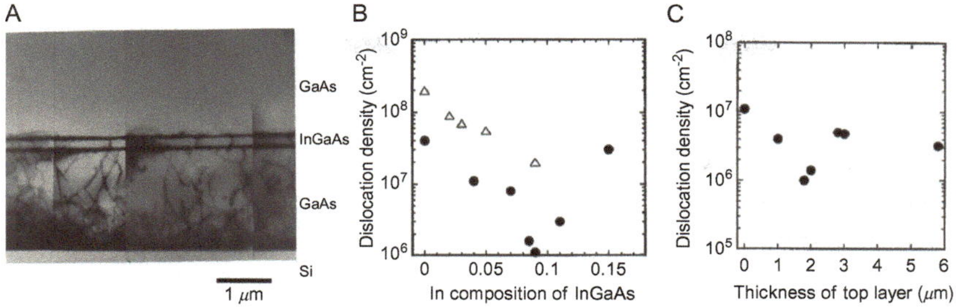

Fig. 10 (A) Cross-sectional TEM of the GaAs-on-Si structure. TD density as a function of (B) In composition in the InGaAs layer and (C) thickness of GaAs overgrown on the strained InGaAs interlayer. *Reproduced from Takano, Y., Hisaka, M., Fujii, N., Suzuki, K., Kuwahara, K., Fuke, S., 1998. Reduction of threading dislocations by InGaAs interlayer in GaAs layers grown on Si substrates. Appl. Phys. Lett. 73, 2917–2919, with the permission of AIP Publishing.*

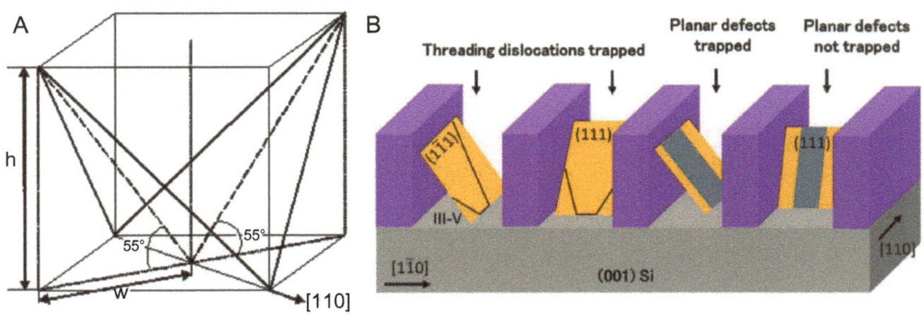

Fig. 11 (A) Schematic showing the crystallography of threading dislocations and their projections onto the (111) plane. (B) The trapping capability of ART for dislocations and planar defects. Panel A: *Reproduced from Bai, J., Park, J.-S., Cheng, Z., Curtin, M., Adekore, B., Carroll, M., Lochtefeld, A., Dudley, M., 2007. Study of the defect elimination mechanisms in aspect ratio trapping Ge growth. Appl. Phys. Lett. 90, 101902, with the permission of AIP Publishing.* Panel B: *Reprinted from Li, Q., Lau, K. M., 2017. Epitaxial growth of highly mismatched III-V materials on (001) silicon for electronics and optoelectronics. Prog. Cryst. Growth Charact. Mater. 63, 105–120, with permission from Elsevier.*

heteroepitaxial films grown on Si, by terminating defects at the sidewalls of patterned dielectric trenches (typically SiO_2) that enclose the growth region (Li and Lau, 2017; Orzali et al., 2015a). Initiated by Ge/Si heteroepitaxy (Langdo et al., 2000), this technique was then applied on III–V/Si growth. As shown in Fig. 11A, the "trapping" of the threading dislocation segments lying in the {111} planes along the ⟨110⟩ directions are based on the crystallographic geometry that line dislocations could project onto the oxide sidewall should the aspect ratio allow (Bai et al., 2007). In principle, the threading dislocations can be terminated by the oxide sidewall when

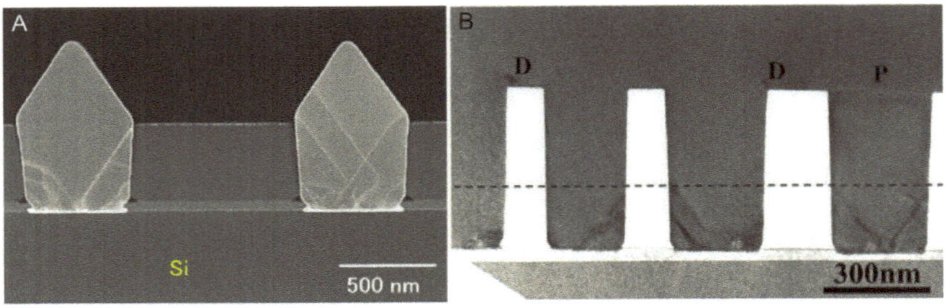

Fig. 12 (A) Cross-sectional dark-field STEM image of GaAs-on-sub-micron-patterned Si by MOCVD, showing the propagation of dislocations and stacking faults. (B) GaAs-on-Si with large aspect ratio (AR = 1.8) to demonstrate a defect-free GaAs on top of the oxide trenches. Panel A: *Reprinted from Li, Q., Lau, K. M., 2017. Epitaxial growth of highly mismatched III-V materials on (001) silicon for electronics and optoelectronics. Prog. Cryst. Growth Charact. Mater. 63, 105–120, with permission from Elsevier.* Panel B: *Reproduced from Li, J., Bai, J., Park, J.-S., Adekore, B., Fox, K., Carroll, M., Lochtefeld, A., Shellenbarger, Z., 2007. Defect reduction of GaAs epitaxy on Si (001) using selective aspect ratio trapping. Appl. Phys. Lett. 91, 021114, with the permission of AIP Publishing.*

the aspect ratio (AR = h/w) is larger than 1.4, determined by the projection angle of ~55° onto the (110) plane. A more detailed explanation is presented in Fig. 11B (Li and Lau, 2017). For most 60° threading dislocations gliding along the ⟨111⟩ planes, they will eventually terminate at the oxide wall if the AR is high enough for the dielectric spacers. Similarly, this approach also works for planar defects parallel to the trenches, while the perpendicular ones are presumably not influenced by the patterned trenches.

Fig. 12 presents the selective-area epitaxy of GaAs on Si with different aspect ratios and various top facets of GaAs nanowires by manipulating the growth conditions. In Fig. 12A, the GaAs was selectively grown on sub-micron-stripe patterned (001) Si by MOCVD, where (111) GaAs planes can be clearly identified. The ART phenomenon was clearly evidenced by the propagation and termination of crystalline defects, although the AR is not sufficiently high for a complete defect necking. Further reducing the pattern size to nanometer-scale with enlarged aspect ratio creates a near defect-free region as the overgrowth proceeds. Still, during the coalescence process, some new dislocations (labeled as "D") and planar defects (labeled as "P") are observed in Fig. 12B, which are not due to the lattice mismatch between GaAs/Si (Li et al., 2007). By carefully tuning the growth conditions, a highly uniform box-shaped GaAs was grown on the shallow trench isolation (STI) patterned 300 mm (001) Si wafers. The original V-shaped {111} Si surfaces is vanished after the two-step GaAs growth, due to reflow of GaAs seed layer at high temperatures, and the highest growth rate on Ga-rich {111} planes (Kunert et al., 2016). This might be more amenable for device implementation.

A more strategic method based on the ART concept is to form the nano-scale V-grooved Si substrates, by KOH or TMAH wet etching the nanopatterned (001) Si substrates (Li et al., 2015; Kunert et al., 2016). Fig. 13A presents the highly ordered GaAs nanowire arrays with the (111) facet planes clearly revealed. With an ultrathin low temperature GaAs stress relaxing layer to accommodate the 4% structural mismatch between GaAs/Si, the GaAs bulk layer possesses a very high crystalline quality. Only a few stacking faults in the {111} planes are observed in the ultrathin GaAs nucleation (Fig. 13B), and almost all of them are terminated by the "tiara"-like undercut structure created during the wet-etching process of the Si patterned with SiO_2. After the SiO_2 dielectric removal, GaAs overgrowth was performed to coalesce the nanowires into 300-nm-thick thin film structures. No planar defects escaped from the top of Si ridge was observed. However, the defect necking along the nanowire direction is much less pronounced,

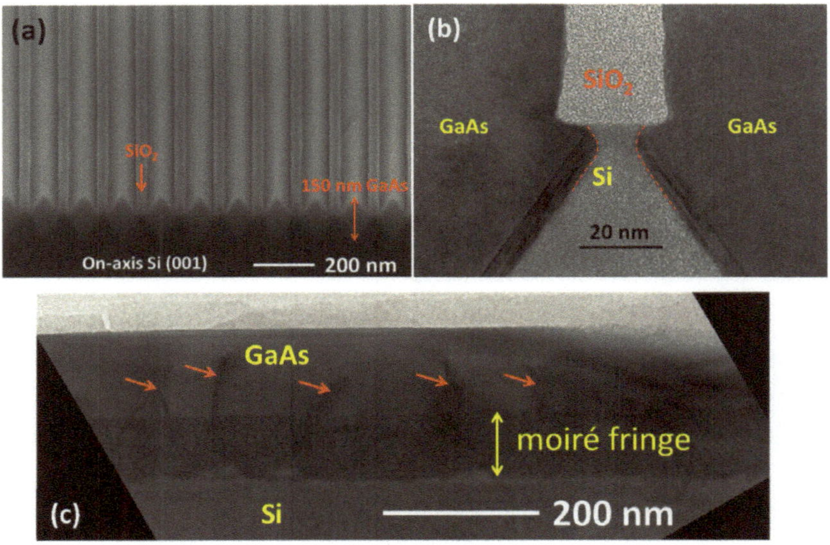

Fig. 13 (A) Tilted SEM image of 150-nm-thick GaAs nanowires separated by SiO_2 dielectrics. (B) Zoomed-in TEM image showing stacking faults trapped by a "tiara-like" structure formed by Si undercuts in GaAs nanowire growth. (C) TEM image of GaAs nanowires parallel to the trench direction. *Reproduced from Li, Q., Ng, K. W., Lau, K. M., 2015. Growing antiphase-domain-free GaAs thin films out of highly ordered planar nanowire arrays on exact (001) silicon. Appl. Phys. Lett. 106, 072105, with the permission of AIP Publishing.*

which can be seen in the dislocation threading upwards in the parallel direction (Fig. 13C). Moiré fringe patterns appear at the V-groove region due to overlapping of lattice constants of Si and GaAs (Li et al., 2015).

Some remaining issues of the SAE method are yet to be tackled, one of which is the prevalent planar defects, especially stacking faults and twin boundaries. Fig. 14A shows the propagation of planar defects. The close-up shown in the scanning transmission electron microscopy (STEM) image (Fig. 14B) can be classified as twins (Orzali et al., 2015b). Twin boundaries act as scattering centers and significantly influence carrier mobility. It is also observed that the meeting point of two twinned lamella facilitates the formation of pits on the surface, and the higher number of the twinned lamellas result in deeper pits. Those twin boundaries mainly originate from the bottom oxide layer with some roughness.

2.2.5 Compositionally graded buffer

Instead of dealing with GaAs/Si mismatch issues in direct epitaxy of GaAs on Si, a natural thinking is to insert or place other semiconductors with lattice constants and thermal expansion coefficients between GaAs and Si as gradual

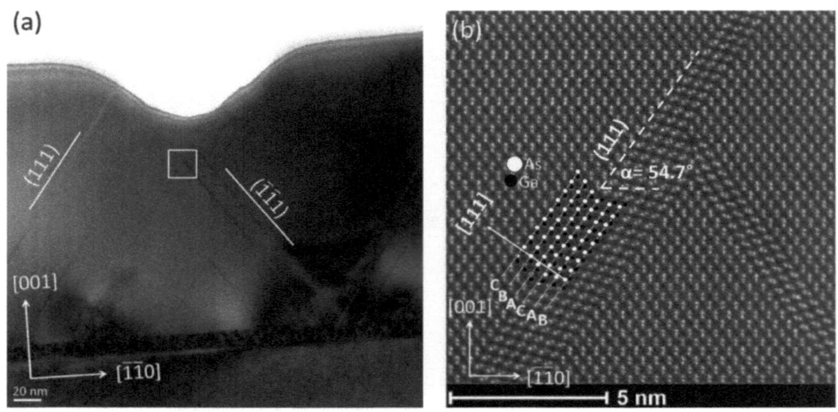

Fig. 14 (A) TEM dark field image of the cross section of GaAs fins parallel to the trenches. (B) High resolution STEM image of the two twinned lamellas. *Reproduced from Orzali, T., Vert, A., O'Brien, B., Herman, J. L., Vivekanand, S., Hill, R. J., Karim, Z., Papa Rao, S. S., 2015a. GaAs on Si epitaxy by aspect ratio trapping: analysis and reduction of defects propagating along the trench direction. J. Appl. Phys. 118, 105307; Orzali, T., Vert, A., Kim, T.-W., Hung, P., Herman, J. L., Vivekanand, S., Huang, G., Kelman, M., Karim, Z., Hill, R. J., PapaRao, S. S., 2015b. Growth and characterization of an In0. 53Ga0. 47As-based metal-oxide-semiconductor capacitor (MOSCAP) structure on 300 mm on-axis Si (001) wafers by MOCVD. J. Cryst. Growth 427, 72–79, with the permission of AIP Publishing.*

transition layers. In recent years, researchers have been trying different kinds of metamorphic buffers between Si and GaAs as crystallographic transitional pathways, i.e., Ge-on-Si intermediate buffer (Sheldon et al., 1985), compositionally graded Ge/Ge_xSi_{1-x} buffer (Ting and Fitzgerald, 2000), and GaAsP or InGaP layers (Komatsu et al., 1997; Takagi et al., 1995). This section focuses on the growth of GaAs on artificial Ge/GeSi/Si substrate by MOCVD for the sake of simplicity, where promising laser results have been obtained based on this structure.

Fig. 15 shows the TEM structure of a Ge film grown on Si, and a GaAs layer grown on top of the Ge/GeSi/Si substrate. Because of the large lattice mismatch between $Si_{0.1}Ge_{0.9}$ and Si, high density of small islands was formed and quickly coalesced into a continuous film as growth proceeds (Chang et al., 2005). With the small mismatch between $Si_{0.1}Ge_{0.9}/Si_{0.05}Ge_{0.95}$ and $Ge/Si_{0.05}Ge_{0.95}$, as well as thermal annealing after each layer growth, the propagating dislocations can be effectively bent and terminated at the interfaces, as depicted in Fig. 15A. Moreover, the resultant surface roughness is greatly improved. Additionally, because of the high quality Ge buffer, the GaAs layer on top of the Ge is quite clean by TEM observation. The weak strain field at the interface of GaAs/Ge, due to 0.01% lattice mismatch, is clearly shown in Fig. 15B. Based on the good quality GaAs/Ge buffer, this structure was used for photonic applications, i.e., InAlGaP LED on Si (Chilukuri et al., 2006), and the GaAs/AlGaAs lasers to be discussed in the following section.

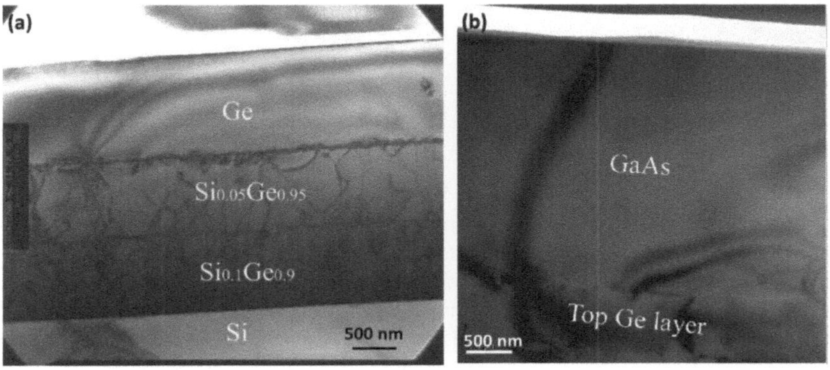

Fig. 15 Cross-sectional TEM images of (A) Ge/SiGe/Si, (B) GaAs on the Ge top layer. *Reprinted by permission from Springer Nature: Chang, E.Y., Yang, T.-H., Luo, G., Chang, C.-Y., 2005. A GeSi-buffer structure for growth of high-quality GaAs epitaxial layers on a Si substrate. J. Electron. Mater. 34, 23–26.*

2.2.6 APB-free GaAs on exact (001) Si surfaces

Although there have been a few reports of growing APB-free GaAs on exact (001) Si substrates, it is not widely applicable to all the MOCVD or MBE growths. The general explanation is attributed to a two-atomic step formation on the silicon vicinal surface (001) toward [110] directions (Bolkhovityanov and Pchelyakov, 2008). A single domain Si surface is thus formed, favoring the passivation of As-As dimers, where the APB-free III–V surface can be obtained upon a certain thickness. In addition, self-annihilation of APBs has also been proposed to suppress APBs (Fang et al., 1990). However, even to date, for conventional III–V growth on planar Si substrate, 4°–6° offcut angle toward [110] directions is presumably required to prevent APB formation, where the substrate mis-orientation can also bring on a lower dislocation density. Recently, notable progress has been made to achieve APB-free GaAs on quasi-normal Ge/Si substrate (Bogumilowicz et al., 2015), and to take a step further, mis-orientation as small as 0.15° in the [110] direction have been demonstrated to be capable of generating APB-free GaAs on nominal 300 mm Si (001) substrates (Alcotte et al., 2016).

Fig. 16A shows high density of randomly oriented APBs on GaAs grown on (001) Si by atomic force microscopy (AFM). After deoxidation using NF_3/NH_3 remote plasma and subsequently annealed in an MOCVD reactor

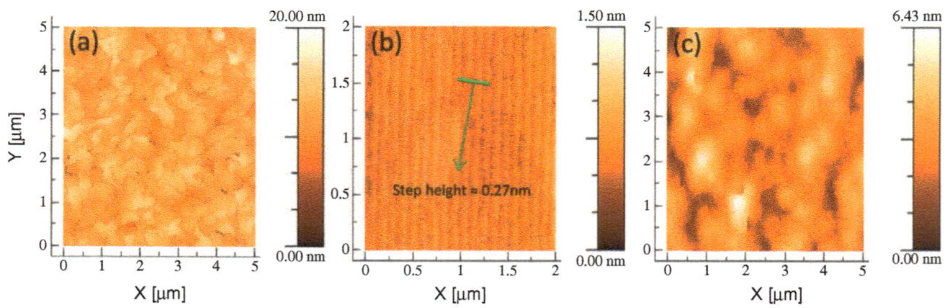

Fig. 16 (A) 5 × 5 µm^2 AFM image of 400 nm thick GaAs growth on un-optimized Si (001) with high density of randomly-oriented APBs: Roughness = 1.6 nm. (B) 2 × 2 µm^2 AFM image of 0.15° Si (001) after optimized preparation (800 °C–950 °C annealing under H_2), showing a double-stepped Si surface. (C) 5 × 5 µm^2 AFM image of APBs-free 150 nm thick GaAs on optimized 0.15° Si (001): Roughness = 0.8 nm. *Alcotte, R., Martin, M., Moeyaert, J., Cipro, R., David, S., Bassani, F., Ducroquet, F., Bogumilowicz, Y., Sanchez, E., Ye, Z., Bao, X. Y., Pin, J. B., Baron, T., 2016. Epitaxial growth of antiphase boundary free GaAs layer on 300 mm Si (001) substrate by metalorganic chemical vapour deposition with high mobility. APL Mater. 4, 046101; licensed under a Creative Commons Attribution (CC BY) license.*

at a high temperature (800–950 °C) for 1–10 min, a (2 × 1) Si surface is reconstructed and predominant double steps are observed on the surface in Fig. 16B. The double-step Si surface promises the APB-free 150 nm GaAs overgrowth. This result coincides with several experimental observations on APB-free III–V-on-Si epitaxy (Li et al., 2013; Orzali et al., 2015a,b; Shi et al., 2017a).

2.3 Epitaxy of InP buffer on Si

By virtue of its excellent compatibility with long wavelength (950–1650 nm) optoelectronic components and high-performance transistors, InP-on-Si has received increasing attention for decades (Lee et al., 1987). Nowadays, 1310 and 1550 nm on-chip lasers for short-reach datacenter communications and metropolitan long-haul telecommunications are mainly fabricated on InP-based lattice-matched III–V alloys. The significance of monolithic integration of photonic and electronic devices has been recognized since 1990, relying on the heteroepitaxial growth of InP on Si (Grundmann et al., 1991). This section thus focuses on the MOCVD growth of InP buffer layer on Si substrates, introducing various strategies to tackle the fundamental issues, such as a large amount of misfit dislocations and planar defects induced by the large lattice mismatch, and the appearance of APBs (Tang et al., 1992). Although less investigated than GaAs/Si epitaxial growth, fruitful results of InP grown on Si have been obtained so far.

2.3.1 Intermediate/compositionally graded buffer

Heteroepitaxy of InP on Si is more challenging than GaAs or GaP on Si because of the larger lattice mismatch (~8%). Despite the successful growth of good quality InP directly on planar Si substrates by carefully optimizing the growth temperature and V/III ratio of the first low temperature InP layer (Grundmann et al., 1991; Yamamoto et al., 1989), the surface roughness, residual stress, as well as crystalline quality of the InP can be further improved by adopting an GaAs or GaP intermediate buffer (Horikawa et al., 1988; Kohama et al., 1989a,b; Seki et al., 1987; Sugo et al., 1990; Tang et al., 2010). Fig. 17A describes the growth sequence of InP on Si with GaAs intermediate buffer by a two-step growth method. The InP epilayer exhibited a mirror-like surface across the 4-in. wafer. 77 K PL spectra of InP layer on different virtual substrates are compared in Fig. 17B. It is noteworthy that in addition to the significantly enhanced PL intensity for InP/GaAs/Si compared to InP/Si reference, the slight red-shift of the peak wavelength suggests a reduced tensile stress by introducing the GaAs intermediate buffer

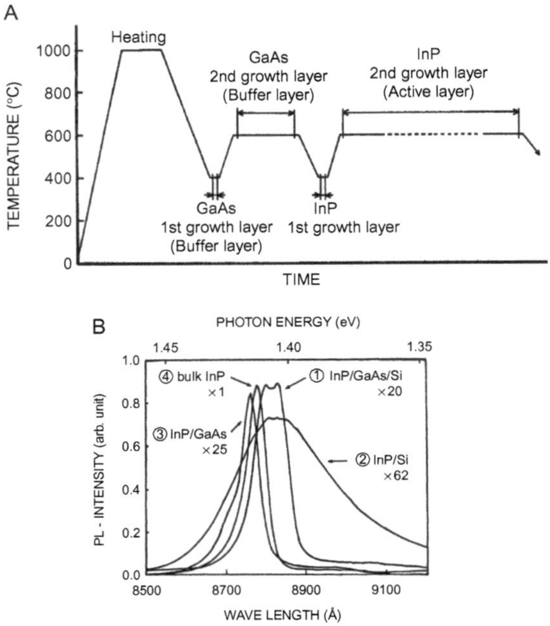

Fig. 17 (A) Growth sequence of InP on Si with GaAs intermediate buffer. (B) 77 K PL spectra of InP on various substrates. *Reproduced from Seki, A., Konushi, F., Kudo, J., Kakimoto, S., Fukushima, T., Koba, M., 1987. MOCVD growth of InP on 4-inch Si substrate with GaAs intermediate layer. Jpn. J. Appl. Phys. 26, L1587, © 1987 IOP Publishing Ltd.*

(Seki et al., 1987). The role of GaP as the intermediate buffer was also explored, suggesting a reduced effect than the GaAs counterpart, in terms of a larger residual stress and an inferior InP buffer quality (Sugo et al., 1990). This is mainly ascribed to the closer lattice constants between GaP and Si, making the division of the structural mismatch between InP and Si non-ideal.

Compositionally graded buffers, such as InGaAs, InAlAs, InAlGaAs, InGaP, and GaAsSb alloys have been studied and developed for high-quality InP buffer grown on GaAs substrates (Chyi et al., 1996; Liu et al., 2007; Quitoriano and Fitzgerald, 2007; Yang et al., 2011). By growing at a high temperature and compositions close to binary compounds, the phase separation in these ternary alloys can be avoided. A low dislocation density of $1.2 \times 10^6/cm^2$ was achieved using the hybrid $In_{0.3}Ga_{0.7}As/In_{0.82}Ga_{0.18}P$ buffer, as shown by the plan-view TEM image in Fig. 18B (Quitoriano and Fitzgerald, 2007). Although this structure possesses strong optical quality, as proven by the comparable RT and LT-PL intensities of the same

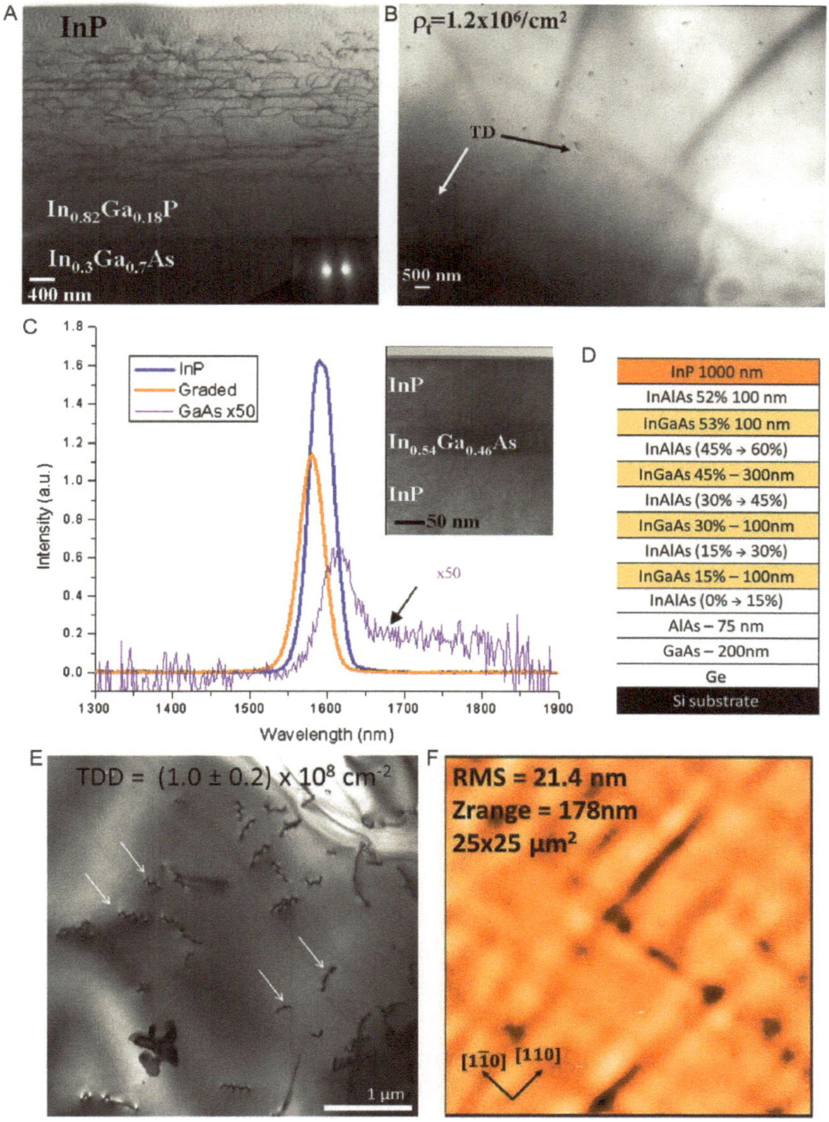

Fig. 18 (A) Cross-sectional and (B) plan-view TEM of InP on hybrid $In_{0.3}Ga_{0.7}As/In_{0.82}Ga_{0.18}P$ buffer. (C) 20 K PL spectra of a 55 nm $In_{0.54}Ga_{0.46}As$ cladded by InP layers on different substrates. Inset shows the zoomed-in TEM of the active region. (D) Schematic of InP on Si using composite buffer with InGaAs interlayers. (E) Plan-view TEM under g = ⟨220⟩ for dislocation counting. (F) 25 × 25 μm^2 AFM image of the sample surface. Cross-hatch pattern is visible along ⟨110⟩ direction. Panel C: *Reproduced from Quitoriano, N.J., Fitzgerald, E.A., 2007. Relaxed, high-quality InP on GaAs by using InGaAs and InGaP graded buffers to avoid phase separation. J. Appl. Phys. 102, 033511, with the permission of AIP Publishing.* Panel F: *Reprinted from Kohen, D., Nguyen, X. S., Made, R. I., Heidelberger, C., Lee, K. H., Lee, K. E. K., Fitzgerald, E. A., 2017. Preventing phase separation in MOCVD-grown InAlAs compositionally graded buffer on silicon substrate using InGaAs interlayers. J. Cryst. Growth 478, 64–70, with permission from Elsevier.*

55 nm $In_{0.54}Ga_{0.46}As$ cladded by InP, the large surface roughness of 61.9 nm prevents their practical applications in device growth and fabrication. Besides this, the compositional grading approach for InP grown on Si are not as effective as that grown on GaAs substrates. Recently, by introducing low temperature InGaAs interlayers in the compositionally graded InAlAs buffer, the phase separation issue is successfully eliminated (Kohen et al., 2017). Yet, the defect density is increased to around $1.0 \times 10^8/cm^2$ and the surface roughness remains a high value (21.4 nm), as shown in Fig. 18E and F.

2.3.2 Thermal annealing

Thermal cycle annealing or post-growth annealing process has not been widely applied in InP-on-Si epitaxy due to the smaller thermal stress induced by a closer thermal expansion coefficient of InP/Si than GaAs/Si (Ababou et al., 1996). Nevertheless, the FWHM of XRD rocking scan can be reduced to 250 arcsec after a long-period annealing for InP/Si (totally 1.5 h at 730–780 °C), as shown in Fig. 19A (Horikawa et al., 1988). For InP/GaAs/Si structures, it was found when the GaAs buffer thickness exceeds 1 μm, the improvement of InP buffer quality is limited, as evidenced by the saturation of XRD FWHM values. With 4 cycles of TCA, the EPD of 6-μm-thick InP buffer was $3 \times 10^7/cm^2$ and the x-ray FWHM was 250 arcsec (Kohama et al., 1989a,b).

2.3.3 Dislocation filters

The dislocation filter layers (DFLs) in the InP/Si system also relies on the interaction between strain field of DFLs and the TDs, similar to GaAs-on-Si heteroepitaxy. SLSs such as In(Ga)AsP/InP, (In)GaP/InP, InGaAs/GaAsP, InGaAs/InP, were proposed to reduce the dislocation density in InP epilayer (Crumbaker et al., 1989; Mori et al., 1993; Samonji et al., 1996; Seki et al., 1988). The InP surfaces turned out to be smoother after SLS insertion (Seki et al., 1988). Nevertheless, the effects of the SLSs in the InP-on-Si system is not as pronounced as what was reported in the GaAs/Si system, and the reduction of threading dislocation density by these techniques was far from sufficient (<30% reduction, and the TD density normally remains on the order of $10^8/cm^2$). It is speculated to be related to the huge misfit dislocation density generated by the InP/Si (Samonji et al., 1996), as well as a severe phase separation in the inserted SLSs (Quitoriano and Fitzgerald, 2007). Some strain-compensated superlattice structures, i.e., InGaAs/InAlAs (Khorenko et al., 2004), and InGaAsP

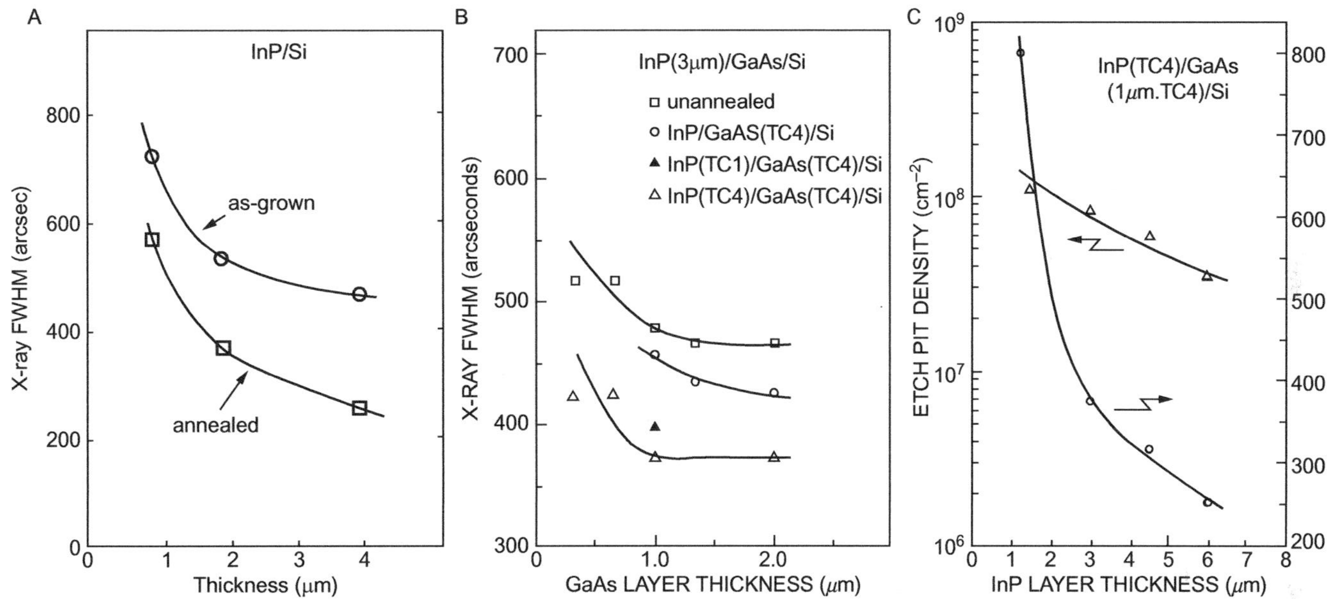

Fig. 19 (A) X-ray FWHM dependence on the thickness of the top InP layer. The annealed samples were treated at 730–780 °C for totally 1.5 h. (B) X-ray FWHM of InP epilayers at different GaAs buffer thicknesses and (C) dependence of x-ray FWHMs and EPDs of InP epilayers on the InP layer thickness. Panel A: *Reprinted from Horikawa, H., Kawai, Y., Akiyama, M., Sakuta, M., 1988. Hetero-epitaxial growth of InP on Si substrates by LP-MOVPE. J. Cryst. Growth 93, 523–526, with permission from Elsevier.* Panel C: *Republished with permission of Electrochemical Society from Kohama, Y., Kadota, Y., Ohmachi, Y., 1989a. Heteroepitaxy of InP on Si substrates by MOCVD. J. Electrochem. Soc. 136, 3853–3856.*

superlattices (Megalini et al., 2018) were also explored, aimed to reduce the threading dislocations more effectively. By growing InP on GaAs-on-V-grooved Si substrates, the lowest reported roughness for InP/Si was obtained at 1.2 nm (Megalini et al., 2018), but no obvious TD reduction can be observed so far.

Another approach developed to filter the TDs is by adopting the self-assembled three-dimensional quantum dots (QDs) as dislocation filter layers (Shi et al., 2017a, 2018). The InAs QDs were first proposed and embedded in the GaAs-on-Si system to impede the propagation of dislocations and to reduce dislocation density (Yang et al., 2007). Compared with conventional two-dimensional SLS, self-assembled QDs formed by Stranski-Krastanow (SK) growth mode provide a larger three-dimensional strain field and a stronger Peach-Koehler force, thus potentially offering more effective dislocation filtering capability. The dislocation filtering effect is examined by cross-sectional TEM, as presented in Fig. 20. With the total InP and GaAs buffer fixed at the same thickness, evidently TD density reduction is observed after the 7-period QDs dislocation filter layer inserted (Fig. 20B). A fourfold reduction of defect density to $\sim 3.2 \times 10^{8}\,\mathrm{cm}^{-2}$ was observed as a result of the interaction of the dislocations with the QDs (Shi et al., 2017a).

2.3.4 InP epitaxially grown on nanopatterned Si

Recently, dominant prevailing growth scheme for InP on Si is the patterned growth by MOCVD, using the ART technique on nano-patterned Si

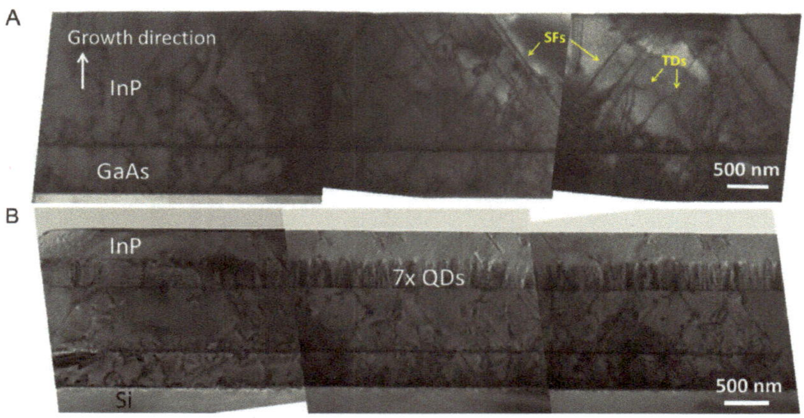

Fig. 20 Large-area cross-sectional TEM images of the InP-on-Si (A) without QDs (total width of 8.94 μm) and (B) with QD filter layers (total width of 8.49 μm). *Reprinted from Shi, B., Li, Q., Lau, K. M., 2017a. Self-organized InAs/InAlGaAs quantum dots as dislocation filters for InP films on (001) Si. J. Cryst. Growth 464, 28–32, with permission from Elsevier.*

substrates. The first attempt of patterned growth of InP-on-Si can be traced back to 1992 (Grundmann et al., 1992). Moreover, APBs can be effectively avoided using this selective area growth method, facilitating the InP on exactly oriented (001) Si substrates.

Starting with the patterned SiO_2 dielectric and nanopatterned Si substrates, the two-step InP epilayer is selectively grown inside the Si trenches. Fig. 21D shows the misfit dislocations and stacking faults generated from the hetero-interface between InP and Si, most of which are confined inside the nanowires by ART. Subsequently, the SiO_2 dielectric is removed and epitaxial lateral overgrowth (ELOG) of the InP nanowires leads to the coalescence of InP nanostructures into a thin film, as shown in Fig. 21A–C. Bending and annihilation of dislocations occur during the overgrowth process, resulting in a significantly reduced defect density in the upper 1 μm InP (Li et al., 2014). To evaluate the defect reduction effect of this growth sequence, an InAlGaAs/InAlAs multi-quantum well (MQW) structure was grown on top of the InP-on-Si virtual substrate, showing comparable PL intensities and linewidths with the MQWs on homoepitaxial InP-on-InP buffer, as evidenced in Fig. 21F (Shi et al., 2015).

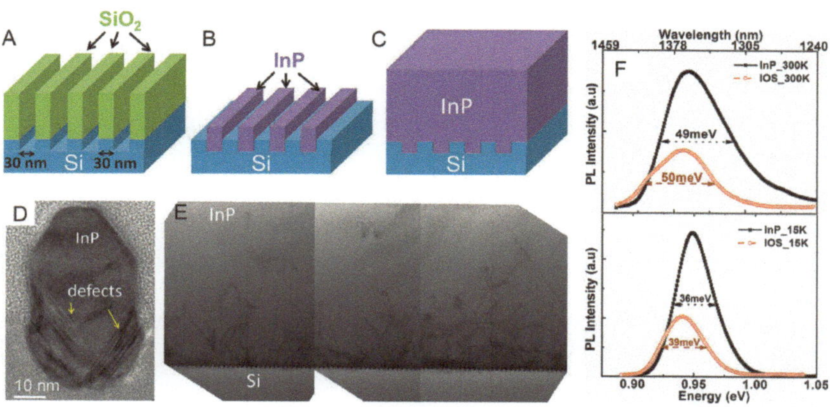

Fig. 21 (A)–(C) Growth procedure of InP on nanostructured Si with a position-controlled seed array. Cross-sectional TEM images of (D) an InP nanowire seed layer and (E) a coalesced InP film observed from the plane perpendicular to the seed stripes. (F) PL spectra recorded at 300 and 15 K for InAlGaAs/InAlAs MQWs grown on the InP-on-nanopatterned-Si and on the InP reference substrate, respectively. Panel E: *Reprinted from Li, Q., Ng, K. W., Lau, K. M., Tang, C. W., Hill, R., Vert, A., 2014. Defect reduction in epitaxial InP on nanostructured Si (001) substrates with position-controlled seed arrays. J. Cryst. Growth 405, 81–86, with permission from Elsevier.* Panel F: *© 2015 IEEE. Reprinted, with permission, from Shi, B., Li, Q., Wan, Y., Ng, K. W., Zou, X., Tang, C. W., Lau, K. M., 2015. InAlGaAs/InAlAs MQWs on Si substrate. IEEE Photon. Technol. Lett. 27, 748–751.*

The surface morphology of the InP film can be affected by the trench width of the patterned Si. In the case of wider trench openings, the high temperature InP coalescence will first occur in a 3D growth mode after nucleation, and finally evolve into a 2D growth mode. However, when the trench width continues to scale down, the growth can only proceed in a 3D mode, resulting in a rough and defective InP epilayer on top (Merckling et al., 2014).

Stacking fault (SF) is an inherent issue for the InP ELOG process on (001) Si (Krost et al., 1994a,b). It has been reported that stacking faults in InP have a rather low generation energy compared to those in other III–V compound semiconductors. Thus, the stacking faults form more easily in InP-on-Si (Tachikawa et al., 1995). As shown in the cross-sectional TEM in Fig. 22, there are two types of SFs in InP, one lying on the $\{1-11\}$ and $\{-111\}$ planes and another on $\{111\}$ and $\{-1-11\}$ planes. The SFs are mostly blocked by the SiO_2 mask but at some points propagating through the mask openings. In addition, SFs are also formed at the interface between the ELOG InP and the SiO_2 mask (Junesand et al., 2014). Meanwhile, TDs are observed to form at the boundaries of coalescence, but can be filtered by the mask, as shown in Fig. 22B (Junesand et al., 2013).

The use of etched $\{111\}$ Si V-grooves in the ART growth process clearly demonstrates its unique advantages. First, the crystallographic alignment between III–V and Si in the V-grooves avoid the formation of APBs. The single step in the "V-shape" (111) Si surface has a height of 0.31 nm and extends over a Si-double layer, eliminating the generation of APB (Li and Lau, 2017; Krost et al., 1994a). Second, III–V nucleation on Si (111) generates fewer defects as compared to nucleation on planer Si (001). Early research has revealed an order of magnitude reduction of dislocation density

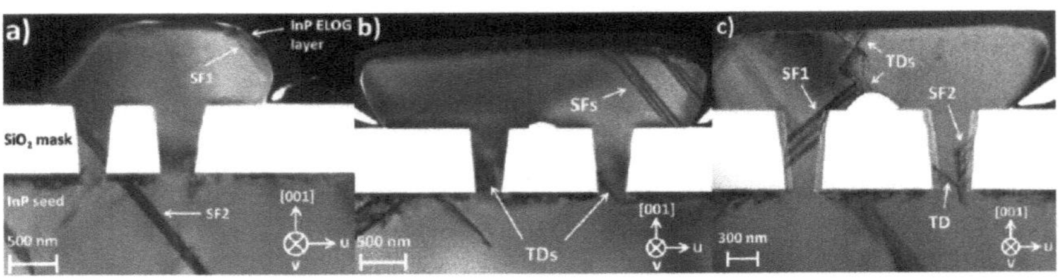

Fig. 22 (A)–(C) Cross-sectional TEM images of sulfur-doped InP ELOG on InP/Si of different sizes. The images were taken in bright field mode under two beam condition tilted off a [100] zone axis. The diffraction vector g was [400] in (A) and (B), and [040] in (C). *Reprinted with permission from Junesand, C., Kataria, H., Metaferia, W., Julian, N., Wang, Z., Sun, Y.-T., Bowers, J. E., Pozina, G., Hultman, L., Lourdudoss, S., 2013. Study of planar defect filtering in InP grown on Si by epitaxial lateral overgrowth. Opt. Mater. Express 3, 1960–1973 © The Optical Society.*

for InP on (111) Si (Krost et al., 1994a,b). The crystallographic improvement is attributed to strain relaxation by the formation of a thin twinned region close to the interface which contains partial dislocations with Burgers vectors parallel to the Si (111) surface. This is in sharp contrast to the heteroepitaxy on the Si (001) surface in which the stress is mainly relaxed by misfits and 60° threading dislocations.

InP on V-grooved Si by selective-area epitaxy is now being intensively studied (Han et al., 2016; Megalini et al., 2017). Fig. 23A shows an SEM image of highly ordered InP nanowires on V-shaped Si. Smooth growth facets consisting of two convex (111) B surfaces and a flat (001) surface are exposed (Han et al., 2016). A majority of the defects generated at the III–V/Si interface are confined within the V-grooved pocket, with a few visible dislocations and stacking faults propagating into the bulk of the InP wire and terminating at the (111) InP surface, as shown in Fig. 23B. The lattice mismatch between InP/Si is accommodated by the formation of a few nanometer-thick stacking faults running parallel to the (111) Si surfaces within the ultra-thin GaAs layer, instead of threading dislocations that would propagate upward. A close-up TEM image of the III–V/Si interface is shown in Fig. 23C.

A novel alternative to the conventional ART technique is referred to the confined epitaxial lateral overgrowth (CELO) (Czornomaz et al., 2015). By forcing the growth direction to turn 90° with respect to the original seed, defect trapping occurs in two directions. For traditional ART, defect generated along the trench can hardly be filtered geometrically. The main idea of this approach is to form a 3D hollow cavity containing a crystalline seed that will be selectively filled by the epitaxial material. All defects will be confined in the seed region by geometry. Due to its binary nature, InP structures were integrated by CELO on Si as a test vehicle to assess the material quality and

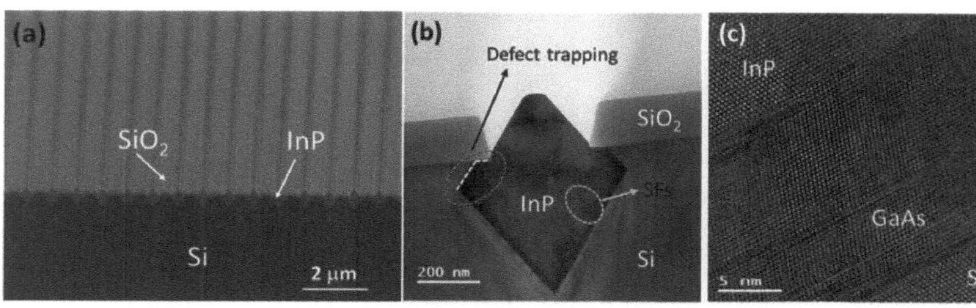

Fig. 23 (A) 70° tilted SEM image of as-grown highly ordered InP nanowires separated by SiO_2 spacers. (B) Cross-sectional TEM image (perpendicular to the trench) of a single semi-rhombic InP nanowire. (C) High resolution TEM image of the III-V/Si interface. *Reproduced from Han, Y., Li, Q., Chang, S.-P., Hsu, W.-D., Lau, K. M. 2016. Growing InGaAs quasi-quantum wires inside semi-rhombic shaped planar InP nanowires on exact (001) silicon. Appl. Phys. Lett. 108, 242105, with the permission of AIP Publishing.*

the defect filtering mechanism. A TEM study of the microcrystalline structure reveals that a high density of defects (mostly SFs) are generated in the seed region, but no evidence of any crystalline disorders can be observed in the fully relaxed InP-on-insulator region. Besides this, a smoother surface can be ex pected without the need for CMP, which offers a simple and highly flexible route for integration of InP-based optoelectronic devices on (001) Si substrates.

Interestingly, the surface morphology of InP on patterned Si substrate can be tailored simply by varying the growth temperature of the InP SAE layers (Lee et al., 2015). Fig. 24 shows the step-by-step morphological evolution of the InP layer on sub-micron patterned Si substrate separated by SiN_x masks. A thin GaP layer was first deposited by a two-step method as the intermediate buffer, followed by the LT-InP deposition. During the LT-InP growth, the sidewalls of InP nearly touched each other due to the faster growth rate of (001) planes. Further growth of HT-InP at 550 °C led to the merge of two (001) planes toward ⟨110⟩ direction, and a flat (001) top surface is thus obtained, as shown in Fig. 24D and E.

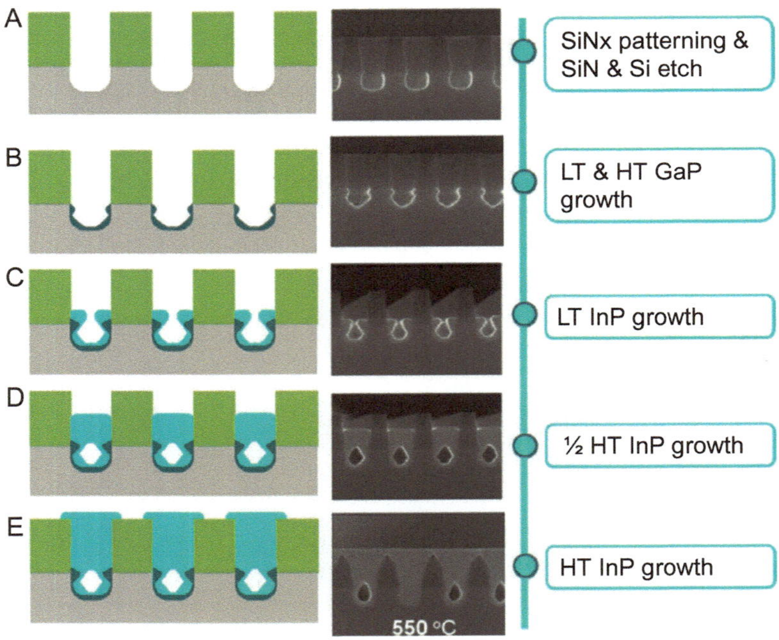

Fig. 24 Evolution of growth behavior of InP layer grown at 550 °C from (A) SiN_x patterning to (E) HT-InP growth. *Reprinted from Lee, S.-M., Cho, Y. J., Park, J.-B., Shin, K. W., Hwang, E., Lee, S., Lee, M.-J., Cho, S.-H., Shin, D. S., Park, J., Yoon, E., 2015. Effects of growth temperature on surface morphology of InP grown on patterned Si (001) substrates. J. Cryst. Growth 416, 113–117, with permission from Elsevier.*

The over-etched Si concaves are critical to reduce the crystalline defects including APBs, TDs, and SFs, through the formation of InP/GaP buffer at the bottom and sidewall of the etched Si. Most defects were trapped by the sidewalls of etched Si concaves and the rest of them were stopped at the sidewalls of SiN_x mask, as shown in Fig. 25A. In the case of HT-InP growth at 650 °C in Fig. 25A, the migration of precursors is dominantly transported on the (001) plane compared to the (111) and (311) plane, since the surface energies of (111) and (311) planes are lower than that of the (001) plane. Therefore, the growth rate of (001) plane is enhanced, resulting in the fast disappearance of (001) plane and leaving the most stable (111) planes. In contrast, the top HT-InP layer grown at 550 °C is equipped with atomically flat surfaces, as shown in Fig. 25C and D. The roughness is only 0.23 nm for the top flat InP (Fig. 25D) without any indication of APBs or dislocations. This is attributed to the different growth rate at different crystal planes and the shorter surface migration length of precursors at lower growth temperatures. After growth of LT-InP layers, the top InP surface is dominated by the (001) plane, and {311} planes appear at the edges of the trenches, as shown in Fig. 25B. The formation of voids at the bottom of the trenches is attributed to the lower growth rate of {111} LT-InP inside the void (Lee et al., 2015).

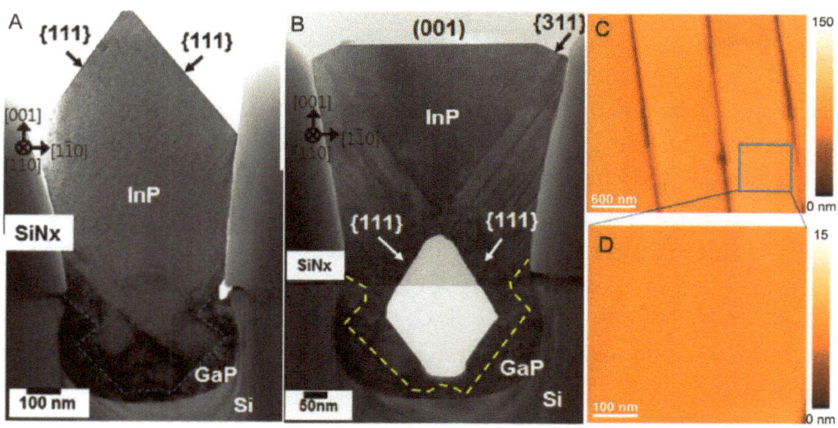

Fig. 25 Cross-sectional TEM images of epitaxially grown InP layer on a patterned Si (001) wafer with SiN_x mask and HT-InP layer grown at (A) 650 °C and (B) 550 °C. (C) and (D) AFM images recorded from the surface of HT-InP grown at 550 °C. *Reprinted from Lee, S.-M., Cho, Y. J., Park, J.-B., Shin, K. W., Hwang, E., Lee, S., Lee, M.-J., Cho, S.-H., Shin, D. S., Park, J., Yoon, E., 2015. Effects of growth temperature on surface morphology of InP grown on patterned Si (001) substrates. J. Cryst. Growth 416, 113–117, with permission from Elsevier.*

2.4 III–V-on-Si from thin film to micro- and nano-scales

Epitaxial III–V nanowires (NWs) on silicon substrate have been extensively investigated since 2004 (Mårtensson et al., 2004). Topologically, these nanowires can be categorized into vertical and planar nanowires, where selective area epitaxy is commonly applied to produce the nanostructures. For the sake of simplicity, readers are referred to several comprehensive reviews and literatures for more comprehensive discussions (Borg et al., 2014; Joyce et al., 2011; Tomioka et al., 2011). The Si-based nanowire lasers will be reviewed in the following section.

3. Epitaxial growth of lasers on silicon emitting at near-infrared wavelengths

While group-IV-based optical building blocks including silicon modulators, micro-ring resonators, waveguides, and germanium photodetectors have demonstrated their techno-economic viability, an efficient Si-based laser source remains the most challenging component due to the inherent indirect bandgap of Si (Liang et al., 2016). Integrating high performance III–V lasers with Si has been regarded as a promising path to resolve this challenge. Two mainstream approaches are devoted to integrate III–V lasers on Si substrates, referring to the hybrid integration and monolithic integration methods. The hybrid integration scheme, relying on the wafer bonding or flip-chip bonding techniques, combines the best properties of III–V and Si, and effectively separates each component without compromises. However, this approach has taken a decade to reach production scale, with stringent requirements on the alignment, making it less cost-effective (Reithmaier and Benyoucef, 2016). The monolithic integration method, as an alternative, is potentially ideal considering the advantages including lower cost in higher volume fabrication, eliminating the requirement of high-precision alignment, and the capability to leverage CMOS processes (Liang and Bowers, 2010). In this section, we will focus on near-infrared III–V lasers monolithically integrated on Si via epitaxial growth, for silicon photonics applications.

3.1 Short-wavelength Si-based lasers

The integration of III–V lasers on Si originates with GaAs/AlGaAs double heterostructure (DH) laser on Si. The first AlGaAs DH laser diode was demonstrated on MBE grown GaAs/Ge/Si wafer in 1984 (Windhorn et al., 1984).

The initial results reveal only one device operating at 77 K under pulsed pumping. In some later work, room-temperature pulsed operation was achieved by increasing the GaAs buffer thickness to 9-μm-thick and improving the laser cavity fabrication technique through chlorine-assisted ion beam etching (Windhorn and Metze, 1985). However, the lasing threshold was beyond 65 kA/cm^2, which requires further improvement.

The first MOCVD grown GaAs/AlGaAs DH laser diode on Si was realized by Sakai et al. (1985). The greatly improved GaAs buffer quality with a GaP intermediate buffer and the adoption of GaP/$GaAs_{0.5}P_{0.5}$ and $GaAs_{0.5}P_{0.5}$/GaAs SLSs enable the room-temperature functioning of the laser diodes (LDs) with a much lower threshold current density of 27 kA/cm^2. The optimization route continued with progressively decreased lasing threshold, yet the devices still operated under pulsed pumping (Fischer et al., 1986b; Van der Ziel et al., 1987b). The first room temperature continuous-wave (cw) operation was then achieved by a combination of MBE grown GaAs-on-Si template and MOCVD grown single GaAs quantum well laser structure (Deppe et al., 1987).

To improve the material quality of the GaAs buffer toward cw operation of the laser, five times of TCA and eight periods of $Al_{0.4}Ga_{0.6}As$-GaAs superlattices were introduced. The 827 nm LDs exhibit a low threshold of 120 mA (13 μm × 375 μm) under cw excitation, which is increased to 150 mA after 20 min, and 178 mA after 90 min. This indicates a rapid degradation for the GaAs/AlGaAs DH laser, ascribed to defects migration and increased heating effect. More concretely, this device degradation involves the growth of dark line defects (DLD) in the ⟨100⟩ direction as well as a gradual darkening of the active area near the mirrors (Van der Ziel et al., 1987a,b). The ⟨100⟩ DLDs are introduced by nonradiative recombination originated at defect sites such as dislocations and stacking faults. They rapidly grow in highly strained regions, resulting in a decrease of internal quantum efficiency η_i and a gradual increase of lasing threshold (Hasegawa et al., 1995). Micro-cracking or cross-hatch pattern is another severe issue to be considered for reliable GaAs/AlGaAs LDs operation (Deppe et al., 1988). By changing the GaAs/Si buffer structure to 10 μm-thick relaxed graded GeSi layer, the surface threading dislocation density for the GaAs/AlGaAs lasers on Si substrates was as low as 2×10^6 cm^{-2}, allowing the continuous room-temperature lasing with a lower threshold current density ($J_{th} = 577$ A/cm^2) at a wavelength of 858 nm (Groenert et al., 2003a,b). The compositional schematic diagram is shown in Fig. 26A. The lower AlGaAs cladding and graded index separate confinement heterostructure

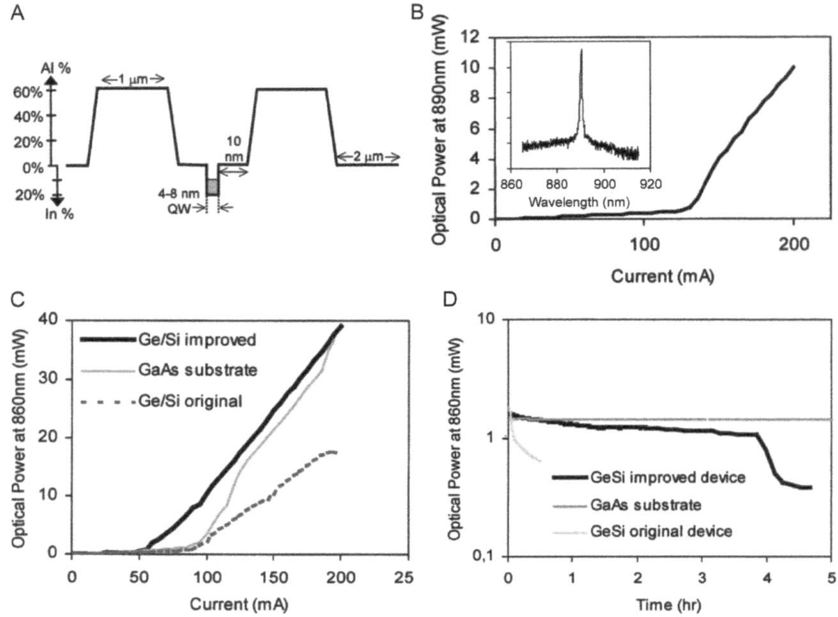

Fig. 26 (A) Compositional schematic diagram of the InGaAs/GaAs strained quantum well lasers grown on the Ge/GeSi/Si substrate. (B) Room temperature cw L-I curve for 5 nm $In_{0.17}Ga_{0.83}As$ quantum well laser on Si. Inset shows the lasing spectrum above threshold. (C) L-I curves for improved GaAs/AlGaAs quantum well lasers fabricated on Ge/GeSi/Si substrates, benchmarked with the previous batch of laser on Si. (D) Device aging test by measuring the output power as a function of time. *Reproduced from Groenert, M. E., Pitera, A. J., Ram, R. J., Fitzgerald, E. A., 2003b. Improved room-temperature continuous wave GaAs/AlGaAs and InGaAs/GaAs/AlGaAs lasers fabricated on Si substrates via relaxed graded GexSi1−x buffer layers. J. Vac. Sci. Technol. B 21, 1064–1069, with the permission of AIP Publishing.*

(GRIN-SCH) prevent optical mode leakage to the higher refractive index Ge/GeSi buffer. Based on the same buffer structure, the lasing wavelength can be extended to 890 nm (inset of Fig. 26B) by changing the active layers to a single 5-nm-thick $In_{0.17}Ga_{0.83}As$/GaAs QW structure. Further optimization of the Ge/GeSi buffer toward a smoother surface by chemical mechanical polishing (CMP), adding an extra Ge cap layer, as well as modifying the FP laser structure to reduce series resistance helped lower the J_{th} to 269 A/cm^2 and increase the lifetime to 4 h (Groenert et al., 2003a,b). The device characteristics are depicted in Fig. 26C and D. Although the QW lasers can work in a room temperature cw mode, the degradation is much faster than the GaAs/AlGaAs counterpart, due to the recombination-enhanced defect reactions with a larger strain (Groenert et al., 2003a,b).

In addition to the GaAs/AlGaAs lasers, some other methodologies were explored to deliver high performance lasers on Si emitting at short wavelengths (below 1 μm), including "proof-of-principle" electrically injected Ga(NAsP)/(BGa)(AsP) QW lasers lattice-matched to (001) silicon substrate (Liebich et al., 2011), optically pumped pure GaAs and InP nanowire lasers (Saxena et al., 2013; Wang et al., 2013, 2015), AlGaAs/GaAs and InGaAs/GaAs core/shell nanowire lasers on Si (Chen et al., 2011; Mayer et al., 2015). Specifically, the nanowire lasers monolithically integrated on Si greatly facilitate low power consumption and dense integration for on-chip interconnections. However, it is challenging to realize nanolasers directly grown on Si due to weak optical confinement and high mode loss into the silicon substrates. By inserting a tailored dielectric interlayer at the NW-Si interface, low-threshold single mode lasing can be achieved in vertical-cavity GaAs/AlGaAs core/shell NW lasers on Si at 10 K under pulsed optical pumping (Mayer et al., 2015). The designed structure and the measurement setup are shown in Fig. 27A and B. Fig. 27C demonstrates the InGaAs/GaAs nanopillars directly grown on silicon, using unique helically propagating cavity modes to strongly confine light within the nanostructure (Chen et al., 2011). The emission wavelength can be tuned by controlling the Indium composition of the InGaAs alloy. A change in Indium composition from 12% to 20% provided a 50 nm tuning range in the output laser wavelength.

Fig. 28A shows another promising method to demonstrate the full compatibility of on-chip lasers with the Si CMOS technology. Leveraging the SAE growth technique in confined regions, an optically pumped InP distributed feedback laser array monolithically grown on V-grooved (001)-silicon operating at room temperature is demonstrated (Wang et al., 2015). Threading dislocations and anti-phase boundaries are suppressed to a <20-nm-thick LT-layer (see Fig. 28B). Moreover, the small lasing wavelength deviation of the precisely controlled in-plane nanolasers in Fig. 28C enables their applications in wavelength-division-multiplexing (WDM).

3.2 Growth of 1.3 μm lasers on Si for datacom applications

Driven by the explosively increased demand for high speed and large capacity communications, one growing interest is to monolithically integrate 1.3 μm lasers on silicon photonic integrated circuits for optical interconnects in datacenters. The active layers have to be designed accordingly, either with $In_xGa_{1-x}As$/In(Ga)P, or $In_xGa_{1-x}As$ /GaAs structures (Chriqui et al., 2003; Huang et al., 2014; Kim et al., 2017; Lu et al., 2017; Shi et al., 2017b). Recent advancements toward 1.3 μm lasers monolithically integrated on

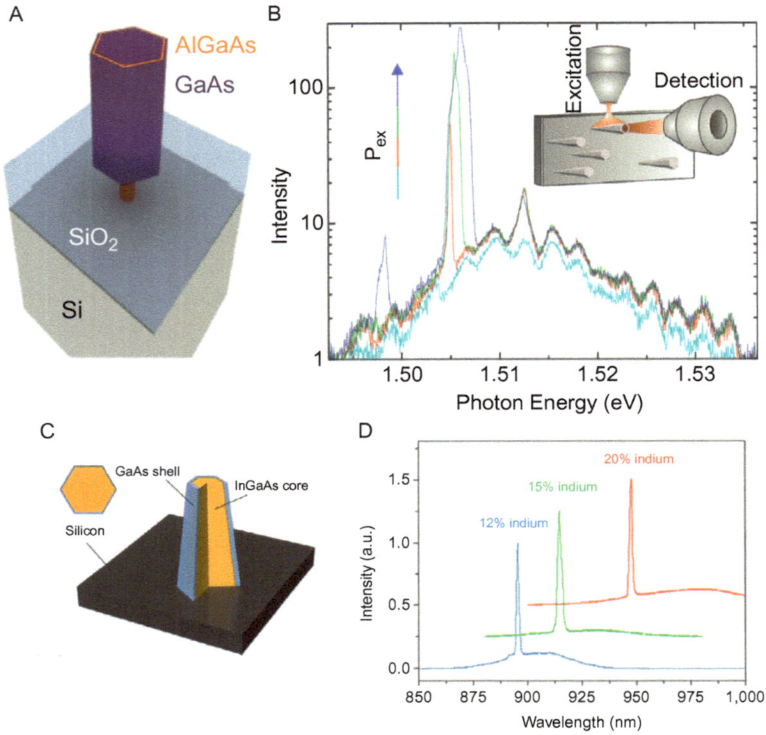

Fig. 27 (A) Schematic of GaAs-AlGaAs nanowire laser on Si with thick SiO_2 dielectric layer. (B) Typical emission spectra recorded from an optically pumped vertical GaAs NW laser excited from the side and detected along the NW-axis at 10 K. The inset depicts schematically the measurement geometry. (C) Schematic of a nanopillar laser monolithically integrated on Si, illustrating its InGaAs core and GaAs shell. (D) Wavelength control of nanopillar lasers by Indium composition variation. Panel B: *Reprinted with permission from Mayer, B., Janker, L., Loitsch, B., Treu, J., Kostenbader, T., Lichtmannecker, S., Reichert, T., Morkötter, S., Kaniber, M., Abstreiter, G., Gies, C., Koblmüller, G., Finley, J. J., 2015. Monolithically integrated high-β nanowire lasers on silicon. Nano Lett. 16, 152–156 Copyright © 2015, American Chemical Society.* Panel D: *Reprinted by permission from Springer Nature: Chen, R., Tran, T.-T. D., Ng, K. W., Ko, W. S., Chuang, L. C., Sedgwick, F. G., Hasnain, C. C., 2011. Nanolasers grown on silicon. Nat. Photonics 5, 170.*

Si are dominated by the InAs/GaAs quantum dots, and their excellent device characteristics (long lifetime, high efficiency, and low threshold) enlighten the promising route to realize practical Si-based on-chip lasers. However, most the results have been obtained via MBE growth (Chen et al., 2016; Liu et al., 2014; Wan et al., 2017), while limited reports for MOCVD growth are available. The main reason involves problems with the growth of the AlGaAs upper cladding layer by MOCVD (Wang et al., 2018).

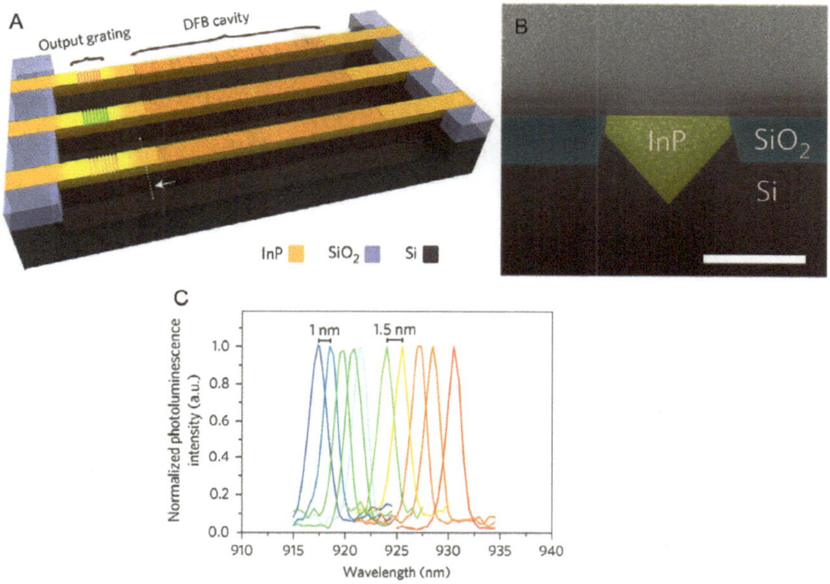

Fig. 28 (A) Schematic of the monolithically integrated InP DFB lasers on Si. (B) False-colored SEM cross-section view of an InP-on-Si waveguide. Scale bar is 500 nm. (C) Lasing spectra from the DFB laser array, demonstrating the capability to control the laser wavelength via the grating design. *Reprinted by permission from Springer Nature: Wang, Z., Tian, B., Pantouvaki, M., W. Guo, P. Absil, Van Campenhout, J., Merckling, C., Van Thourhout, D., 2015. Room-temperature InP distributed feedback laser array directly grown on silicon. Nat. Photonics 9, 837.*

The blue-shift in emission wavelength of QDs by post-growth annealing is due to the enhanced interdiffusion of indium adatoms in the QD layer (Tatebayashi et al., 2005). By replacing the conventional p-AlGaAs upper cladding with the low-temperature (550 °C) GaInP layer, room temperature electrically pumped cw InAs/GaAs QD laser grown on Si by MOCVD was demonstrated (Wang et al., 2018). A more comprehensive review on the development of high performance QD lasers on Si is available (Norman et al., 2018), which will not be discussed in detail in this chapter.

The first room-temperature cw $In_{0.75}Ga_{0.25}As_{0.5}P_{0.5}$/InP buried ridge waveguide laser integrated on Si substrates was grown by two-step low-pressure MOCVD (Razeghi et al., 1988). Emitting at 1.3 μm, these QW lasers exhibited a threshold current density of 4 kA/cm^2. A preliminary aging test has been performed in cw operation mode at room temperature, showing near 5% increase of the injected current after 5 h aging without any decrease of differential quantum efficiency. A more reliable 1.3 μm LDs heteroepitaxially grown on Si was delivered by further lowering the threading dislocation density in the InP buffer (Yamada et al., 1995). The high quality LDs can operate stably over 800 h, with only 10% increase in the driving current. This longer lifetime is attributed to the high quality 13-μm-thick InP buffer on 2-μm GaAs/Si, with an extremely low EPD of $5 \times 10^6\,cm^{-2}$. The successful realization of MOCVD grown strained InGaAs-GaAs QW laser structures integrated on Si substrates via strained relaxed Ge/GeSi buffer layers was reported in 2003 (Chriqui et al., 2003). Room-temperature pulsed lasing at 1.04 μm was realized on 6° offcut Si substrates with 11-μm-thick Ge/GeSi compositionally graded buffer. The large J_{th} (2.0 kA/cm^2) is ascribed to the inferior quality of laser facet since the cleaving is not straightforward on offcut Si. One decade later, Huang et al. demonstrated 1.03 μm InGaAs/GaAs QW laser on exact Si (001) in pulsed mode at room temperature by MBE growth (Huang et al., 2014), utilizing the high-quality GaP on (001) Si with atomic level control (Volz et al., 2011). Yet, the devices were still beyond practical applications due to the large threshold current density of 5.6 kA/cm^2.

Recently, monolithically integrated nanowire lasers are intensively investigated as a promising candidate to miniaturize the Si-based lasers emitting at longer wavelengths, since the nanoscale interfaces between nanowires and substrates effectively relax the strain without the need of buffer layers (Kim et al., 2017). Fig. 29A shows the cross-sectional STEM image of the InGaAs/GaAs nano-ridge lasers. The nano-ridges extended considerably above the Si substrate, enabling the maximum mode confinement without Si recess step to avoid substrate leakage loss (Shi et al., 2017b). Exploiting the ART technique, unwanted defects are confined to a narrow trench defined on the silicon substrate. The InGaAs/GaAs nanoridges were then fabricated into DFB lasers. Under pulsed optical pumping, single mode lasing with side mode suppression over 28 dB was realized, and precise control of the emission wavelength over 60 nm was demonstrated in Fig. 29B.

Silicon-on-insulator (SOI) platform was naturally ideal for the integration of III–V nanowire lasers on photonic circuits. For example, InGaAs/InGaP core/shell nanowire arrays were selectively grown on a pre-patterned SOI substrate to form a photonic crystal nanobeam cavity (Kim et al., 2017). Superior optical field confinement was therefore achieved by the nanobeam cavities composed of vertically standing nanowires (Fig. 30B). Moreover, as seen in Fig. 30A, the direct coupling between the nanowire array lasers with SOI waveguides was confirmed by measuring the light emission from the output grating couplers. The coupling efficiency of nanowire array lasers to the waveguides can be over 50% while maintaining a high quality factor

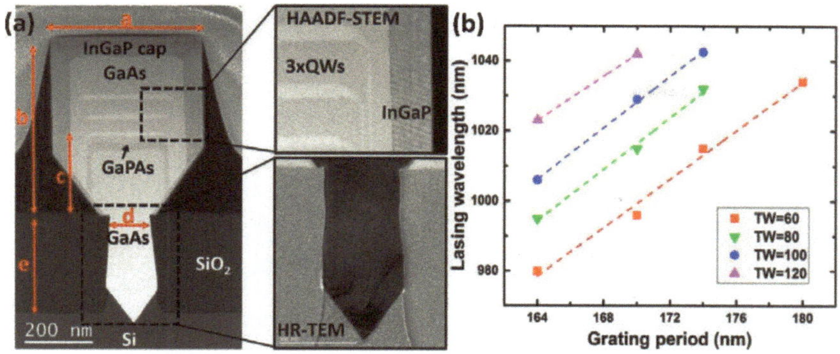

Fig. 29 (A) STEM images of the GaAs nano-ridge waveguide. The upper one shows zoomed-in QWs, barriers, and InGaP passivation layer, while the lower high-resolution TEM shows that defects are trapped in the V-shaped trench. (B) Lasing wavelength versus grating period for trench sizes varying from 60 to 120 nm. *Reprinted with permission from Shi, Y., Wang, Z., Van Campenhout, J., Pantouvaki, M., Guo, W., Kunert, B., Van Thourhout, D., 2017b. Optical pumped InGaAs/GaAs nano-ridge laser epitaxially grown on a standard 300-mm Si wafer. Optica 4, 1468–1473 © The Optical Society.*

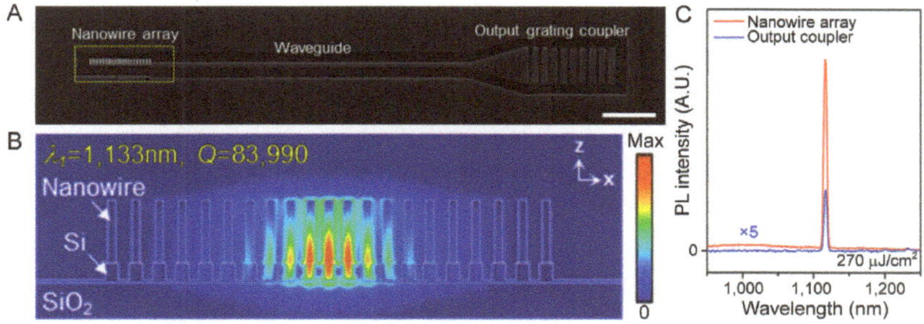

Fig. 30 (A) 30° tilted SEM images of a nanowire array cavity integrated with a waveguide and an output coupler. Scale bar is 5 μm. (B) Electric field profile of the fundamental cavity mode. (C) PL spectra measured on top of a nanowire array (red line) and on top of an output coupler (blue line) above the lasing threshold, indicating that the cavity peaks are effectively coupled to the waveguide. *Reprinted with permission from Kim, H., Lee, W.-J., Farrell, A. C., Morales, J. S., Senanayake, P., Prikhodko, S. V., Ochalski, T. J., Huffaker, D. L., 2017. Monolithic InGaAs nanowire array lasers on silicon-on-insulator operating at room temperature. Nano Lett. 17, 3465–3470 Copyright © 2017, American Chemical Society.*

(Fig. 30C), by designing the cavity structure. This bottom-up approach to form nanowire-based photonic crystals on SOI substrates offers an attractive degree of freedom in designing novel photonic devices in a low-cost and high-volume pathway, showing the potential toward the fulfillment of nanophotonic integrated circuits on Si.

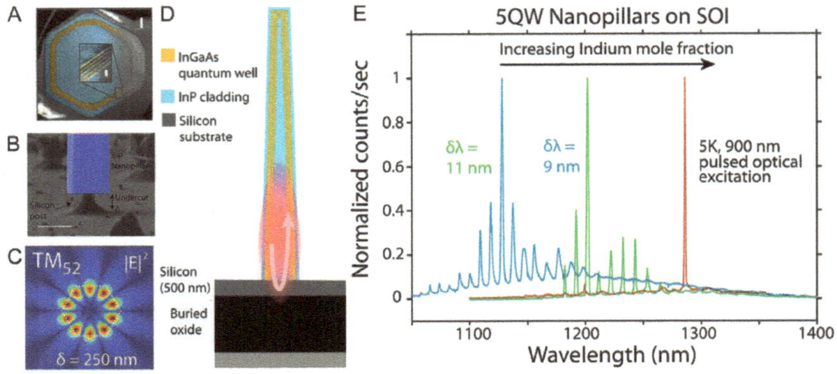

Fig. 31 (A) Cross-sectional TEM of a nanopillar with five QWs grown conformally on the InP core, with an inset showing the individual QWs. Scale bar is 100 nm (inset: 20 nm). (B) SEM image of a nanopillar with a selective silicon undercut using $SF_6 + O_2$ plasma. Scale bar is 1 μm. (C) Electric field profile of the helically propagating modes with an undercut δ = 250 nm. (D) Schematic of nanopillar lasers monolithically integrated on a SOI substrate. (E) Tuning the Indium composition of the QW can be used to shift the lasing wavelength from ~1100 to ~1300 nm. *Reprinted with permission from Lu, F., Bhattacharya, I., Sun, H., Tran, T.-T. D., Ng, K. W., Malheiros-Silveira, G. N. Hasnain, C. C., 2017. Nanopillar quantum well lasers directly grown on silicon and emitting at silicon-transparent wavelengths. Optica 4, 717–723 © The Optical Society.*

Fig. 31A and D show the schematic of InGaAs/InP MQW nanopillar lasers directly grown on SOI substrate (Lu et al., 2017). The thin Si waveguide layer (500 nm) enhances the reflection from the bottom facet of the pillar, leading to multiple longitudinal modes lasing. The lasing wavelength can be extended to 1286 nm by increasing the composition of the InGaAs well, measured under pulsed optical pumping at 5 K. An alternative approach to increase the nanopillar/Si interface reflection is to create an abrupt structural discontinuity at the III–V/Si interface, formed by a selective undercut etching of the silicon substrate, as depicted in Fig. 31B. The small air gap at the nanopillar base provides an enhanced light reflection for the helically propagating modes, with a quality factor Q exceeding 1000. Benefiting from this, the nanopillars can subsequently lase at room-temperature under pulsed optical pumping.

3.3 Long-wavelength 1.55 μm lasers integrated on Si

The full deployment of Si-photonics at 1.55 μm telecom band is appealing to long-haul and free space communication systems, LiDAR, and sensing applications (Castellano et al., 2017). This spectral range however is out-of-reach with GaAs quantum-dot lasers and naturally calls for InP-based alloys.

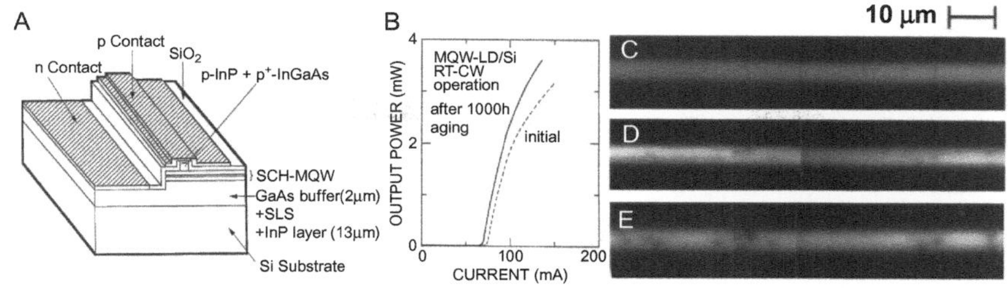

Fig. 32 (A) Schematic diagram of an InGaAs/InGaAsP MQW ridge waveguide LD on Si. (B) L-I characteristics before and after aging in MQW LD on Si. Typical EL images of the DH LDs during the aging test: (C) 0 h, (D) 1000–2000 h, and (E) over 5000 h. Panel B: *Reproduced from Sugo, M., Mori, H., Itoh, Y., Sakai, Y., Tachikawa, M., 1991. 1.5 μm-long-wavelength multiple quantum well laser on a Si substrate. Jpn. J. Appl. Phys. 30, 3876–3878 © 1991 IOP Publishing Ltd.* Panel E: *Reproduced from Sasaki, T., Mori, H., Tachikawa, M., Yamada, T., 1998. Aging tests of InP-based laser diodes heteroepitaxially grown on Si substrates. J. Appl. Phys. 84, 6725–6728, with the permission of AIP Publishing.*

The first room-temperature cw InP-based LDs on a (001) Si substrate emitting at 1.54 μm was demonstrated by employing a hybrid MOCVD and vapor mixing epitaxy (VME) method (Sugo et al., 1991). The laser structure on Si is illustrated in Fig. 32A, with six pairs of 7.5 nm InGaAs well/10 nm InGaAsP barrier MQW on 13 μm InP/5-period InGaAsP/InP SLS/2 μm GaAs buffer layer. A threshold current of 55 mA was measured for the ridge waveguide LD with 6-μm stripe width and 300-μm cavity length ($J_{th} = 3.1$ kA/cm^2). The even slightly decreased lasing threshold after 1000 h aging is attributed to recombination-enhanced annealing effect, which is shown in Fig. 32B. A further aging test of the same device indicates no degradation even after 2000 h (Sugo et al., 1992). Continuing to lower the TD density from above 10^7 cm^{-2} to 5×10^6 cm^{-2}, the maximum operation temperature of the InP-based laser was increased to 80 °C and the lifetime was extended to over 7000 h at 50 °C (Yamada et al., 1997). To study the device characteristics evolution during aging, electroluminescence (EL) images were compared during the aging tests (Sasaki et al., 1998). It was discovered that the degradation mainly occurs in the active region and its vicinity. The EL images are shown in Fig. 32C–E. Although no defects exist in the initial image (Fig. 32C), dark spot defects (DSDs) appeared as the aging proceeded (Fig. 32D). However, no more DSDs were generated once they reached a certain amount (Fig. 32E). For this scenario, the DSDs neither change in size nor develop into dark line defects (DLDs), which accelerates device degradation. The absence of DLDs agrees with the phenomenon that InP-based LDs seldom suffer rapid degradation caused by the dislocation climbing. This also explains the more stable operation of InP-based LDs/Si than GaAs-based ones.

However, the aggressively thick InP buffer is not cost-effective nor practical. The desirable approach is to bring about decent laser performances with a thin III–V buffer layer. Recently, taking advantage of ART concept introduced previously, the 1.5-μm-thick InP epilayer was directly grown on pre-patterned V-shaped (001) Si substrate, eliminating the adoption of thick GaAs intermediate buffer (Zhu et al., 2018a). On the basis of the superior InP-on-Si virtual substrate, an InGaAs/InAlGaAs MQW laser was grown atop, as shown schematically in Fig. 33A. The as-fabricated FP ridge waveguide (RWG) laser exhibited a threshold current density of 3.3 kA/cm^2, a high total output power of 180 mW, and a maximum operation temperature up to 60 °C under pulsed current injection (see temperature-dependent LI curves in Fig. 33B). The extracted characteristic temperatures T_0 are 133 and 46.3 K in the range of 20–40 °C and 40–60 °C, respectively

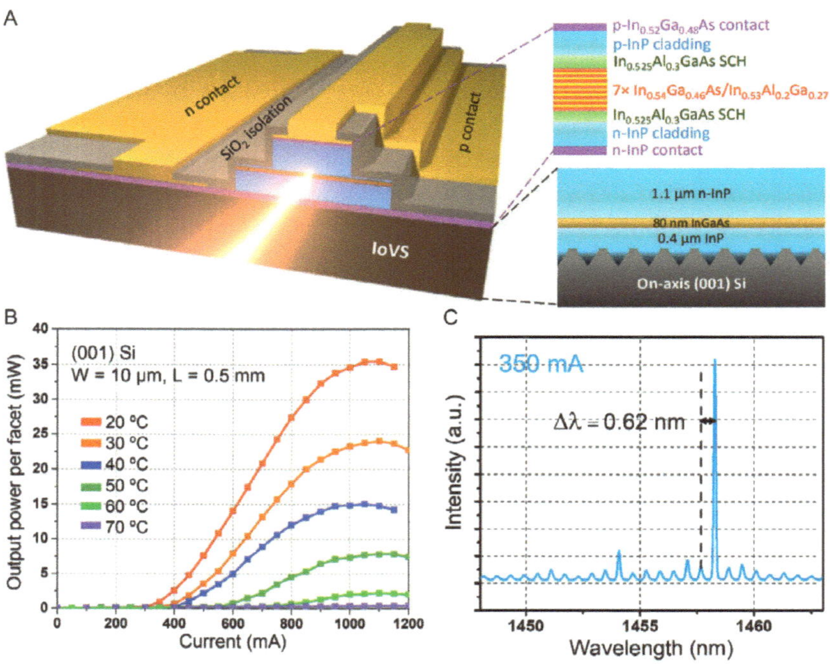

Fig. 33 (A) Schematic diagram of the whole LD epitaxial structure on (001) V-grooved silicon substrate. (B) Measured LI curves of a 10 μm × 0.5 mm laser on silicon as a function of the heat sink temperature. (C) Zoomed-in emission spectrum at 350 mA current injection, showing lasing peak at 1.46 μm and FP oscillations in the cavity with a FSR = 0.62 nm. *Reprinted with permission from Zhu, S., Shi, B., Li, Q., Lau, K. M., 2018a. Room-temperature electrically-pumped 1.5 μm InGaAs/InAlGaAs laser monolithically grown on on-axis (001) Si. Opt. Express 26, 14514–14523 © The Optical Society.*

(Zhu et al., 2018a). Multiple longitudinal mode lasing was observed at high injection levels with the primary lasing peak centered at 1.46 μm (Fig. 33C). This Si-based long-wavelength light source offers a new promising path to fulfill InP/Si monolithic integration for Si photonics.

The emergence of 1.55 μm QD lasers opens up a new avenue for the lasers epitaxially grown on Si substrates. The dislocation diluting property of QDs promises the potential of high efficiency and thermally stable on-chip lasers (Zhu et al., 2017). The first realization of 1.55 μm band QD laser directly grown on (001) Si was based on the combination of QD active materials and the microdisk laser (MDL) structures (Shi et al., 2017c). The initial attempt demonstrated a low temperature (4.5 K) cw operation of optically pumped microdisk lasers with a high threshold of 1.6 mW. Further improving the material gain of multiple QD stacks, as well as shrinking the microdisk size to subwavelength scale, room-temperature lasing was achieved with an ultra-low threshold power of 2.73 μW (Shi et al., 2017d). The two-step grown 1.5-μm InP buffer on GaAs/Si is shown in Fig. 34A, with a less defective top InP epilayer identified. The active region, composed of 7-layer InAs/InAlGaAs QDs, is revealed by cross-sectional TEM and AFM images in Fig. 34C and D. Only a few large clusters are formed in addition to the QDs, thanks to a relatively smooth InP/Si surface morphology, with a low roughness of 1.6 nm (Fig. 34E). Single mode lasing was therefore achieved at a wavelength of 1563 nm, due to the smooth and steep sidewall of the fabricated subwavelength microdisks. The good circularity of the microdisk structure is presented in Fig. 34F. Moreover, these microlasers can successfully operate at elevated temperatures over 60 °C, indicating the feasibility of directly grown on-chip light sources for low-cost, large-scale, and high-density photonic integrated circuits (Shi et al., 2017d).

The first electrically injected 1.5 μm InAs quantum dot laser diode was epitaxially grown on the on-axis (001) V-groove patterned Si substrates subsequently (Zhu et al., 2018b). Fig. 35A shows a SEM image of the complete growth structure after the Fabry-Perot ridge waveguide laser fabrication. The 2-μm-thick GaAs intermediate buffer helps relieve the lattice mismatch between InP and Si, and three sets of $In_{0.6}Ga_{0.4}As$/InP SLS was inserted in the InP buffer to filter most of the dislocations, leading to a reduced surface defect density of $1.5 \times 10^8\,cm^{-2}$. The laser consisting of 5-period optimized InAs/InAlGaAs quantum dots was grown on top of the InP buffer. The room temperature PL spectra of single sheet and five-layer QDs reveal a multimodal dot distribution due to the QDs inhomogeneity.

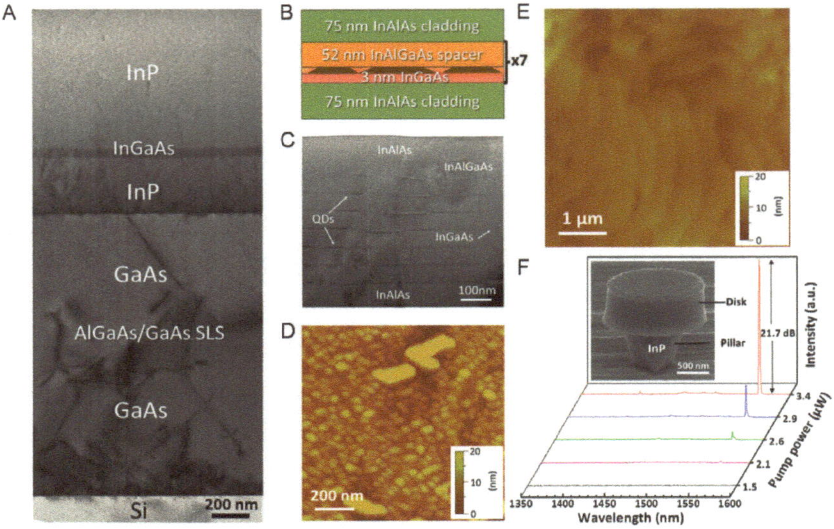

Fig. 34 (A) Cross-sectional TEM image of the InP-on-Si template. (B) Schematic diagram of the microdisk laser structure. (C) Zoomed-in TEM of the microdisk laser on Si, highlighting the InAs/InGaAs/InAlGaAs DWELLs. (D) Representative surface morphology of uncapped InAs QDs on top of the InP-on-Si substrate, with a few large clusters observed. (E) Typical $5\,\mu m^2 \times 5\,\mu m^2$ AFM image of the InP thin film grown on Si, showing a step-flow morphology with a low roughness of 1.6 nm. (F) Emission spectra as pulse pumping progressively increases to demonstrate single-mode lasing at 1563 nm. Inset shows a SEM image of the good device topology of the fabricated subwavelength MDL. *Reproduced from Shi, B., Zhu, S., Li, Q., Tang, C. W., Wan, Y., Hu, E. L., Lau, K. M., 2017d. 1.55 μm room-temperature lasing from subwavelength quantum-dot microdisks directly grown on (001) Si. Appl. Phys. Lett. 110, 121109, with the permission of AIP Publishing.*

Yet, the QDs remain coherent during the stacking process, showing a dot density around $4.5 \times 10^{10}/cm^2$, as characterized by AFM scan in the inset of Fig. 35B. These electrical laser diodes can sustain up to 80 °C under pulsed current injection (see Fig. 35C). Meanwhile, comparing with the QW lasers on Si (Zhu et al., 2018a), the Si-based QD lasers exhibit lower threshold densities at various cavity lengths, as shown in Fig. 35D. This result therefore verifies the suitability of adopting QDs as the gain medium for III–V lasers directly grown on Si. The dependence of lasing thresholds on the cavity length is due to smaller losses for longer cavities (Zhu et al., 2018b).

In addition to the InP-based material system, GaSb and Ge are studied as alternatives to enrich the 1.55 μm laser on Si with their unique advantages. GaSb-based semiconductor alloys are grown on Si by MBE in a particular relaxation mode: strain relaxation occurs by the formation of 90° lomer

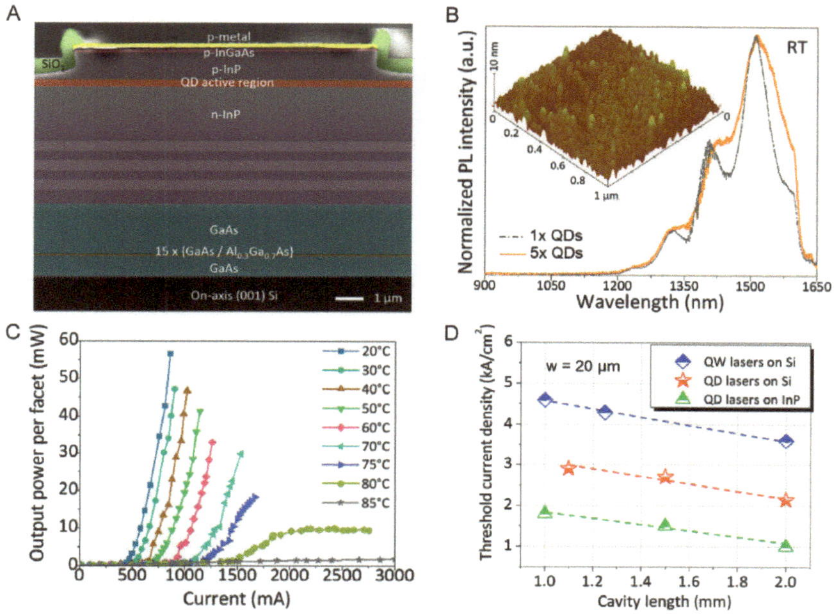

Fig. 35 (A) Cross-sectional SEM image of the electrical 1.5 μm QD LD on Si after FP laser fabrication. (B) Room temperature PL spectra of single sheet and five layers of InAs/InAlGaAs QDs. Inset shows the 3D AFM surface morphology of an uncapped QD layer on top of the five-period QD stacks. (C) Temperature-dependent L-I curves from 20 °C to 85 °C. (D) Dependence of threshold current density on the cavity length for QD and QW lasers on InP and Si. *Reproduced from Zhu, S., Shi, B., Li, Q., Lau, K. M., 2018b. 1.5 μm quantum-dot diode lasers directly grown on CMOS-standard (001) silicon. Appl. Phys. Lett. 113, 221103, with the permission of AIP Publishing.*

dislocations arranged in a two-dimensional network, and these dislocations are confined at the III-Sb/Si interface. This particular relaxation mode leads to a very low surface defect density ($\sim 8 \times 10^5\,cm^{-2}$) of the III-Sb layers grown on Si without resorting to thick buffer layers (Balakrishnan et al., 2006). In fact, this material system is more suitable for mid-infrared optoelectronics where high-performance laser diodes have been produced.

The first 1.54 μm GaSb/AlGaSb MQW laser on miscut Si substrate was realized by MBE, under pulsed operation at 77 K (Jallipalli et al., 2007). By designing a new active region with a higher strain and a stronger carrier confinement (4-nm-wide GaInSb QWs confined by AlGaAsSb barriers), the lasers work reluctantly in a room temperature pulsed condition (Cerutti et al., 2010). Further on, after well addressing the issues including structural defects arising from strain relaxation of III-Sb on Si (optimized band structure shown in Fig. 36A), and the inferior as-cleaved facet quality, the second generation of lasers can achieve room temperature cw lasing on 6° offcut Si,

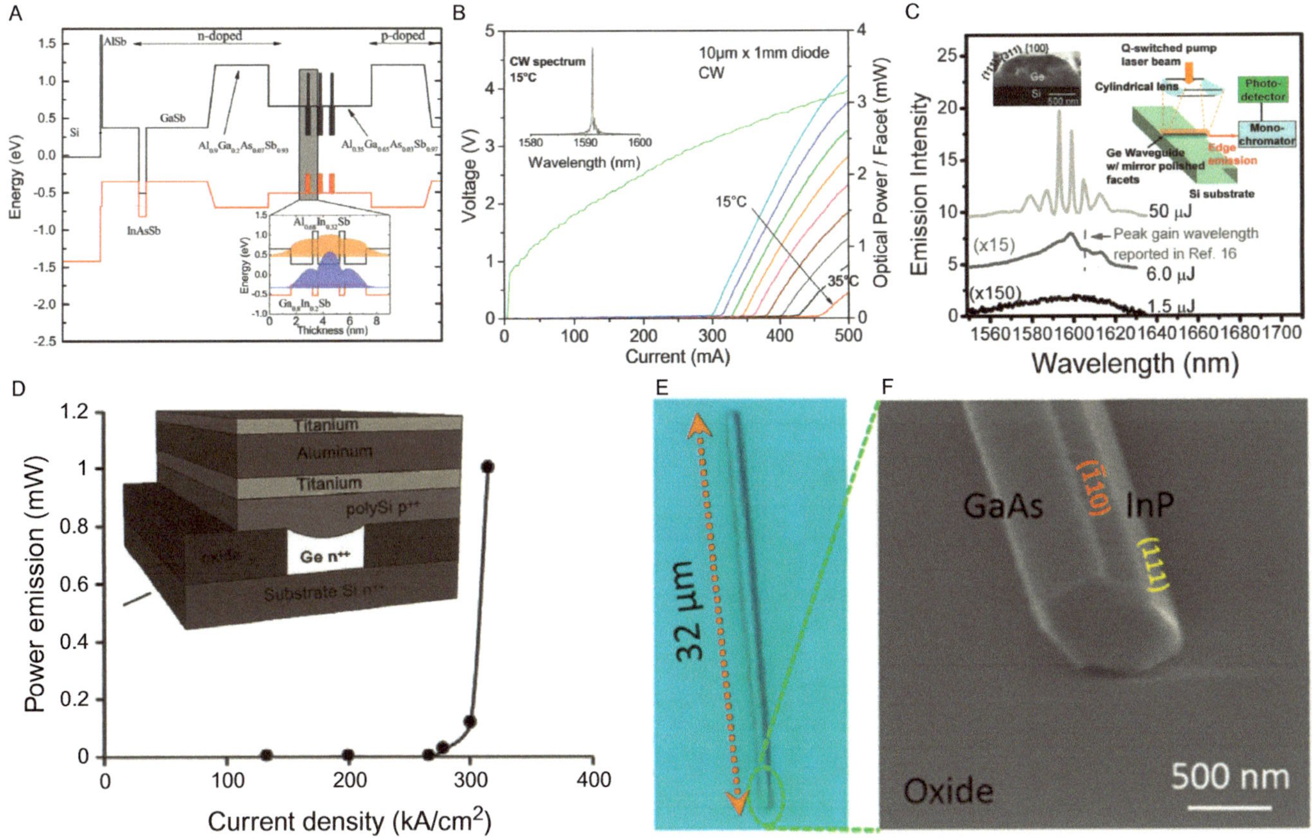

Fig. 36 See figure legend on next page.

as shown in Fig. 36B (Castellano et al., 2017). It is noted that all the GaSb/Si lasers are currently obtained by MBE, while MOCVD grown GaSb-based laser on Si substrates remains a severe challenge.

The pseudo-direct gap properties and compatibility of Ge with Si CMOS technology stimulated the research interest toward Ge-on-Si lasers. Traditionally considered an indirect gap material, Ge can be band engineered to behave like a direct-gap material by introducing a large tensile strain and a heavy n-type doping to compensate the energy difference between the direct and indirect conduction valleys (Liu et al., 2007). Fig. 36C shows an optically pumped Ge-on-Si edge emitting laser first demonstrated at room temperature with a gain spectrum of 1590–1610 nm (Liu et al., 2010). Applying the delta-doping during epitaxial growth of Ge (n-type doping level $> 4 \times 10^{19}\,cm^{-3}$), electrical Ge-on-Si laser was accomplished at room temperature under pulsed operation (Camacho-Aguilera et al., 2012). Although the threshold is quite large (Fig. 36D), lasing was observed in the range from 1520 to 1700 nm.

Another interesting approach to achieve monolithically integrated 1550 nm laser on Si without growing a thick buffer is the use of InGaAs/InP nanowire lasers. Fig. 36E and F depicts a transferred InGaAs/InP nanowire originally grown on V-grooved Si (001) substrate by SAE, onto the SiO_2/Si

Fig. 36 (A) Band structure of the whole GaSb-on-Si laser heterostructure. Inset: The GaInSb/AlInSb composite QWs with the confined electron and hole levels and their wave functions. (B) L-I-V curves taken from a 10 μm × 1 mm ridge LD under cw operation between 15 °C and 35 °C. Inset shows laser emission spectrum at 15 °C. (C) Edge-emission spectra of a Ge-on-Si waveguide with polished facets under 1064 nm pulsed excitation. The inset shows a cross-sectional SEM picture of the Ge waveguide and a schematic drawing of the experimental setup for optical pumping. (D) L-I curve for a 270 μm long waveguide device at 15 °C under pulsed current injection. Inset shows Ge-on-Si waveguide laser. (E) Microscopic image of the transferred InP/InGaAs nanoridge. (F) Zoomed-in SEM image illustrating the high end-facet quality. Panel B: *Castellano, A., Cerutti, L., Rodriguez, J., Narcy, G., Garreau, A., Lelarge, F., Tournié, E., 2017. Room-temperature continuous-wave operation in the telecom wavelength range of GaSb-based lasers monolithically grown on Si. APL Photonics 2, 061301; licensed under a Creative Commons Attribution (CC BY) license.* Panel C: *Reprinted with permission from Liu, J., Sun, X., Camacho-Aguilera, R., Kimerling, L. C., Michel, J., 2010. Ge-on-Si laser operating at room temperature. Opt. Lett. 35, 679–681 © The Optical Society.* Panel D: *Reprinted with permission from Camacho-Aguilera, R., Cai, Y., Patel, N., Bessette, J. T., Romagnoli, M., Kimerling, L. C., Michel, J., 2012. An electrically pumped germanium laser. Opt. Express 20, 11316–11320 © The Optical Society.* Panel F: *Reproduced from Han, Y., Li, Q., Zhu, S., Ng, K. W., Lau, K. M., 2017. Continuous-wave lasing from InP/InGaAs nanoridges at telecommunication wavelengths. Appl. Phys. Lett. 111, 212101, with the permission of AIP Publishing.*

substrate to avoid mode leakage through the Si substrate (Han et al., 2018). Room temperature pulsed lasing with emission wavelengths covering the entire O-, E-, S- and C-band with ultra-low lasing thresholds has been demonstrated. These results pave the way toward telecom InP/InGaAs nano-laser arrays on CMOS standard Si or SOI substrates.

4. Conclusion

In conclusion, the epitaxial growth of III–V compound semiconductors such as GaAs and InP on Si substrate is first discussed, focusing on the introduction of different defect engineering techniques using the MOCVD system. Some novel methods and recent progresses within this research regime are presented. Based on the advanced III–V/Si platform, lasers monolithically integrated on the on-axis or offcut Si substrates are comprehensively reviewed. Specifically, flourishing efforts are being devoted to improve the 1.3 and 1.55 μm QD lasers on (001) Si by MOCVD, for high density inter/intra-chip data transmission. Monolithically integrated III–V nanowires on Si offers the strategic solution to minimize power consumption for dense silicon photonic integrations. However, prior to achieve electrical-driven nanolasers, it is necessary to realize optically pumped room-temperature cw nanowire lasers directly grown on Si substrate, and to extend the emission wavelength to Si-transparent windows.

Acknowledgments

The authors acknowledge fruitful discussions and contributions from Si Zhu, Qiang Li, Yu Han, Chak Wah Tang and Evelyn L. Hu.

References

Ababou, Y., Desjardins, P., Chennouf, A., Leonelli, R., Hetherington, D., Yelon, A., L'Espérance, G., Masut, R.A., 1996. Structural and optical characterization of InP grown on Si (111) by metalorganic vapor phase epitaxy using thermal cycle growth. J. Appl. Phys. 80, 4997–5005.

Adomi, K., Strite, S., Morkoç, H., 1990. Antiphase-domain-free GaAs grown on pseudomorphic Si (100) surfaces by molecular beam epitaxy. Appl. Phys. Lett. 56, 469–471.

Akahori, K., Wang, G., Okumura, K., Soga, T., Jimbo, T., Umeno, M., 2001. Improvement of the MOCVD-grown InGaP-on-Si towards high-efficiency solar cell application. Sol. Energy Mater. Sol. Cells 66, 593–598.

Akiyama, M., Kawarada, Y., Kaminishi, K., 1984. Growth of single domain GaAs layer on (100)-oriented Si substrate by MOCVD. Jpn. J. Appl. Phys. 23, L843.

Akiyama, M., Nishi, S., Kaminishi, K., 1986. Growth and properties of single domain GaAs, AlGaAs and their heterostructures on Si by MOCVD and MBE. Surf. Sci. 174, 19–30.

Alcotte, R., Martin, M., Moeyaert, J., Cipro, R., David, S., Bassani, F., Ducroquet, F., Bogumilowicz, Y., Sanchez, E., Ye, Z., Bao, X.Y., Pin, J.B., Baron, T., 2016. Epitaxial growth of antiphase boundary free GaAs layer on 300 mm Si (001) substrate by metalorganic chemical vapour deposition with high mobility. APL Mater. 4, 046101.

Bai, J., Park, J.-S., Cheng, Z., Curtin, M., Adekore, B., Carroll, M., Lochtefeld, A., Dudley, M., 2007. Study of the defect elimination mechanisms in aspect ratio trapping Ge growth. Appl. Phys. Lett. 90, 101902.

Balakrishnan, G., Huang, S., Khoshakhlagh, A., Jallipalli, A., Rotella, P., Amtout, A., Krishna, S., Haines, C.P., Dawson, L.R., Huffaker, D.L., 2006. Room-temperature optically-pumped GaSb quantum well based VCSEL monolithically grown on Si (100) substrate. Electron. Lett. 42, 350–352.

Bogumilowicz, Y., Hartmann, J., Cipro, R., Alcotte, R., Martin, M., Bassani, F., Moeyaert, J., Baron, T., Pin, J.B., Bao, X., Ye, Z., Sanchez, E., 2015. Anti-phase boundaries-free GaAs epilayers on "quasi-nominal" Ge-buffered silicon substrates. Appl. Phys. Lett. 107, 212105.

Bolkhovityanov, Y.B., Pchelyakov, O.P., 2008. GaAs epitaxy on Si substrates: modern status of research and engineering. Physics-Uspekhi 51, 437–456.

Borg, M., Schmid, H., Moselund, K.E., Signorello, G., Gignac, L., Bruley, J., Breslin, C., Kanungo, P.D., Werner, P., Riel, H., 2014. Vertical III–V nanowire device integration on Si (100). Nano Lett. 14, 1914–1920.

Camacho-Aguilera, R., Cai, Y., Patel, N., Bessette, J.T., Romagnoli, M., Kimerling, L.C., Michel, J., 2012. An electrically pumped germanium laser. Opt. Express 20, 11316–11320.

Castellano, A., Cerutti, L., Rodriguez, J., Narcy, G., Garreau, A., Lelarge, F., Tournié, E., 2017. Room-temperature continuous-wave operation in the telecom wavelength range of GaSb-based lasers monolithically grown on Si. APL Photonics 2, 061301.

Cerutti, L., Rodriguez, J.-B., Tournie, E., 2010. GaSb-based laser, monolithically grown on silicon substrate, emitting at 1.55 μm at room temperature. IEEE Photon. Technol. Lett. 22, 553–555.

Chang, E.Y., Yang, T.-H., Luo, G., Chang, C.-Y., 2005. A GeSi-buffer structure for growth of high-quality GaAs epitaxial layers on a Si substrate. J. Electron. Mater. 34, 23–26.

Chen, R., Tran, T.-T.D., Ng, K.W., Ko, W.S., Chuang, L.C., Sedgwick, F.G., Hasnain, C.C., 2011. Nanolasers grown on silicon. Nat. Photonics 5, 170.

Chen, S., Li, W., Wu, J., Jiang, Q., Tang, M., Shutts, S., Elliott, S.N., Sobiesierski, A., Seeds, A.J., Ross, I., Smowton, P.M., Liu, H., 2016. Electrically pumped continuous-wave III–V quantum dot lasers on silicon. Nat. Photonics 10, 307.

Chilukuri, K., Mori, M.J., Dohrman, C.L., Fitzgerald, E.A., 2006. Monolithic CMOS-compatible AlGaInP visible LED arrays on silicon on lattice-engineered substrates (SOLES). Semicond. Sci. Technol. 22, 29.

Choi, C., Otsuka, N., Munns, G., Houdre, R., Morkoc, H., Zhang, S., Klein, M.V., 1987. Effect of insitu and exsitu annealing on dislocations in GaAs on Si substrates. Appl. Phys. Lett. 50, 992–994.

Chriqui, Y., Saint-Girons, G., Bouchoule, S., Moison, J.-M., Isella, G., Von Kaenel, H., Sagnes, I., 2003. Room temperature laser operation of strained InGaAs/GaAs QW structure monolithically grown by MOVCD on LE-PECVD Ge/Si virtual substrate. Electron. Lett. 39, 1658–1660.

Chyi, J.I., Shieh, J.L., Pan, J.W., Lin, R.M., 1996. Material properties of compositional graded $In_x Ga_{1-x}As$ and $In_xAl_{1-x}As$ epilayers grown on GaAs substrates. J. Appl. Phys. 79, 8367–8370.

Crumbaker, T., Lee, H., Hafich, M., Robinson, G., 1989. Growth of InP on Si substrates by molecular beam epitaxy. Appl. Phys. Lett. 54, 140–142.

Czornomaz, L., Uccelli, E., Sousa, M., Deshpande, V., Djara, V., Caimi, D., Rossell, M.D., Erni, R., Fompeyrine, J., 2015. Confined epitaxial lateral overgrowth (CELO): a novel concept for scalable integration of CMOS-compatible InGaAs-on-insulator MOSFETs on large-area Si substrates. In: VLSI Technology Symposium, pp. T172–T173.

Deppe, D., Holonyak Jr., N., Nam, D., Hsieh, K., Jackson, G., Matyi, R., Shichijo, H., Epler, J.E., Chung, H.F., 1987. Room-temperature continuous operation of p-n $Al_xGa_{1-x}As$-GaAs quantum well heterostructure lasers grown on Si. Appl. Phys. Lett. 51, 637–639.

Deppe, D., Hall, D., Holonyak Jr., N., Matyi, R., Shichijo, H., Epler, J.E., 1988. Effects of microcracking on $Al_xGa_{1-x}As$-GaAs quantum well lasers grown on Si. Appl. Phys. Lett. 53, 874–876.

El-Masry, N., Tarn, J., Humphreys, T., Hamaguchi, N., Karam, N., Bedair, S., 1987. Effectiveness of strained-layer superlattices in reducing defects in GaAs epilayers grown on silicon substrates. Appl. Phys. Lett. 51, 1608–1610.

Fang, S., Adomi, K., Iyer, S., Morkoc, H., Zabel, H., Choi, C., Otsuka, N., 1990. Gallium arsenide and other compound semiconductors on silicon. J. Appl. Phys. 68, R31–R58.

Fischer, R., Morkoc, H., Neumann, D., Zabel, H., Choi, C., Otsuka, N., Longerbone, M., Erickson, L.P., 1986a. Material properties of high-quality GaAs epitaxial layers grown on Si substrates. J. Appl. Phys. 60, 1640–1647.

Fischer, R., Kopp, W., Morkoc, H., Pion, M., Specht, A., Burkhart, G., Appelman, H., McGougan, D., Rice, R., 1986b. Low threshold laser operation at room temperature in GaAs/(Al, Ga) As structures grown directly on (100) Si. Appl. Phys. Lett. 48, 1360–1361.

Freundlich, A., Grenet, J., Neu, G., Leycuras, A., Verie, C., Gibart, P., Landa, G., Carles, R., 1988. Influence of MOVPE growth parameters on the structural and optical properties of GaAs on Si (100). J. Cryst. Growth 93, 487–493.

Groenert, M.E., Leitz, C.W., Pitera, A.J., Yang, V., Lee, H., Ram, R.J., Fitzgerald, E.A., 2003a. Monolithic integration of room-temperature cw GaAs/AlGaAs lasers on Si substrates via relaxed graded GeSi buffer layers. J. Appl. Phys. 93, 362–367.

Groenert, M.E., Pitera, A.J., Ram, R.J., Fitzgerald, E.A., 2003b. Improved room-temperature continuous wave GaAs/AlGaAs and InGaAs/GaAs/AlGaAs lasers fabricated on Si substrates via relaxed graded Ge_xSi_{1-x} buffer layers. J. Vac. Sci. Technol. B 21, 1064–1069.

Grundmann, M., Krost, A., Bimberg, D., 1991. Low-temperature metalorganic chemical vapor deposition of InP on Si (001). Appl. Phys. Lett. 58, 284–286.

Grundmann, M., Krost, A., Bimberg, D., Ehrmann, O., Cerva, H., 1992. Maskless growth of InP stripes on patterned Si (001): defect reduction and improvement of optical properties. Appl. Phys. Lett. 60, 3292–3294.

Han, Y., Li, Q., Chang, S.-P., Hsu, W.-D., Lau, K.M., 2016. Growing InGaAs quasi-quantum wires inside semi-rhombic shaped planar InP nanowires on exact (001) silicon. Appl. Phys. Lett. 108, 242105.

Han, Y., Ng, W.K., Ma, C., Li, Q., Zhu, S., Chan, C.S., Lennon, S., Taylor, R.A., Wong, K.S., Lau, K.M., 2018. Room-temperature InP/InGaAs nano-ridge lasers grown on Si and emitting at telecom bands. Optica 5 (8), 918–923.

Hasegawa, Y., Egawa, T., Jimbo, T., Umeno, M., 1995. Influences of dark line defects on characteristics of AlGaAs/GaAs quantum well lasers grown on Si substrates. Jpn. J. Appl. Phys. 34, 2994.

Hayafuji, N., Ochi, S., Miyashita, M., Tsugami, M., Murotani, T., Kawagishi, A., 1988. Effectiveness of AlGaAs/GaAs superlattices in reducing dislocation density in GaAs on Si. J. Cryst. Growth 93, 494–498.

Hayafuji, N., Miyashita, M., Nishimura, T., Kadoiwa, K., Kumabe, H., Murotani, T., 1990. Effect of employing positions of thermal cyclic annealing and strained-layer superlattice on defect reduction in GaAs-on-Si. Jpn. J. Appl. Phys. 29, 2371.
Heck, M.J., Davenport, M.L., Bowers, J.E., 2013. Progress in Hybrid-Silicon Photonic Integrated Circuit Technology. SPIE.
Horikawa, H., Kawai, Y., Akiyama, M., Sakuta, M., 1988. Hetero-epitaxial growth of InP on Si substrates by LP-MOVPE. J. Cryst. Growth 93, 523–526.
Huang, X., Song, Y., Masuda, T., Jung, D., Lee, M., 2014. InGaAs/GaAs quantum well lasers grown on exact GaP/Si (001). Electron. Lett. 50, 1226–1227.
Itoh, Y., Nishioka, T., Yamamoto, A., Yamaguchi, M., 1988. GaAs heteroepitaxial growth on Si for solar cells. Appl. Phys. Lett. 52, 1617–1618.
Jallipalli, A., Kutty, M., Balakrishnan, G., Tatebayashi, J., Nuntawong, N., Huang, S., Dawson, L.R., Huffaker, D.L., Mi, Z., Bhattacharya, P., 2007. 1.54 μm GaSb/AlGaSb multi-quantum-well monolithic laser at 77 K grown on miscut Si substrate using interfacial misfit arrays. Electron. Lett. 43, 1198–1199.
Joyce, H.J., Gao, Q., Tan, H.H., Jagadish, C., Kim, Y., Zou, J., Smith, L.M., Jackson, H.E., Yarrison-Rice, J.M., Parkinson, P., Johnston, M.B., 2011. III–V semiconductor nanowires for optoelectronic device applications. Prog. Quantum Electron. 35, 23–75.
Junesand, C., Kataria, H., Metaferia, W., Julian, N., Wang, Z., Sun, Y.-T., Bowers, J.E., Pozina, G., Hultman, L., Lourdudoss, S., 2013. Study of planar defect filtering in InP grown on Si by epitaxial lateral overgrowth. Opt. Mater. Express 3, 1960–1973.
Junesand, C., Gau, M.-H., Sun, Y.-T., Lourdudoss, S., Lo, I., Jimenez, J., Postigo, P., Morales, F.M., Hernandez, J., Molina, S., Abdessamad, A., Pozina, G., Hultman, L., Pirouz, P., 2014. Defect reduction in heteroepitaxial InP on Si by epitaxial lateral overgrowth. Mater. Express 4, 41–53.
Jung, D., Callahan, P.G., Shin, B., Mukherjee, K., Gossard, A.C., Bowers, J.E., 2017. Low threading dislocation density GaAs growth on on-axis GaP/Si (001). J. Appl. Phys. 122, 225703.
Kakinuma, H., Ueda, T., Gotoh, S., Yamagishi, C., 1999. Reduction of threading dislocations in GaAs on Si by the use of intermediate GaAs buffer layers prepared under high V–III ratios. J. Cryst. Growth 205, 25–30.
Kawanami, H., Hatayama, A., Nagai, K., Hayashi, Y., 1987. The growth of single domain GaAs films on double domain Si (001) substrates by molecular beam epitaxy. Jpn. J. Appl. Phys. 26, L173.
Khorenko, V., Mofor, A., Bakin, A., Neumann, S., Guttzeit, A., Wehmann, H.-H., Prost, W., Schlachetzki, A., Tegude, F.-J., 2004. Buffer optimization for InP-on-Si [001] quasi-substrates. In: Indium Phosphide and Related Materials, pp. 118–121.
Kim, H., Lee, W.-J., Farrell, A.C., Morales, J.S., Senanayake, P., Prikhodko, S.V., Ochalski, T.J., Huffaker, D.L., 2017. Monolithic InGaAs nanowire array lasers on silicon-on-insulator operating at room temperature. Nano Lett. 17, 3465–3470.
Kohama, Y., Kadota, Y., Ohmachi, Y., 1989a. Heteroepitaxy of InP on Si substrates by MOCVD. J. Electrochem. Soc. 136, 3853–3856.
Kohama, Y., Kadota, Y., Ohmachi, Y., 1989b. InP grown on Si substrates with GaP buffer layers by metalorganic chemical vapor deposition. Jpn. J. Appl. Phys. 28, 1337.
Kohen, D., Nguyen, X.S., Made, R.I., Heidelberger, C., Lee, K.H., Lee, K.E.K., Fitzgerald, E.A., 2017. Preventing phase separation in MOCVD-grown InAlAs compositionally graded buffer on silicon substrate using InGaAs interlayers. J. Cryst. Growth 478, 64–70.
Komatsu, Y., Hosotani, K., Fuyuki, T., Matsunami, H., 1997. Heteroepitaxial growth of InGaP on Si with InGaP/GaP step-graded buffer layers. Jpn. J. Appl. Phys. 36, 5425.
Kroemer, H., 1987. Polar-on-nonpolar epitaxy. J. Cryst. Growth 81, 193–204.

Krost, A., Schnabel, R., Heinrichsdorff, F., Rossow, U., Bimberg, D., Cerva, H., 1994a. Defect reduction in GaAs and InP grown on planar Si (111) and on patterned Si (001) substrates. J. Cryst. Growth 145, 314–320.

Krost, A., Heinrichsdorff, F., Bimberg, D., Cerva, H., 1994b. InP on Si (111): accommodation of lattice mismatch and structural properties. Appl. Phys. Lett. 64, 769–771.

Kunert, B., Guo, W., Mols, Y., Tian, B., Wang, Z., Shi, Y., Van Thourhout, D., Pantouvaki, M., Van Campenhout, J., Langer, R., Barla, K., 2016. III/V nano ridge structures for optical applications on patterned 300 mm silicon substrate. Appl. Phys. Lett. 109, 091101.

Langdo, T., Leitz, C., Currie, M., Fitzgerald, E., Lochtefeld, A., Antoniadis, D., 2000. High quality Ge on Si by epitaxial necking. Appl. Phys. Lett. 76, 3700–3702.

Lee, M., Wuu, D., Tung, H., 1987. Heteroepitaxial growth of InP directly on Si by low pressure metalorganic chemical vapor deposition. Appl. Phys. Lett. 50, 1725–1726.

Lee, S.-M., Cho, Y.J., Park, J.-B., Shin, K.W., Hwang, E., Lee, S., Lee, M.-J., Cho, S.-H., Shin, D.S., Park, J., Yoon, E., 2015. Effects of growth temperature on surface morphology of InP grown on patterned Si (001) substrates. J. Cryst. Growth 416, 113–117.

Li, Q., Lau, K.M., 2017. Epitaxial growth of highly mismatched III-V materials on (001) silicon for electronics and optoelectronics. Prog. Cryst. Growth Charact. Mater. 63, 105–120.

Li, J., Bai, J., Park, J.-S., Adekore, B., Fox, K., Carroll, M., Lochtefeld, A., Shellenbarger, Z., 2007. Defect reduction of GaAs epitaxy on Si (001) using selective aspect ratio trapping. Appl. Phys. Lett. 91, 021114.

Li, Q., Zhou, X., Tang, C.W., Lau, K.M., 2013. Material and device characteristics of metamorphic $In_{0.53}Ga_{0.47}As$ MOSHEMTs grown on GaAs and Si substrates by MOCVD. IEEE Trans. Electron Devices 60, 4112–4118.

Li, Q., Ng, K.W., Lau, K.M., Tang, C.W., Hill, R., Vert, A., 2014. Defect reduction in epitaxial InP on nanostructured Si (001) substrates with position-controlled seed arrays. J. Cryst. Growth 405, 81–86.

Li, Q., Ng, K.W., Lau, K.M., 2015. Growing antiphase-domain-free GaAs thin films out of highly ordered planar nanowire arrays on exact (001) silicon. Appl. Phys. Lett. 106, 072105.

Liang, D., Bowers, J.E., 2010. Recent progress in lasers on silicon. Nat. Photonics 4, 511.

Liang, D., Huang, X., Kurczveil, G., Fiorentino, M., Beausoleil, R., 2016. Integrated finely tunable microring laser on silicon. Nat. Photonics 10, 719.

Liebich, S., Zimprich, M., Beyer, A., Lange, C., Franzbach, D., Chatterjee, S., Hossain, N., Sweeney, S.J., Volz, K., Kunert, B., Stolz, W., 2011. Laser operation of Ga (NAsP) lattice-matched to (001) silicon substrate. Appl. Phys. Lett. 99, 071109.

Liu, J., Sun, X., Pan, D., Wang, X., Kimerling, L.C., Koch, T.L., Michel, J., 2007. Tensile-strained, n-type Ge as a gain medium for monolithic laser integration on Si. Opt. Express 15, 11272–11277.

Liu, J., Sun, X., Camacho-Aguilera, R., Kimerling, L.C., Michel, J., 2010. Ge-on-Si laser operating at room temperature. Opt. Lett. 35, 679–681.

Liu, A.Y., Zhang, C., Norman, J., Snyder, A., Lubyshev, D., Fastenau, J.M., Liu, A.W.K., Gossard, A.C., Bowers, J.E., 2014. High performance continuous wave 1.3 μm quantum dot lasers on silicon. Appl. Phys. Lett. 104, 041104.

Lu, F., Bhattacharya, I., Sun, H., Tran, T.-T.D., Ng, K.W., Malheiros-Silveira, G.N., Hasnain, C.C., 2017. Nanopillar quantum well lasers directly grown on silicon and emitting at silicon-transparent wavelengths. Optica 4, 717–723.

Lum, R., Klingert, J., Davidson, B., Lamont, M., 1987. Improvements in the heteroepitaxy of GaAs on Si. Appl. Phys. Lett. 51, 36–38.

Mahajan, S., 2000. Defects in semiconductors and their effects on devices. Acta Mater. 48, 137–149.

Mårtensson, T., Svensson, C.P.T., Wacaser, B.A., Larsson, M.W., Seifert, W., Deppert, K., Gustafsson, A., Wallenberg, L.R., Samuelson, L., 2004. Epitaxial III – V nanowires on silicon. Nano Lett. 4, 1987–1990.

Mayer, B., Janker, L., Loitsch, B., Treu, J., Kostenbader, T., Lichtmannecker, S., Reichert, T., Morkötter, S., Kaniber, M., Abstreiter, G., Gies, C., Koblmüller, G., Finley, J.J., 2015. Monolithically integrated high-β nanowire lasers on silicon. Nano Lett. 16, 152–156.

Megalini, L., Bonef, B., Cabinian, B.C., Zhao, H., Taylor, A., Speck, J.S., Bowers, J.E., Klamkin, J., 2017. 1550-nm InGaAsP multi-quantum-well structures selectively grown on v-groove-patterned SOI substrates. Appl. Phys. Lett. 111, 032105.

Megalini, L., Šuran Brunelli, S.T., Charles, W.O., Taylor, A., Isaac, B., Bowers, J.E., Klamkin, J., 2018. Strain-compensated InGaAsP superlattices for defect reduction of InP grown on exact-oriented (001) patterned Si substrates by metal organic chemical vapor deposition. Materials 11, 337.

Merckling, C., Waldron, N., Jiang, S., Guo, W., Collaert, N., Caymax, M., Vancoille, E., Barla, K., Thean, A., Heyns, M., Vandervorst, W., 2014. Heteroepitaxy of InP on Si (001) by selective-area metal organic vapor-phase epitaxy in sub-50 nm width trenches: the role of the nucleation layer and the recess engineering. J. Appl. Phys. 115 023710.

Mori, H., Sugo, M., Itoh, Y., 1993. Heteroepitaxy of InP on Si substrates. Adv. Mater. 5, 208–209.

Norman, J.C., Jung, D., Wan, Y., Bowers, J.E., 2018. Perspective: the future of quantum dot photonic integrated circuits. APL Photonics 3, 030901.

Orzali, T., Vert, A., O'Brien, B., Herman, J.L., Vivekanand, S., Hill, R.J., Karim, Z., Papa Rao, S.S., 2015a. GaAs on Si epitaxy by aspect ratio trapping: analysis and reduction of defects propagating along the trench direction. J. Appl. Phys. 118, 105307.

Orzali, T., Vert, A., Kim, T.-W., Hung, P., Herman, J.L., Vivekanand, S., Huang, G., Kelman, M., Karim, Z., Hill, R.J., PapaRao, S.S., 2015b. Growth and characterization of an $In_{0.53}Ga_{0.47}As$-based metal-oxide-semiconductor capacitor (MOSCAP) structure on 300 mm on-axis Si (001) wafers by MOCVD. J. Cryst. Growth 427, 72–79.

Quitoriano, N.J., Fitzgerald, E.A., 2007. Relaxed, high-quality InP on GaAs by using InGaAs and InGaP graded buffers to avoid phase separation. J. Appl. Phys. 102 033511.

Razeghi, M., Defour, M., Blondeau, R., Omnes, F., Maurel, P., Acher, O., Brillouet, F., Fan, J.C.C., Salerno, J., 1988. First cw operation of a $Ga_{0.25}In_{0.75}As_{0.5}P_{0.5}$-InP laser on a silicon substrate. Appl. Phys. Lett. 53, 2389–2390.

Reithmaier, J.P., Benyoucef, M., 2016. III-V/Si integration for photonics. ECS Trans. 72, 171–179.

Richardson, C.J., Lee, M.L., 2016. Metamorphic epitaxial materials. MRS Bull. 41, 193–198.

Sakai, S., Soga, T., Takeyasu, M., Umeno, M., 1985. AlGaAs/GaAs DH lasers on Si substrates grown using super lattice buffer layers by MOCVD. Jpn. J. Appl. Phys. 24, L666.

Samonji, K., Yonezu, H., Takagi, Y., Iwaki, K., Ohshima, N., Shin, J., Pak, K., 1996. Reduction of threading dislocation density in InP-on-Si heteroepitaxy with strained short-period superlattices. Appl. Phys. Lett. 69, 100–102.

Sasaki, T., Mori, H., Tachikawa, M., Yamada, T., 1998. Aging tests of InP-based laser diodes heteroepitaxially grown on Si substrates. J. Appl. Phys. 84, 6725–6728.

Saxena, D., Mokkapati, S., Parkinson, P., Jiang, N., Gao, Q., Tan, H.H., Jagadish, C., 2013. Optically pumped room-temperature GaAs nanowire lasers. Nat. Photonics 7, 963.

Seki, A., Konushi, F., Kudo, J., Kakimoto, S., Fukushima, T., Koba, M., 1987. MOCVD growth of InP on 4-inch Si substrate with GaAs intermediate layer. Jpn. J. Appl. Phys. 26, L1587.

Seki, A., Konushi, F., Kudo, J., Koba, M., 1988. Epitaxial growth of InP on Si by OMVPE—defect reduction in epitaxial InP using $InAs_xP_{1-x}$/InP superlattices. J. Cryst. Growth 93, 527–531.
Sheldon, P., Yacobi, B., Jones, K., Dunlavy, D., 1985. Growth and characterization of GaAs/Ge epilayers grown on Si substrates by molecular beam epitaxy. J. Appl. Phys. 58, 4186–4193.
Shi, B., Li, Q., Wan, Y., Ng, K.W., Zou, X., Tang, C.W., Lau, K.M., 2015. InAlGaAs/InAlAs MQWs on Si substrate. IEEE Photon. Technol. Lett. 27, 748–751.
Shi, B., Li, Q., Lau, K.M., 2017a. Self-organized InAs/InAlGaAs quantum dots as dislocation filters for InP films on (001) Si. J. Cryst. Growth 464, 28–32.
Shi, Y., Wang, Z., Van Campenhout, J., Pantouvaki, M., Guo, W., Kunert, B., Van Thourhout, D., 2017b. Optical pumped InGaAs/GaAs nano-ridge laser epitaxially grown on a standard 300-mm Si wafer. Optica 4, 1468–1473.
Shi, B., Zhu, S., Li, Q., Wan, Y., Hu, E.L., Lau, K.M., 2017c. Continuous-wave optically pumped 1.55 μm InAs/InAlGaAs quantum dot microdisk lasers epitaxially grown on silicon. ACS Photonics 4, 204–210.
Shi, B., Zhu, S., Li, Q., Tang, C.W., Wan, Y., Hu, E.L., Lau, K.M., 2017d. 1.55 μm room-temperature lasing from subwavelength quantum-dot microdisks directly grown on (001) Si. Appl. Phys. Lett. 110, 121109.
Shi, B., Li, Q., Lau, K.M., 2018. Epitaxial growth of high quality InP on Si substrates: the role of InAs/InP quantum dots as effective dislocation filters. J. Appl. Phys. 123, 193104.
Shi, B., Wang, L., Taylor, A.A., Suran Brunelli, S., Zhao, H., Song, B., Klamkin, J., 2019. MOCVD grown low dislocation density GaAs-on-V-groove patterned (001) Si for 1.3 μm quantum dot laser applications. Appl. Phys. Lett. 114, 172102.
Shinohara, M., 1988. Dislocation-free GaAs epitaxial growth with the use of modulation-doped AlAs-GaAs superlattice buffer layers. Appl. Phys. Lett. 52, 543–545.
Soga, T., Hattori, S., Sakai, S., Takeyasu, M., Umeno, M., 1984. MOCVD growth of GaAs on Si substrates with AlGaP and strained superlattice layers. Electron. Lett. 20, 916–918.
Soga, T., Jimbo, T., Umeno, M., 1990. Low etch pit density GaAs on Si grown by metalorganic chemical vapor deposition. Appl. Phys. Lett. 56, 1433–1435.
Sugo, M., Uchida, N., Yamamoto, A., Nishioka, T., Yamaguchi, M., 1989. Residual strains in heteroepitaxial III-V semiconductor films on Si (100) substrates. Jpn. J. Appl. Phys. 65, 591–595.
Sugo, M., Yamaguchi, M., Al-Jassim, M., 1990. Heteroepitaxial growth of InP on Si substrates. J. Cryst. Growth 99, 365–370.
Sugo, M., Mori, H., Itoh, Y., Sakai, Y., Tachikawa, M., 1991. 1.5 μm-long-wavelength multiple quantum well laser on a Si *substrate*. Jpn. J. Appl. Phys. 30, 3876–3878.
Sugo, M., Mori, H., Sakai, Y., Itoh, Y., 1992. Stable cw operation at room temperature of a 1.5-μm wavelength multiple quantum well laser on a Si substrate. Appl. Phys. Lett. 60, 472–473.
Tachikawa, M., Mori, H., 1990. Dislocation generation of GaAs on Si in the cooling stage. Appl. Phys. Lett. 56, 2225–2227.
Tachikawa, M., Mori, H., Sugo, M., Itoh, Y., 1993. Continuous GaAs film growth on epitaxial Si surface in initial stage of GaAs/Si heteroepitaxy. Jpn. J. Appl. Phys. 32, L1252.
Tachikawa, M., Yamada, T., Sasaki, T., Mori, H., Kadota, Y., 1995. Laser-diode-quality InP/Si grown by hydride vapor phase epitaxy. Jpn. J. Appl. Phys. 34, L657.
Takagi, Y., Yonezu, H., Kawai, T., Hayashida, K., Samonji, K., Ohshima, N., Pak, K., 1995. Suppression of threading dislocation generation in GaAs-on-Si with strained short-period superlattices. J. Cryst. Growth 150, 677–680.
Takano, Y., Hisaka, M., Fujii, N., Suzuki, K., Kuwahara, K., Fuke, S., 1998. Reduction of threading dislocations by InGaAs interlayer in GaAs layers grown on Si substrates. Appl. Phys. Lett. 73, 2917–2919.

Tang, G.-P., Lubnow, A., Wehmann, H.-H., Zwinge, G., Schlachetzki, A., 1992. Antiphase-domain-free InP on (100) Si. Jpn. J. Appl. Phys. 31, L1126.

Tang, C.W., Zhong, Z., Lau, K.M., 2010. Hetero-epitaxy of III-V compounds by MOCVD on silicon substrates. ECS Trans. 28, 227–231.

Tatebayashi, J., Ishida, M., Hatori, N., Ebe, H., Sudou, H., Kuramata, A., Sugawara, M., Arakawa, Y., 2005. Lasing at 1.28 μm of InAs-GaAs quantum dots with AlGaAs cladding layer grown by metal-organic chemical vapor deposition. IEEE J. Sel. Top. Quantum Electron. 11, 1027–1034.

Ting, S., Fitzgerald, E., 2000. Metal-organic chemical vapor deposition of single domain GaAs on Ge/Ge_xSi_{1-x}/Si and Ge substrates. J. Appl. Phys. 87, 2618–2628.

Tomioka, K., Tanaka, T., Hara, S., Hiruma, K., Fukui, T., 2011. III–V nanowires on Si substrate: selective-area growth and device applications. IEEE J. Sel. Top. Quantum Electron. 17, 1112–1129.

Uen, W.-Y., Li, Z.-Y., Huang, Y.-C., Chen, M.-C., Yang, T.-N., Lan, S.-M., Wu, C.-H., Hong, H.-F., Chi, G.-C., 2006. Heteroepitaxial growth of GaAs on Si by MOVPE using a-GaAs/a-Si double-buffer layers. J. Cryst. Growth 295, 103–107.

Van der Ziel, J., Dupuis, R., Logan, R., Pinzone, C., 1987a. Degradation of GaAs lasers grown by metalorganic chemical vapor deposition on Si substrates. Appl. Phys. Lett. 51, 89–91.

Van der Ziel, J., Dupuis, R., Logan, R., Mikulyak, R., Pinzone, C., Savage, A., 1987b. Low threshold pulsed and continuous laser oscillation from AlGaAs/GaAs double heterostructures grown by metalorganic chemical vapor deposition on Si substrates. Appl. Phys. Lett. 50, 454–456.

Volz, K., Beyer, A., Witte, W., Ohlmann, J., Németh, I., Kunert, B., Stolz, W., 2011. GaP-nucleation on exact Si (0 0 1) substrates for III/V device integration. J. Cryst. Growth 315, 37–47.

Wan, Y., Norman, J., Li, Q., Kennedy, M.J., Liang, D., Zhang, C., Huang, D., Zhang, Z., Liu, A.Y., Torres, A., Jung, D., Gossard, A.C., Hu, E.L., Lau, K.M., Bowers, J.E., 2017. 1.3 μm submilliamp threshold quantum dot micro-lasers on Si. Optica 4, 940–944.

Wang, Z., Tian, B., Paladugu, M., Pantouvaki, M., Le Thomas, N., Merckling, C., Guo, W., Dekoster, J., Van Campenhout, J., Absil, P., Van Thourhout, D., 2013. Polytypic InP nanolaser monolithically integrated on (001) silicon. Nano Lett. 13, 5063–5069.

Wang, Z., Tian, B., Pantouvaki, M., Guo, W., Absil, P., Van Campenhout, J., Merckling, C., Van Thourhout, D., 2015. Room-temperature InP distributed feedback laser array directly grown on silicon. Nat. Photonics 9, 837.

Wang, J., Hu, H., Yin, H., Bai, Y., Li, J., Wei, X., Liu, Y., Huang, Y., Ren, X., Liu, H., 2018. 1.3 μm InAs/GaAs quantum dot lasers on silicon with GaInP upper cladding layers. Photonics Res. 6, 321–325.

Windhorn, T., Metze, G., 1985. Room-temperature operation of GaAs/AlGaAs diode lasers fabricated on a monolithic GaAs/Si substrate. Appl. Phys. Lett. 47, 1031–1033.

Windhorn, T., Metze, G., Tsaur, B.Y., Fan, J.C., 1984. AlGaAs double-heterostructure diode lasers fabricated on a monolithic GaAs/Si substrate. Appl. Phys. Lett. 45, 309–311.

Yamada, T., Tachikawa, M., Sasaki, T., Mori, H., Kadota, Y., Yamamoto, M., 1995. Stable CW operation of 1.3 μm double-heterostructure laser heteroepitaxially grown on Si. Electron. Lett. 31, 455–457.

Yamada, T., Tachikawa, M., Sasaki, T., Mori, H., Kadota, Y., 1997. 7000 h continuous wave operation of multiple quantum well laser on Si at 50°C. Appl. Phys. Lett. 70, 1614–1615.

Yamaguchi, M., 1991. Dislocation density reduction in heteroepitaxial III-V compound films on Si substrates for optical devices. J. Mater. Res. 6, 376–384.

Yamaguchi, M., Tachikawa, M., Itoh, Y., Sugo, M., Kondo, S., 1990. Thermal annealing effects of defect reduction in GaAs on Si substrates. J. Appl. Phys. 68, 4518–4522.
Yamamoto, A., Uchida, N., Yamaguchi, M., 1989. Optimization of InP/Si heteroepitaxial growth conditions using organometallic vapor phase epitaxy. J. Cryst. Growth 96, 369–377.
Yang, V., Groenert, M., Leitz, C., Pitera, A., Currie, M., Fitzgerald, E.A., 2003. Crack formation in GaAs heteroepitaxial films on Si and SiGe virtual substrates. J. Appl. Phys. 93, 3859–3865.
Yang, J., Bhattacharya, P., Mi, Z., 2007. High-performance $In_{0.5}Ga_{0.5}As$/GaAs quantum-dot lasers on silicon with multiple-layer quantum-dot dislocation filters. IEEE Trans. Electron Devices 54, 2849–2855.
Yang, L., Bulsara, M.T., Lee, K.E., Fitzgerald, E.A., 2011. Compositionally-graded InGaAs-InGaP alloys and GaAsSb alloys for metamorphic InP on GaAs. J. Cryst. Growth 324, 103–109.
Zamir, S., Meyler, B., Salzman, J., 2001. Thermal microcrack distribution control in GaN layers on Si substrates by lateral confined epitaxy. Appl. Phys. Lett. 78, 288–290.
Zhou, Z., Yin, B., Michel, J., 2015. On-chip light sources for silicon photonics. Light Sci. Appl. 4, e358.
Zhu, S., Shi, B., Li, Q., Wan, Y., Lau, K.M., 2017. Parametric study of high-performance 1.55 μm InAs quantum dot microdisk lasers on Si. Opt. Express 25, 31281–31293.
Zhu, S., Shi, B., Li, Q., Lau, K.M., 2018a. Room-temperature electrically-pumped 1.5 μm InGaAs/InAlGaAs laser monolithically grown on on-axis (001) Si. Opt. Express 26, 14514–14523.
Zhu, S., Shi, B., Li, Q., Lau, K.M., 2018b. 1.5 μm quantum-dot diode lasers directly grown on CMOS-standard (001) silicon. Appl. Phys. Lett. 113, 221103.

Further reading

Han, Y., Li, Q., Zhu, S., Ng, K.W., Lau, K.M., 2017. Continuous-wave lasing from InP/InGaAs nanoridges at telecommunication wavelengths. Appl. Phys. Lett. 111, 212101.
Liu, W., Lubyshev, D., Fastenau, J., Wu, Y., Bulsara, M., Fitzgerald, E., Urteaga, M., Ha, W., Bergman, J., Brar, B., Hoke, W.E., LaRoche, J.R., Herrick, K.J., Kazior, T.E., Clark, D., Smith, D., Thompson, R.F., Drazek, C., Daval, N., 2009. Monolithic integration of InP-based transistors on Si substrates using MBE. J. Cryst. Growth 311, 1979–1983.

CHAPTER EIGHT

Nano-ridge laser monolithically grown on (001) Si

Dries Van Thourhout[a], Yuting Shi[a], Marina Baryshnikova[b], Yves Mols[b], Nadezda Kuznetsova[b], Yannick De Koninck[b], Marianna Pantouvaki[b], Joris Van Campenhout[b], Robert Langer[b], Bernardette Kunert[b,*]
[a]Photonics Research Group, Ghent University-imec, Ghent, Belgium
[b]Imec, Leuven, Belgium
*Corresponding author: e-mail address: bernardette.kunert@imec.be

Contents

1. Introduction

The monolithic integration of III–V devices directly on silicon (Si) is the most promising approach to meet the high requirements of cost-efficiency and scalability in co-integration with silicon photonics. The obvious challenges are linked to the large lattice mismatch between Si and most III–V compound materials of interest, which leads to the formation of strain induced misfit defects in the III–V heterolayer. A well-known method to achieve a high III–V crystal quality on Si is selective area growth (SAG) in narrow trenches fabricated on (001) Si substrates. Strain induced dislocation defects are confined inside these narrow trenches close to the III–V/Si interface by so-called aspect ratio trapping (ART). The successful restriction of misfit defects in a highly confined pattern has been demonstrated for various III–V material systems (Kunert et al., 2018b) and

Semiconductors and Semimetals, Volume 101
ISSN 0080-8784
https://doi.org/10.1016/bs.semsem.2019.07.002

its application is explored for very different laser integration approaches (Han et al., 2018; Megalini et al., 2017; Tian et al., 2017; Wang et al., 2018; Wirths et al., 2018).

In this chapter we would like to report about a new nano-ridge laser concept, which is based on ART but adds an important aspect called nano-ridge engineering (NRE). A pronounced growth out of the trenches together with an engineered nano-ridge shape leads to an enlarged III–V nano-ridge volume, which has been optimized for laser diode integration. In this NRE integration concept the defective III–V/Si interface is locally separated from the active device region, which is essential to achieve sufficient device lifetime.

After an introduction to ART and NRE, the monolithic growth by metal organic vapor phase epitaxy (MOVPE) of GaAs-based nano-ridge laser structures on 300 mm (∅) on-axis Si substrates is explained. Nanometer scaled distributed feedback (DFB) lasers are fabricated and optically pumped laser operation is demonstrated at room-temperature, which emphasizes the large potential of this laser integration approach. In a concluding discussion the advantages of the NRE laser approach for the co-integration with silicon photonic is discussed.

2. Aspect ratio trapping (ART)

Selective area growth (SAG) in narrow trenches is an advantageous integration approach for the monolithic deposition of lattice mismatched III–V materials on Si substrates. The Si wafer is patterned with a dielectric mask, such as silicon dioxide (SiO_2), which contains open stripes exposing the Si surface only in predefined trenches. A high aspect ratio (AR) of the trench pattern is the key requirement for an efficient reduction of the defect density in the III–V heterolayer by ART. The AR is defined by the trench depth d over the trench width w (d/w). The lattice mismatch between III–V compound materials and Si will induce plastic relaxation to release the strain in the heterolayer during the monolithic deposition. For example, dislocation half-loops nucleate at the III–V surface during growth, which are line defects. These threading dislocation (TD) loops are most likely based on 60° misfit defects and, thus, will glide down toward the III–V/Si interface along {111} planes. In case the {111} plane is parallel to the trench orientation, the {111} plane will hit the oxide sidewall due to the angle of 54.7° between this {111} plane and the wafer surface. Hence both TD arms of the half-loop are trapped inside the trench when the AR is sufficient. If the dislocation loop glides along a {111} plane positioned perpendicular to the mask sidewalls, the TD arms are also trapped because the TD loop extends in width during motion until the arms hit the oxide sidewalls. For a high enough AR ($\gg$1.43) all threading dislocations can be entirely pinned at the mask sidewall inside the trench if the applied growth conditions ensure the nucleation of glissile threading dislocations moving along {111} planes.

Planar defects (PD), such as stacking faults and twins, are two-dimensional and run as well along the different {111} planes. In a trench pattern only one kind of PD defect is trapped, the one which lies in {111} planes parallel to the trench orientation. Any PD in a plane perpendicular to the trench orientation is not trapped and will completely penetrate through the overgrown III/V material. Fig. 1 contains a sketch of these four defect cases.

In conclusion for ART in trenches: The formation of dislocation defects to release the mismatch induced strain is unavoidable, but misfit defect trapping is very efficient for the III–V deposition in narrow trenches with high AR under optimized growth conditions. It could finally enable the integration of completely misfit defect-free III–V devices on Si. PDs are only confined in one case out of two; hence, it is important to minimize the nucleation of PDs from the beginning. Next to optimizing the III–V seed deposition, which has an impact on the PD density, the cleaning of the patterned wafers as well as the pre-treatment of the Si surface before epitaxial growth are crucial process steps to achieve a high crystal quality.

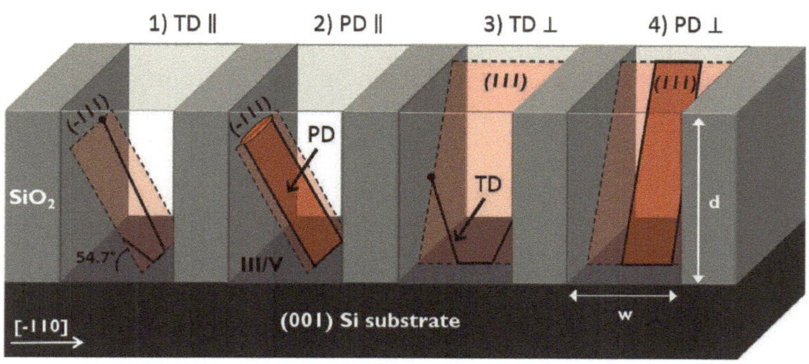

Fig. 1 Sketch of ART in trenches concerning threading dislocations (TDs) and planar defects (PDs). If a TD glides in a {111} plane parallel to the trench orientation, both TD arms will hit the oxide sidewalls for a sufficient AR due to the inclined angle between the {111} plane and the mask sidewall (1). For the same reason also a PD lying in this {111} plane will be trapped (2). TD in {111} planes perpendicular to the trench orientations are pinned as the dislocation half-loop expands in width while gliding (3). The only defect not being trapped in a trench is a PD in {111} planes perpendicular to the mask sidewalls (4). The AR is defined by the trench depth over the trench width (*d/w*).

The difference in lattice parameter between III–V and Si induces the biggest challenge in heteroepitaxy, but both classes of semiconductors also differ in the crystal basis. Si, which crystalizes in diamond structure, consists only of Si atoms while the zincblende structure of III–V compound semiconductors contains group-III and group-V atoms. This crystal difference leads to the formation of anti-phase disorder such as anti-phase domains (APDs) in the III–V film when deposited on a (001) Si surface with monoatomic surface steps. In contrast to this, the monolithic growth of III–V materials on a (111) Si plane has no risk of anti-phase disorder formation.

In an ingenious approach, this fact is used for the growth in narrow trenches fabricated on (001) Si substrates along ⟨110⟩ directions (Paladugu et al., 2012). If the trench bottom is processed toward a V-shape exposing two {111} Si facets, the monolithic growth of III–V material doesn't induce any anti-phase disorder. This is a clear advantage for the integration on on-axis (001) Si substrates because no special Si surface treatment, such as a high temperature bake, is necessary to reduce the presence of monoatomic steps and, hence, APDs (Kunert et al., 2018b; Volz et al., 2015). The atomic model in Fig. 2 (left) illustrates that a monoatomic step on a {111} plane doesn't initiate anti-phase disorder.

The patterning process based on shallow trench isolation (STI), commonly applied in the Si industry, is well established to fabricate nano-meter scaled structures on 300 mm (∅) Si. In this process flow, Si is partly removed and replaced by SiO_2 to fabricate Si stripes embedded in oxide. Then Si is etched away to create trenches with a V-shaped bottom. Such a V-profile defined by two {111} facets, as shown in Fig. 2 (right), can be easily

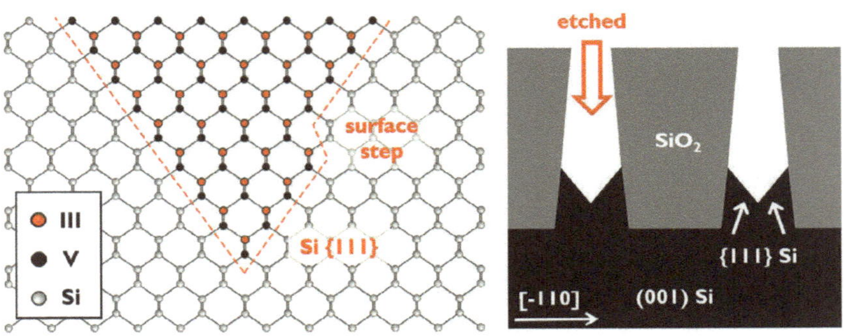

Fig. 2 Left: Atom model for the growth of III–V material on a V-shaped Si surface with two {111} facets. Right: Sketch of the trench formation based on the shallow trench isolation (STI) process and applying a wet-etch step.

wet-etched with potassium hydroxide (KOH), ammonium hydroxide (NH_4OH), or tetramethylammonium hydroxide (TMAH).

Another advantage of the III–V growth on {111} is that relaxation and thus the nucleation of misfit defects seems to be enhanced (Guo et al., 2017). It was reported for the growth of GaAs in V-shaped trenches that for certain MOVPE conditions a periodic and dense array of 60° misfit dislocations along both {111} Si facets is formed (Guo et al., 2017; Kunert et al., 2018b). In such a case the lattice mismatch is mainly accommodated via this 60° misfit array. However, any residual strain can easily initiate additional TDs inside the GaAs crystal, which have to be trapped by the oxide sidewalls to achieve a defect-free top region. In Figs. 3 and 4 the crystal property of GaAs deposited in 40, 100 and 300 nm wide trenches is investigated by transmission electron microscopy (TEM). The oxide mask was fabricated based on the STI process and is about 300 nm thick, which implies an AR (w/d) of 7.5, 3 and 1, respectively. In these growth experiments, not only the trenches were filled with GaAs, but the deposition was also continued out of the trenches to form box-shaped nano-ridges. These GaAs nano-ridges were then capped with a thin InGaP layer. A tilted cross-section scanning electron microscopy (SEM) image of such a GaAs/InGaP nano-ridge on top of a 100 nm wide trench is shown in Fig. 3A. The box-shape formation based on NRE will be explained in more detail in the next section. Here the TEM characterization was focused on the GaAs material inside the trenches

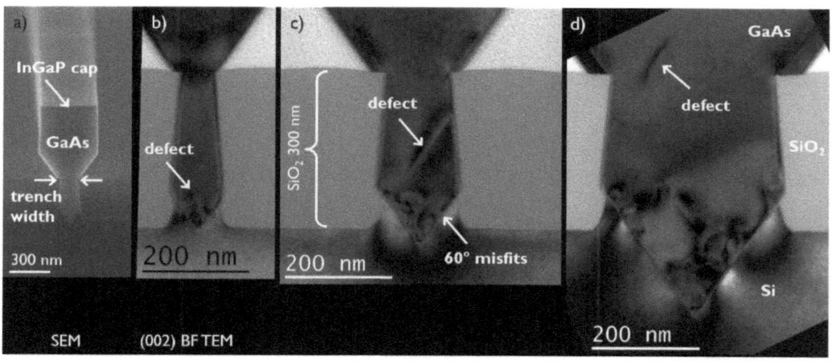

Fig. 3 (A) Tilted cross-section scanning electron microscopy (SEM) image of a GaAs nano-ridge with an InGaP cap layer. The trench width is defined by the trench opening close to the mask surface. (B–D) are (002) two-beam bright field transmission electron microscopy (BF TEM) images of GaAs inside 40, 100 and 300 nm wide trenches (Kunert et al., 2018b). BF TEM images are very sensitive to strain fields; hence, the periodic 60° misfit array is visible at the GaAs/Si interface due to the induced strain fields.

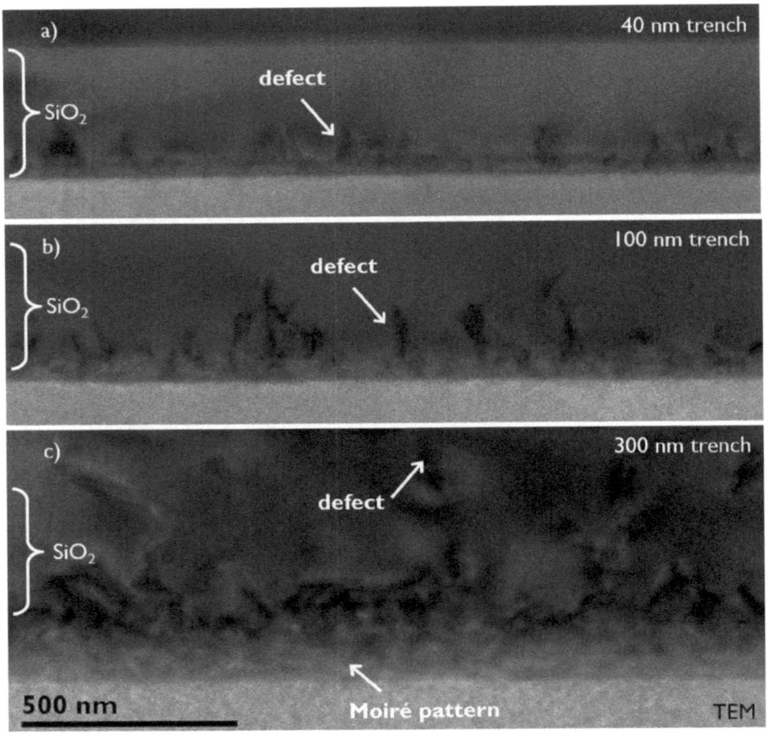

Fig. 4 (A–C) are TEM images taken along the trench orientation of the GaAs samples discussed in Fig. 3, therefore, the TEM samples were prepared along the trenches (Kunert et al., 2018b). The corresponding trench widths are 40 (A), 100 (B) and 300 nm (C). The white brackets indicate the SiO_2 region in each case. The Moiré pattern at the trench bottom is caused by the overlap of the GaAs and Si crystal in the V-shape section, which is especially recognizable in image (C).

of different widths. Fig. 3B–D is (002) two-beam bright field (BF) TEM images in cross-section of GaAs inside 40, 100 and 300 nm wide trenches. The periodic arrangement of 60° misfit defects along the GaAs/Si interface is visible in all three cases by the dark/light contrast fluctuation induced by the defect strain fields. Some TDs were nucleated in the GaAs bulk material due to some residual strain. As these line defects induce a strong strain field, their presence is clearly recognizable by the blurry lines and contrast shadow. For 40 and 100 nm wide trenches (AR $\gg$ 1.43), all defects are confined inside the trench close to the III–V/Si interface whereas the top GaAs region is free of misfit defects. In the case of the 300 nm wide trench, where the AR of 1 is obviously below 1.43, also the upper GaAs material contains a threading dislocation.

This correlation between AR and defect trapping becomes more conclusive in Fig. 4, which shows the corresponding TEM images for a sample preparation direction along the trench. The thickness of the SiO_2 is indicated by the bracket. Due to the etched V-shape, the bottom tip reaches into the Si beyond the mask depth in case of the 100 and 300 nm wide trenches. TDs are visible as blurry lines. In the case of the 40 and 100 nm wide trenches in Fig. 4A and B, all misfit defects are trapped inside the trench. The TD lines are obviously more confined and hence shorter in length for growth inside 40 nm wide trenches. Only for the 300 nm wide trench, Fig. 4C, defect lines extend into the top GaAs region outside the mask pattern due to the low AR.

The TEM comparison in Figs. 3 and 4 emphasizes the efficiency of ART to restrict misfit defects inside narrow trenches. However, this trapping effect is not just guaranteed by the choice of a trench width with sufficiently high AR. It is important to initiate glissile TD gliding along the {111} planes. If TD are pinned at the growth surface or don't glide appropriately, even misfit defects can penetrate through the full III–V structure (Kunert et al., 2018b). Furthermore, it is of key importance that the III–V material is becoming fully relaxed inside the trench before growing out of the trench. Any remaining mismatch strain, which builds-up with layer thickness, can induce the nucleation of new misfit defects even outside the trench.

Box-shaped GaAs nano-ridges were investigated applying electron channeling contrast imaging (ECCI). No indications of misfit defects were found for the growth in narrow trenches ($\leq$150 nm trench width). The quantified defect density of $3 \times 10^6\,cm^{-2}$ reported in Kunert et al. (2018b) is limited by the investigated III–V surface area and hence could be much lower. Additional defect revealing techniques are currently investigated to confirm that no misfit defects are overlooked applying ECCI and to increase the inspection area. As discussed before, twins and stacking faults in a {111} plane perpendicular to the trench sidewalls are not trapped. In contrast to the TD density, the density of PD increases with decreasing trench width. This tendency was observed for different material systems selectively grown in trenches (Kunert et al., 2018b; Schulze et al., 2018) and indicates that the trench cleaning and/or pre-treatment of the Si surface before epitaxial growth strongly affects the PD density. For GaAs nano-ridges a PD density of 0.14–0.45 μm^{-1} (PD counted per micrometer trench length) was reported based on ECCI going from 500 to 100 nm wide trenches.

To conclude, the final defect density in the heterolayer is determined by the MOVPE growth conditions but also by the quality of the structured SiO_2/Si wafer and Si-surface preparation before deposition. Hence, it is important to optimize the full integration process flow starting with the fabrication of the patterned wafers.

3. Nano-ridge engineering (NRE)

A high III–V crystal quality can be achieved applying ART, where choosing very narrow trenches is beneficial. But the application of such highly confined trenches restricts also the total III–V volume, even considering the fabrication of very deep trenches. A very innovative way to still profit from high AR for defect trapping but at the same time to increase the III–V volume is the growth out of the trench together with a defined shape formation. This approach is called NRE. The nano-ridge shape on top of the trench line is engineered by the applied MOVPE growth conditions. The four cross-section SEM images in Fig. 5 illustrate the possibility of NRE. GaAs nano-ridges were grown on patterned Si wafers using growth parameters varied over a wide range. This had led to the formation of very distinct nano-ridge shapes, such as elongated rectangles (A), different diamond profiles (B and D) or triangles (D). Looking at these series of experiments, it becomes obvious that the defect-free III–V volume is clearly increased on top of the narrow trenches.

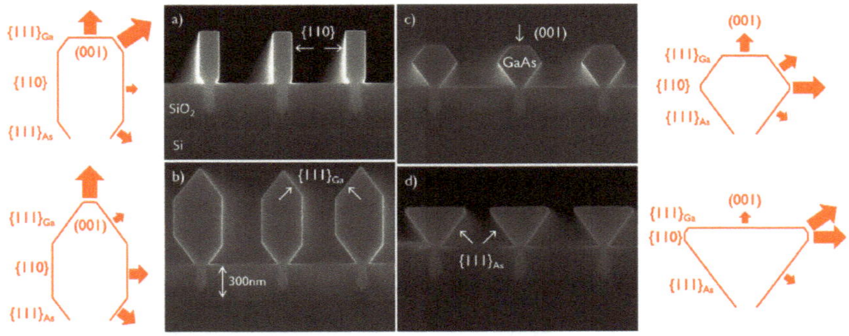

Fig. 5 (A–D) are cross-section SEM images of GaAs nano-ridges deposited under very different MOVPE growth conditions (Kunert et al., 2018a). The simplified sketch next to each SEM image compares the growth rates on the different facets indicated by the size of the red arrow. The facets with the highest growth rate (largest arrow) disappear quickly during deposition and the nano-ridge shape is defined by the facets with low growth rate.

Adjusting the MOVPE parameters, it is possible to manipulate the growth rate hierarchy of the different nano-ridge facets due to their unique surface properties. The facets with the highest growth rates will disappear quickly whereas the nano-ridge shape is defined by the facets with low growth rates. The correlation of the growth rate and shape evolution is indicated by the sketches next to the SEM images in Fig. 5 (Kunert et al., 2018a).

The manipulation of the growth rate to define the final III–V structure was reported before for homoepitaxial SAG of InP and GaAs in a micro-meter scaled pattern (Gil-Lafon et al., 2001; Sun et al., 2001). If the correlation between the growth rates on the different facets and the deposition conditions is known, e.g., the growth rate polar diagram (Jones et al., 1995), a Wulff construction allows to predict the final ridge shape (Gil-Lafon et al., 2001).

The novelty here is the combination of NRE with ART in III–V heteroepitaxy on Si and the application of trench dimensions in the nano-meter range. The pronounced growth of defect-free III–V material out of the trench to increase the material volume opens up the possibility of new III–V integration concepts on Si substrates. Especially the box-shaped nano-ridges are of great interest due to the broadened (001) crystal plane. Fig. 6 underlines the uniform nano-shape structure and ridge line in case of the deposition of a GaAs p-n-junction. Fig. 6A and B is cross-section SEM images of these nano-ridge heterostructures deposited on trench patterns with different separations. Fig. 6C is a tilted SEM image in lower magnification of the nano-ridge array in (B). The GaAs inside and on top of the trench is n-doped with Si whereas the upper GaAs region is p-doped with carbon (C). Both junction layers are separated by a thin InGaP layer as a contrast marker for the SEM inspections. The nano-ridge arrays are very uniform for both trench separations. The ridge size is larger in (B) in comparison to (A) due to loading effect typical in SAG as more material from the mask surface contributes to the growth of each nano-ridge.

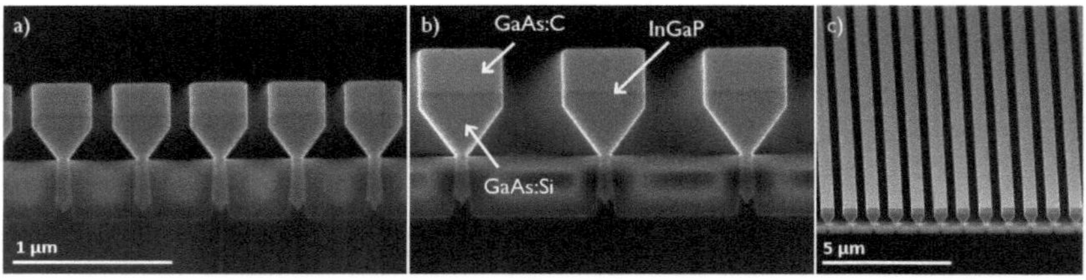

Fig. 6 (A–B) are cross-section SEM images of a nano-ridge heterostructure based on different mask layouts. (C) is a tilted cross-section SEM image of the nano-ridges in (B) in lower magnification. The heterostructure consists of a Si- and C-doped GaAs region, which is separated by a thin InGaP layer to provide a better image contrast.

4. Nano-ridge DFB-laser

Nano-ridge dimensions of several hundred nano-meters together with the uniform growth along the line pattern lead straightforwardly to the design of a nano-ridge laser where the nano-ridge acts as a waveguide. The GaAs nano-ridge on top of the trench is fully relaxed and provides the lattice constant of GaAs for the monolithic deposition of different heterolayers. In the first laser design a three times InGaAs/GaAs multi-quantum-well (MQW) structure was chosen as an active material system. The indium (In)-concentration of about 20% leads to a QW emission wavelength of 1025 nm (Kunert et al., 2016a,b). The InGaAs QWs are pseudomorphically grown on the GaAs nano-ridge surface and, hence, are compressively strained. The QWs are separated by GaAs barriers. Most InGaAs QW material is deposited on the top (001) nano-ridge surface to confine electron-hole pairs and to emit photons. However, due to the three-dimensional structure of the nano-ridges, the three times InGaAs/GaAs multi-layer stack is also grown on the two {110} sidewalls. As the thickness of these InGaAs side layers is only between 1 and 2 nm, no electron-hole pairs are bonded and no contribution to light emission is expected. The GaAs sidewall growth, especially from the last thick barrier layer, is as a matter of fact desired because it ensures a carrier confinement inside the QWs also toward the lateral sidewalls. Due to the accumulated compressive strain of the MQW stack, a thin GaPAs layer with 11% phosphor (P) is deposited before integrating the active material stack. This layer is inserted for strain compensation purposes as GaPAs is deposited under tension on GaAs. The final heterolayer of the laser device is an InGaP cap grown around the nano-ridge, which is lattice matched to GaAs with an In-concentration of 51%. This cap layer provides an additional carrier confinement toward the GaAs barrier and QWs. Another function of the InGaP layer is to suppress any carrier losses due to GaAs surface recombination as no GaAs surface is exposed. Fig. 7 holds a high-angle annular dark field (HAADF) scanning TEM image of the nano-ridge laser grown on top of 100 nm wide trenches, which reveals the different heterolayers. The inset is a magnification of the QW region and shows more clearly the sidewall growth. The total height and width of the nano-ridge laser in Fig. 7 is about 600 and 460 nm, respectively.

The hetero-offsets in the laser stack and the corresponding strain states of the different heterolayers become obvious looking at the summary in Fig. 8A, which shows the band gap of different III–V materials versus their

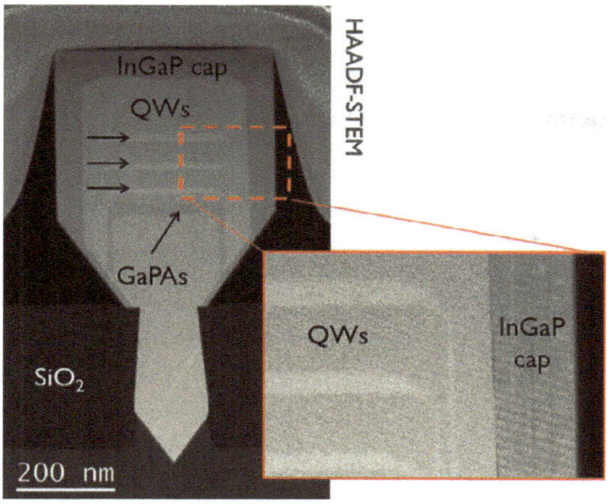

Fig. 7 The figure is a high-angle annular dark field scanning TEM image of the complete nano-ridge laser grown on top of a 100nm wide trench line (Shi et al., 2017). The different heterolayers are clearly visible due to the image contrast. The inset shows a magnification of the QW layers and InGaP side passivation.

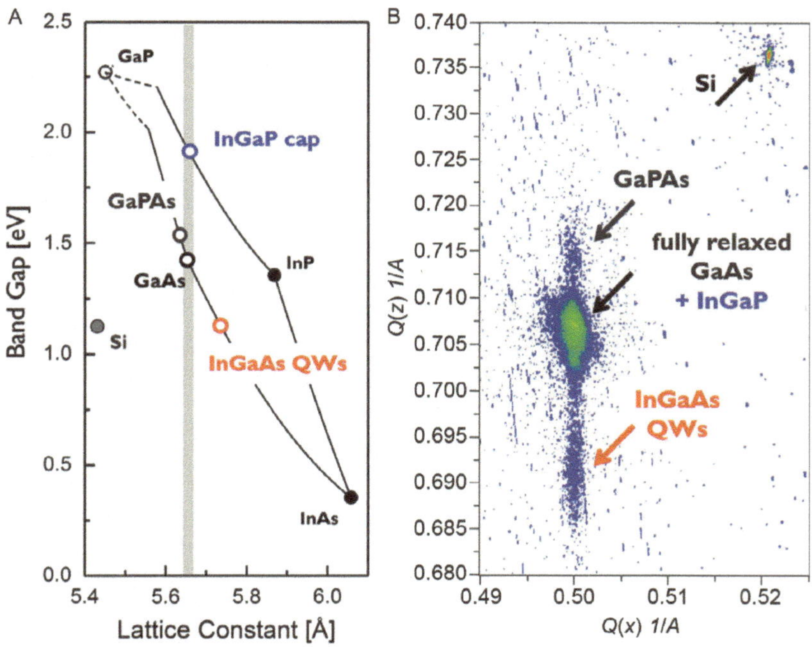

Fig. 8 (A) Plot of the band gap versus the lattice constant for different III-V compound materials. (B) Reciprocal space map (RSM) around the (224) reflex of the nano-ridge laser discussed in Fig. 7. (B).

lattice constant. InGaAs with 20% In has in comparison to GaAs a lower band gap and a larger lattice constant (red) whereas the incorporation of P into GaAs leads to an increase in band gap and reduction in lattice constant (gray). The combination of both material systems inside a multi-layer stack offers the possibility of strain compensation for well-defined layer thicknesses and strain states. In our first nano-ridge laser approach just a thin GaPAs layer was inserted under the MQW stack, mainly to investigate the growth characteristics on the different facets and to provide an additional carrier confinement toward the trench due to the slight increase in band gap. The InGaP cap lattice matched to GaAs has a quite large band gap which is very beneficial to act as good carrier confinement layer. Fig. 8B is a reciprocal space map (RSM) around the (224) reflex of the nano-ridge structure discussed in Fig. 7. This high-resolution X-ray diffraction method provides information about the lattice constant out-of-plane in growth direction (z) and in-plane as a function of the X-ray incident direction (x or y). The RSM in Fig. 8B was measured with an X-ray incident direction parallel to the trench orientation, hence, the in-plane lattice constant parallel (x) to the trench orientation was measured.

Next to the Si peak in Fig. 8B, the peak with the highest intensity is related to the GaAs nano-ridge. The extracted lattice constant is the one of free-standing GaAs, hence the GaAs nano-ridge is fully relaxed. Because the InGaAs QWs are compressively strained to GaAs, the diffraction pattern is aligned vertically under the GaAs peak. In the same way, the GaPAs related diffraction peak is lined up above the GaAs peak position due to the tensile strain. The broadening in vertical direction is caused by the limited QW and GaPAs layer thicknesses. The InGaP layer is not visible as a separate peak in Fig. 8B as the composition is adjusted to be lattice matched to GaAs. The complete vertical alignment of the diffraction peaks of all heterolayers verifies the perfect pseudomorphic growth of the entire device stack.

The growth of the nano-ridge laser is carried out by MOVPE on 300 mm patterned Si wafers using commercial metal-organic precursors. The pattern is defined by a specific die-layout, which is repeated and distributed over the full wafer. Each die contains various trench structures with different trench widths, lengths and orientations, which allows a structural comparison study based on a single growth experiment. The left top-view SEM image of Fig. 9 shows several vertical nano-ridges grouped always into 10 lines having equal trench width and trench separation. This top-view image emphasizes that the ridge lines of the complete laser device stack are very collinear. This region was cleaved to carry out XSEM inspections. SEM images of

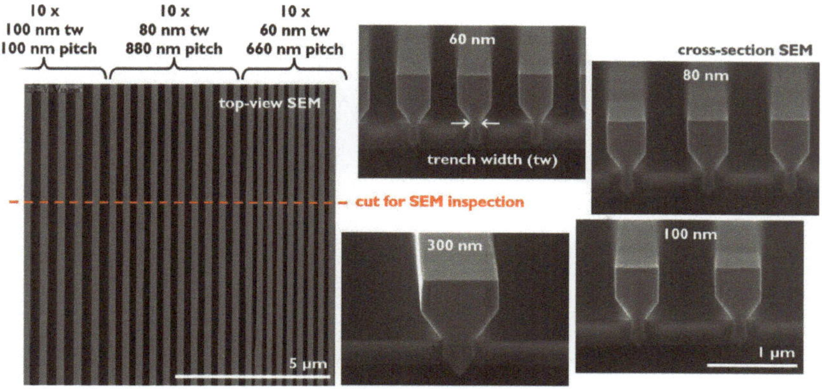

Fig. 9 Left is a top-view SEM image of several nano-ridge lasers, grouped into 10 lines based on the same trench width and separation. The red dashed line indicates the cleave edge for cross-section SEM inspection. On the right side are four tilted cross-section SEM images of the nano-ridge laser based on 60, 80, 100 and 300 nm wide trenches.

the nano-ridge laser based on 60, 80, 100 and 300 nm wide trenches are shown in the right part of Fig. 9. The InGaP cap layer is clearly visible but also the QWs are weakly recognizable in the XSEM images. The total nano-ridge size differs as a function of the trench width and separation, but the principle layout is the same in all cases. Nano-ridge lasers based on 60, 80, 100 and 120 nm wide trenches were processed into distributed feedback lasers, which is described in the next section.

5. Optical characterization of nano-ridge DFB-lasers

In first instance we characterized the photoluminescence (PL) of the nano-ridge devices including three compressively strained InGaAs quantum wells but without an InGaP cap layer. The devices where pumped with a continuous wave pump laser with wavelength 532 nm and intensity 270 W/cm^2. The measured PL spectra, shown in Fig. 10 (Kunert et al., 2016b), exhibit a clear emission peak between 1010 and 1040 nm for all structures. For the narrowest (<40 nm) and wider (>150 nm) trench widths a clear decrease in the strength of the PL signal can be observed. For the wider trenches, this decrease can be correlated to a reduced AR and hence an increase of the defect density in the nano-ridges. In this specific sample without an InGaP passivation layer, the PL intensity also declines due to carrier losses caused by GaAs surface recombination and its impact likely increases for smaller nano-ridge dimensions.

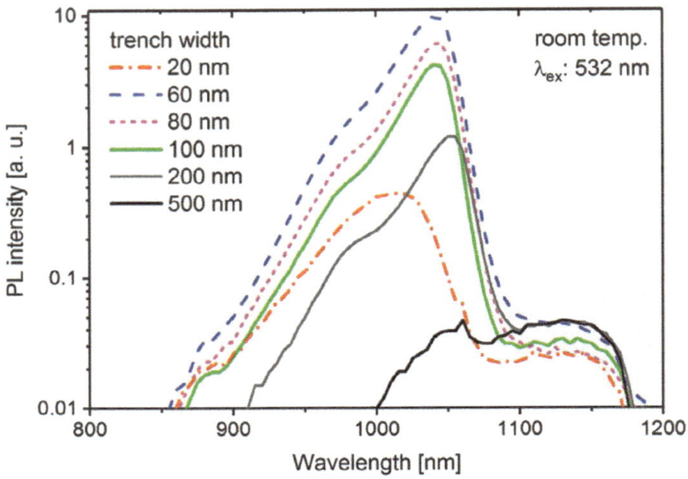

Fig. 10 Normalized room-temperature photoluminescence spectra of nano-ridges with different trench width (Kunert et al., 2016b).

In addition to the basic PL characterization, we also used a variable stripe length (VSL) method to extract the optical gain of GaAs nano-ridges over a wide range of pumping conditions (Shi et al., 2018). For the nano-ridge with a trench width of 100 nm shown in Fig. 7 we found a modal gain up to 450 cm^{-1}. Given the QW confinement factor (CF) in the order of 9%, this is equivalent with a material gain above 4000 cm^{-1}, on par with similar QWs grown on GaAs substrates. This confirms the excellent quality of the grown material.

In a next step, we investigated through simulations if the nano-ridges can support a low loss optical mode. As explained above, the NRE approach allows to enlarge the volume of III–V material grown above the patterned substrate. Compared to earlier nano-lasers (Tian et al., 2017; Wang et al., 2015), this has the advantage that leakage of the optical mode toward the high refractive index silicon substrate can be suppressed, without complex post-processing steps. Fig. 11 shows the lowest order transverse electric (TE)-like optical mode in nano-ridge devices with trench size 60, 80, 100 and 120 nm. The dimensions for the simulated devices were obtained from actual fabricated structures. In all cases the optical mode is strongly confined in the nano-ridge and the calculated leakage loss is less than 5 dB/cm. The cumulative confinement in the three InGaAs QW's gradually increases from 8.8% to 9.6%, values comparable with classical laser structures now used in commercial applications. Due to the high index contrast, the

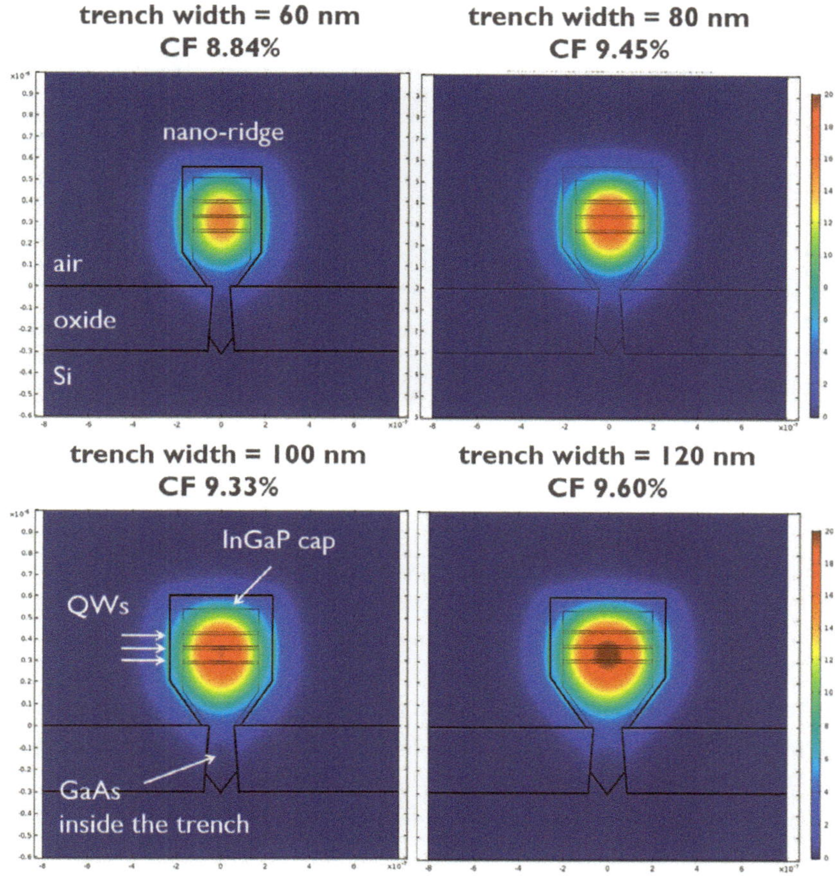

Fig. 11 Lowest order TE-like optical mode in nano-ridges with different trench sizes, calculated using Finite Difference Eigenmode simulation. The increase in nano-ridge volume with increasing trench width results in an increase of the calculated effective index for these modes. We find $n_{eff} = 3.022$, 3.122, 3.167 and 3.210, respectively. The confinement factor (CF) also changes with trench width.

nano-ridges actually support up to four guided modes (for 100 nm trench width), with the lowest order mode having a transverse magnetic (TM)-like character (Shi et al., 2017). However, given that the simulated leakage loss for this mode is considerably higher than for the TE-like mode and given that the gain of the compressively strained InGaAs QWs is expected to be considerably higher for the in-plane polarized TE-like optical mode (Chuang, 2009), only the latter is considered in the actual laser design.

Fig. 12 shows a schematic picture and a top-view SEM of the fabricated DFB-laser. A first order grating is etched in the top layer of the nano-ridge.

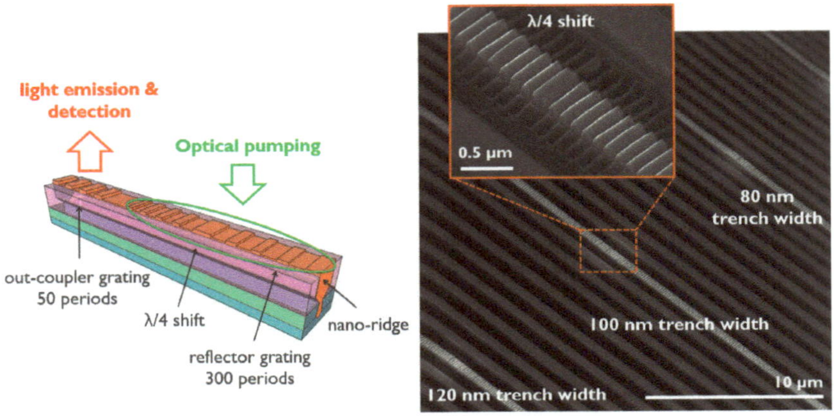

Fig. 12 Left: Schematic design of the optically pumped DFB-laser combined with second order grating for out-coupling. Right: Tilted top-view SEM from fabricated DFB-lasers (Shi et al., 2017). The devices where patterned sufficiently far away from each other so that it was possible to address them individually with a patterned pump spot.

The grating period (170 nm) was optimized using 3D Finite Difference Time Domain simulations such that its reflectivity overlaps with the PL peak of the material (1020 nm). The etching depth was restricted to 110 nm, to avoid etching the QW layers. In this way, a 300-period grating provides a reflectivity of 90% (coupling strength $\kappa L = 3.3$). The total laser cavity consists of two 300-period sections separated by a $\lambda/4$ shift section in the center (to suppress multi-mode lasing). Twenty micro-meter away from the laser cavity, a second order grating (period 340 nm) was defined to couple out the laser light orthogonally to the substrate, making it possible to characterize the devices on a standard micro-PL setup (Fig. 12).

The devices where characterized on the same setup as used for the PL-measurement described above. However, in this case the pump source was a Nd: YAG pulsed laser (7 ns pulse width, 938 Hz repetition rate, 532 nm wavelength). The pump power was controlled using a half-wave plate and a polarizer, for details see Shi et al. (2017). The effective pump power is calculated by measuring the pump intensity and simulating how much of the incident light is absorbed in the nano-ridge. All measurements were carried out at room-temperature. Fig. 13A and B shows the measured output power, as function of the pump power (L-L curve) on a linear and logarithmic scale, respectively. A clear threshold behavior can be seen. Fitting to a standard rate-equations model (Shore and Ogura, 1992) allows to extract a threshold pump power $P_{th} = 37$ mW (equivalent with a pump intensity of $34\,kW/cm^2$) and a spontaneous emission coupling factor $\beta = 0.008$.

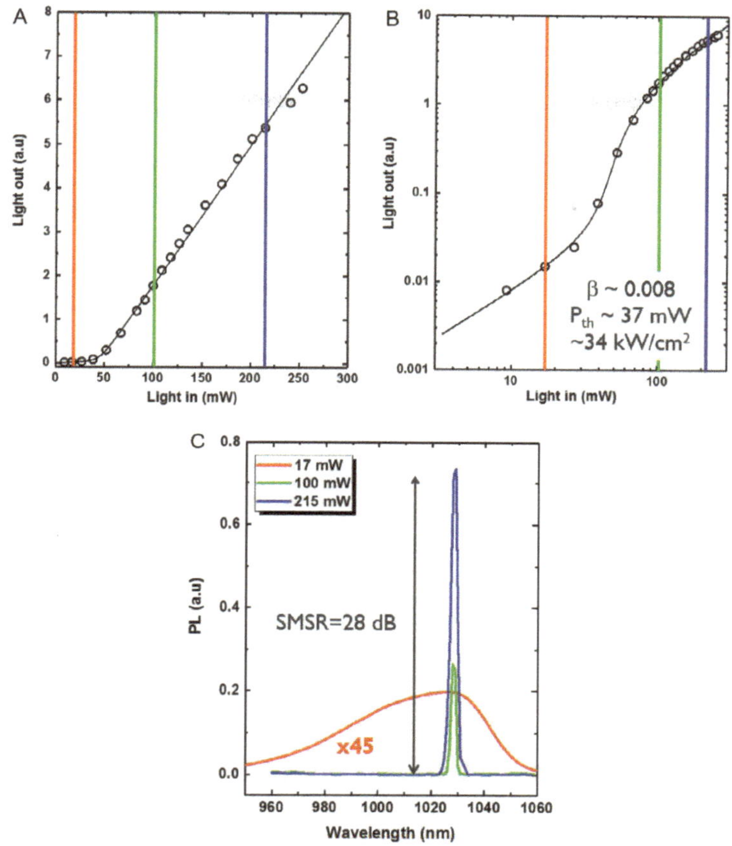

Fig. 13 Measured (black dots) and simulated (solid line) L-L curve for a DFB-laser with trench width 100 nm and grating period 170 nm on (A) linear and (B) logarithmic scale. (C) Spectra under different excitation powers.

Fig. 13C shows the broad photoluminescence spectrum below threshold (red) and the narrow lasing spectra at two different pump powers (green and blue) above threshold. The clamping of the spontaneous emission spectrum above threshold and the high side mode suppression (up to 28 dB) are further signatures of laser operation.

Fig. 14A shows the above threshold lasing spectra of DFB-lasers fabricated in nano-ridges based on different trench widths and fixed grating period of 170 nm. It is obvious that the lasing wavelength increases monotonously with trench width. As explained above, an increasing trench width results in a larger volume for the nano-ridge and hence an increased effective refractive index for the optical mode (Fig. 11). This in turn leads to a shift

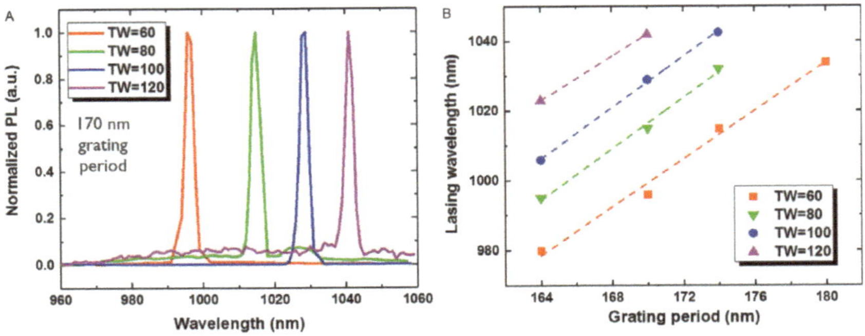

Fig. 14 (A) Normalized lasing spectra for DFB-lasers with 170 nm period, fabricated in nano-ridges with trench width 60, 80, 100 and 120 nm. (B) Lasing wavelength versus grating period for nano-ridge lasers with different trench width.

of the Bragg-wavelength, explaining the observed behavior. Fig. 14B shows the results from a full study whereby also the period of the grating was varied. These results show that the lasing wavelength of the device can be accurately controlled, either through the grating period or through the trench width. This opens prospects for using the devices in wavelength division multiplexing (WDM) applications whereby multiple devices with different wavelength are used to increase the total bandwidth of an optical link.

6. Integration with silicon photonics

The major advantage of the ART plus NRE integration process, besides the high quality of the grown-out material, is the associated SAG: the III–V material is only deposited where it is desired. Other areas, which possibly already contain predefined photonic or electronic devices can be masked with SiO_2 and will remain untouched. Compared to other hetero-epitaxy approaches, which rely on a thick buffer to suppress defects, this promises a more straightforward integration scheme with, e.g., silicon waveguides and hence the possibility to build upon the massive earlier developments in the domain of silicon photonics. Fig. 15A shows an exemplary scenario with a silicon waveguide placed directly next to the nano-ridge. If the dimensions of the silicon waveguide are chosen such that its effective index is phase matched with that of the GaAs nano-ridge, all light is coupled from the nano-ridge laser to the passive waveguide over a length less than 45 μm (see Fig. 15B). An alternative option is to include the silicon waveguide in the STI layer, diagonally below the III–V nano-ridge, as was proposed in Shi et al. (2019). This has the advantage that no additional

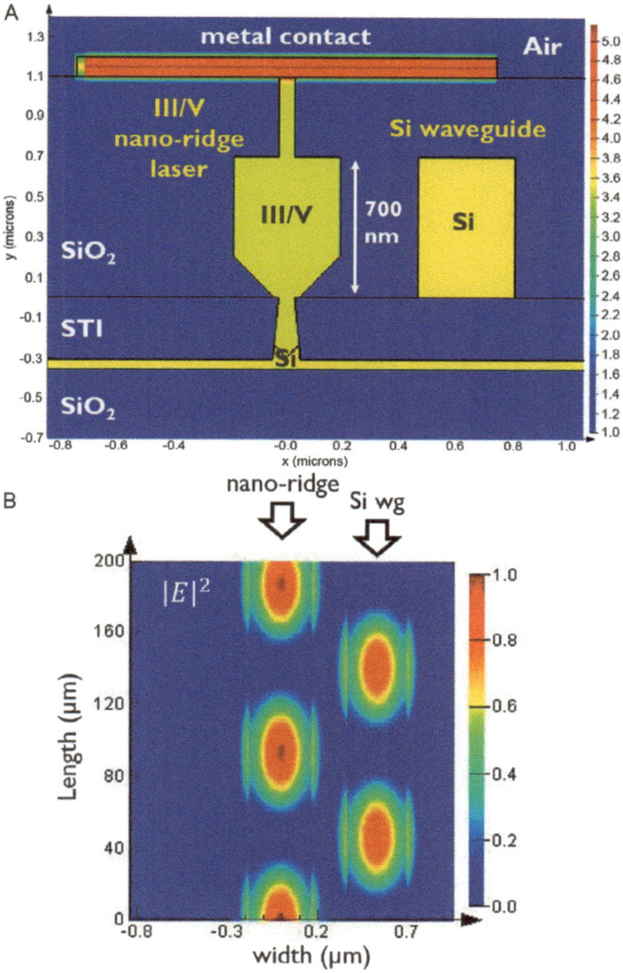

Fig. 15 Schematic view of a nano-ridge laser sideways coupled to a silicon waveguide (A) and a simulation of the optical field coupling between a III/V nano-ridge and a silicon waveguide (wg) in top-view, $|E|^2$ versus length and width (B).

processing steps are required to define the waveguide. The drawback of this so-called directional coupler design is its relative intolerance to fabrication variations. An adiabatic coupler design (Shi et al., 2019) can overcome this, at the cost of a slight increase in the coupler length. Once the III–V nano-ridge gain section is coupled to a silicon waveguide, one can conceive more complex external cavity laser designs: standard filter elements such as DBR-gratings, ring resonators or even complex multi-channel devices such as arrayed waveguide gratings can be integrated with the waveguides to provide the wavelength selective feedback needed for realizing widely tunable lasers or multi-wavelength lasers.

Co-integration with other active devices such as silicon pn-modulators, Ge-detectors or even silicon transistors needs a careful design of the integration flow and the temperature budget of each step needs to be taken into account. High temperature steps have to be carried out before the III–V SAG-process, while back-end processes such as metallization have to be performed afterward.

7. Next steps

The next challenge now is to demonstrate electrical injection. Directly depositing a metal contact on top of the nano-ridge would lead to unacceptably high waveguide losses (>200 dB/cm). As an alternative, an etched or regrown fin on top of the nano-ridge as shown in Fig. 15A allows to keep the metal contact and the optical mode separated, while still allowing for efficient current injection. The bottom trench can then be used for current extraction.

The proof of sufficient device lifetime is the final milestone for this new integration route. Since the NRE approach leads to a clear local separation of the defective III–V/Si interface region (inside narrow trenches) from the active device region (grown-out of the trenches), we expect no degrading impact of the misfit defects on the nano-ridge device performance. However, there are still a lot of open questions, which have to be explored in the near future. And beyond laser devices, there is also huge potential of his novel NRE integration approach for other optoelectronic applications.

References

Chuang, S.L., 2009. Physics of photonic devices. In: Boreman, G. (Ed.), Wiley Series in Pure and Applied Optics. Wiley.

Gil-Lafon, E., Napierala, J., Castelluci, D., Pimpinelli, A., Cadoret, R., Gérard, B., 2001. Selective growth of GaAs by HVPE: keys for accurate control of the growth morphologies. J. Cryst. Growth 222 (3), 482–496. https://doi.org/10.1016/S0022-0248(00)00961-1.

Guo, W., Mols, Y., Belz, J., Beyer, A., Volz, K., Schulze, A., Langer, R., Kunert, B., 2017. Anisotropic relaxation behavior of InGaAs/GaAs selectively grown in narrow trenches on (001) Si substrates. J. Appl. Phys. 122 (2), 025303. https://doi.org/10.1063/1.4991481.

Han, Y., Qiang, L., Ng, K.W., Si, Z., Lau, K.M., 2018. InGaAs/InP quantum wires grown on silicon with adjustable emission wavelength at telecom bands. Nanotechnology 29 (22), 225601. IOP Publishing. https://doi.org/10.1088/1361-6528/aab53b.

Jones, S.H., Salinas, L.S., Jones, J.R., Mayer, K., 1995. Crystallographic orientation dependence of the growth rate for GaAs low pressure organometallic vapor phase Epitaxy. J. Electron. Mater. 24 (1), 5–14. https://doi.org/10.1007/BF02659719.
Kunert, B., Guo, W., Mols, Y., Langer, R., Barla, K., 2016a. Integration of III/V hetero-structures by selective area growth on Si for Nano- and optoelectronic. ECS Trans. 75 (8), 409–419.
Kunert, B., Guo, W., Mols, Y., Tian, B., Wang, Z., Shi, Y., Van Thourhout, D., et al., 2016b. III/V Nano ridge structures for optical applications on patterned 300 mm silicon substrate. Appl. Phys. Lett. 109 (9). 091101. https://doi.org/10.1063/1.4961936.
Kunert, B., Langer, R., Pantouvaki, M., Van Campenhout, J., Van Thourhout, D., 2018a. Gaining an edge with nano-ridges. Comp. Semicond. 24 (05), 36–41.
Kunert, B., Mols, Y., Baryshniskova, M., Waldron, N., Schulze, A., Langer, R., 2018b. How to control defect formation in monolithic III/V hetero-epitaxy on (100) Si? A critical review on current approaches. Semicond. Sci. Technol. 33 (9), 93002. IOP Publishing. https://doi.org/10.1088/1361-6641/aad655.
Megalini, L., Bonef, B., Cabinian, B.C., Zhao, H., Taylor, A., Speck, J.S., Bowers, J.E., Klamkin, J., 2017. 1550-nm InGaAsP multi-quantum-well structures selectively grown on v-groove-patterned SOI substrates. Appl. Phys. Lett. 111 (3), 2–7. https://doi.org/10.1063/1.4994318.
Paladugu, M., Merckling, C., Loo, R., Richard, O., Bender, H., Dekoster, J., Vandervorst, W., Caymax, M., Heyns, M., 2012. Site selective integration of III–V materials on Si for nanoscale logic and photonic devices. Cryst. Growth Design 12 (10), 4696–4702. https://doi.org/10.1021/cg300779v.
Schulze, A., Strakos, L., Vystavel, T., Loo, R., Pacco, A., Collaert, N., Vandervorst, W., Caymax, M., 2018. Non-destructive characterization of extended crystalline defects in confined semiconductor device structures. Nanoscale 10, 7058–7066. https://doi.org/10.1039/C8NR00186C.
Shi, Y., Wang, Z., Van Campenhout, J., Pantouvaki, M., Guo, W., Kunert, B., Van Thourhout, D., 2017. Optical pumped InGaAs/GaAs nano-ridge laser Epitaxially grown on a standard 300-mm Si wafer. Optica 4 (12), 1468. https://doi.org/10.1364/OPTICA.4.001468.
Shi Y., Kunert B., Baryshnikova M., Pantouvaki M., Van Campenhout J. and Van Thourhout D., Optical gain characterization of nano-ridge amplifiers epitaxially grown on a standard Si wafer, Conference on Lasers and Electro-Optics, OSA Technical Digest (online) (Optical Society of America, 2018), paper JTu2A.21. https://doi.org/10.1364/cleo_at.2018.jtu2a.21.
Shi, Y., Kunert, B., Van Campenhout, J., Van Thourhout, D., 2019. Novel adiabatic tapered coupler for small III-V lasers grown on SOI wafer. In: European Conference on Integrated Optics W.Po1.23, Papers are available from https://www.ecio-conference.org/previous-editions/.
Shore, K.A., Ogura, M., 1992. Threshold characteristics of microcavity semiconductor lasers. Opt. Quant. Electron. 24 (2), S209–S213.
Sun, Y.T., Rodríguez Messmer, E., Söderström, D., Jahan, D., Lourdudoss, S., 2001. Temporally resolved selective area growth of InP in the openings off-oriented from [110] direction. J. Cryst. Growth 225 (1), 9–15. https://doi.org/10.1016/S0022-0248(01)00970-8.
Tian, B., Wang, Z., Pantouvaki, M., Absil, P., Van Campenhout, J., Merckling, C., Van Thourhout, D., 2017. Room temperature O-band DFB laser array directly grown on (001) silicon. Nano Lett. 17 (1), 559–564. https://doi.org/10.1021/acs.nanolett.6b04690.
Volz, K., Stolz, W., Dadgar, A., Krost, A., 2015. 31 Growth of III/Vs on silicon: nitrides, phosphides, Arsenides and Antimonides. In: Handbook of Crystal Growth. Elsevier, pp. 1249–1300. https://doi.org/10.1016/C2013-0-09792-7.

Wang, Z., Tian, B., Pantouvaki, M., Guo, W., Absil, P., Van Campenhout, J., Merckling, C., Van Thourhout, D., 2015. Room-temperature InP distributed feedback laser array directly grown on silicon. Nat. Photonics 9 (12), 837–842, Nature Publishing Group. https://doi.org/10.1038/nphoton.2015.199.

Wang, M., Li, Z., Zhou, X., Li, Y., Wang, P., Yu, H., Wang, W., Pan, J., 2018. InGaAs/InP multi-quantum-well nanowires directly grown on SOI substrates and optical property characterizations. In: Conference on Lasers and Electro-Optics. OSA, Washington, D.C.JTh2A.76https://doi.org/10.1364/CLEO_AT.2018.JTh2A.76.

Wirths, S., Mayer, B.F., Schmid, H., Sousa, M., Gooth, J., Riel, H., Moselund, K.E., 2018. Room-temperature lasing from monolithically integrated GaAs microdisks on silicon. ACS Nano 12, 2169–2175. https://doi.org/10.1021/acsnano.7b07911.

CHAPTER NINE

Quantum dot microcavity lasers on silicon substrates

Yating Wan, Justin Norman, John Bowers*
Institute for Energy Efficiency, University of California Santa Barbara, Santa Barbara, CA, United States
*Corresponding author: e-mail address: bowers@ece.ucsb.edu

Contents

1. Introduction

Traditional copper networking cables have been replaced with fiber optic cables for transcontinental distances down to meters through much

Semiconductors and Semimetals, Volume 101
ISSN 0080-8784
https://doi.org/10.1016/bs.semsem.2019.05.002

more efficient and cost-effective light transmission. Highly integrated 100-Gbps silicon photonics transceiver are now being shipped in volume and 400 G transceivers are expected to reach volume production in Q4 2019 (Jones et al., 2019). While the revolution of photonic integration was started on InP, and are boosted with hybrid wafer bonding, higher production volumes requires low-cost, large-area Si substrates containing high gain III-V epitaxial materials (Bowers et al., 2019). In contrast to conventional long-haul fiber optics, datacom applications focus on power efficiency, reliability, high temperature operation, volume manufacturability and overall system cost in order to be technically and economically competitive (Cheng et al., 2018). Many traditional CMOS foundries are now exploring epitaxial growth of III-V on Si as a "More-than-Moore" technology to combat the gate length scaling limits of transistors. The same technology is also being leveraged for photonics, with recent advances in using QDs to mitigate crystal defects during III/V-on-Si growth (Liu and Bowers, 2018). InAs QD lasers grown on Si substrates are already showing exceptional performance, exceeding what has been achieved through heterogeneous integration, and reliability is reaching a point of commercial viability (Norman et al., 2018). Formerly, the use of miscut Si substrates to avoid antiphase boundary (APB) formation would hamper full compatibility with CMOS processing, but recent breakthroughs with microelectronics-standard nominal (001) Si substrates have demonstrated comparable or even superior performance (Li and Lau, 2017). A number of breakthroughs, including sub-milliamp threshold (Wan et al., 2017a), 175 mW single-facet output powers (Jung et al., 2017a), 119 °C continuous wave temperature (Liu et al., 2014), near zero linewidth enhancement factor (Zhang et al., 2018a), isolator-free stability at optical feedback levels of up to 90% (Duan et al., 2019), and >100,000 h lifetimes at 35 °C (Jung et al., 2017b), have now made the III/V-on-Si epitaxial growth an appealing platform for photonic integration, from both a performance and cost standpoint, warranting discussion for more advanced photonic integration (Wang et al., 2017).

For photonic integration, the forces that drove the evolution of discrete optical modules to compact elements on a single chip will continue to drive improvements in cost, size, weight, and power (cSWaP). By confining light to small volumes with resonant recirculation thus eliminating the need for gratings or Fabry-Perot (*F-P*) facets, microcavity lasers have generated great interest for compact on-chip light sources (No, 2019). While the best edge-emitting lasers have thresholds on the order of tens of mA, microcavity lasers have achieved below-ten-mA thresholds by significantly improving the spontaneous emission efficiency and modifying the electromagnetic field distribution. Through light confinement by total internal reflection at a curved boundary, high-quality modes support light propagation close to the boundary of the cavity periphery and can be coupled to a bus waveguide for wavelength division multiplexing applications (Yao et al., 2018). Since the first demonstration of microcavity lasers with four cleaved facets forming a closed loop (Scifres et al., 1976), continuous research has been conducted with a target for low power consumption and small footprint. However, sidewall scattering and recombination hinder performance when the size of the resonator is reduced to tens of microns. For microring semiconductor lasers, to obtain minimal bending losses down to a radii of 5 μm or smaller, the waveguide must be etched at least 1 μm below the active region (De La Rue et al., 2014). The necessity of deeply etched waveguide structure, with the active region exposed to the sidewall, exacerbates the issue with nonradiative sidewall-recombination, resulting in higher thresholds and enhanced device degradation.

Combining the advantages of QDs with microcavity structures has opened a new size regime for microcavity lasers (Kryzhanovskaya et al., 2014). The same property that mitigates the sensitivity to dislocations during III/V-on-Si epitaxial growth, also reduces the sensitivity to nonradiative recombinations at sidewalls of the microresonator. In quantum well (QW) lasers, carrier in-plane diffusion length reaches several microns and can readily reach the device sidewalls. Consequently, radiative recombination efficiency drops drastically with the shrinkage of the cavity size. However, in QD lasers, the in-plane diffusion length is reduced to ~0.5 μm. This enhanced carrier localization gives rise to a weaker surface recombination effects and enables deeply etched process through the active layer without imposing an obvious penalty on the lasing threshold (Moore et al., 2006). This provides notable advantages in scaling to small dimensions, shrinking on-chip lasers with size as small as 1 μm in diameter (Kryzhanovskaya et al., 2015a,b; Wan et al., 2016a), and sub-milliamp thresholds (Wan et al., 2017b).

2. Fundamentals of semiconductor QDs and historical review of QD lasers on Si

First proposed in 1982 (Arakawa and Sakaki, 1982), QD lasers have made advances and outperformed bulk, QW and quantum wire lasers

(Bimberg and Pohl, 2011). QDs can be formed through a self-assembly process using InAs on (In,Ga,Al)(As,P) layers, and are promising to make lasers with low threshold current densities, temperature insensitive behavior, high power operation, isolator-free stability, suppressed sidewall recombination rates, and mitigated sensitivity to defects in epitaxial integration on Si (Norman et al., 2019). Each of these advantages can be attributed to the discrete density of states and three-dimensional carrier confinement (Bimberg et al., 2000).

2.1 Principle of operation: Quantization of the density of states

In conventional double heterostructure lasers, the active region is comprised of a low band-gap material with a thickness of 0.1–0.3 μm that is surrounded on either side with a larger band-gap material. This double heterostructure helps to effectively trap electrons and holes within the active region, thereby reducing the required threshold current density for lasing. The electron and the hole in the exciton are normally confined to an Exciton Bohr Radius. However, when the thickness of the active layer reduces to 50–100 Å, which is comparable to the de Broglie wavelength, the carriers are forced to exist in bound states with discrete allowed energies, as opposed to the continuous band of states associated with typical bulk semiconductors (Coldren et al., 2012). Modification of the quantum confinement results in redistribution of the allowed electronic states over energy and momentum, i.e., changing the shape of the density of states. This energy quantization effect was first utilized to increase the density of states near the band edge in QWs and effectively lower the lasing threshold (Dingle and Henry, 1976). When the active layer has quantum confinement in three dimensions, as in the case of QDs, a complete localization of electrons and holes ideally gives rise to a sequence of delta-function-like states possessing atom-like degeneracy. In practice, these delta-function-like states show a small finite line width from phonon scattering, as shown in Fig. 1.

2.2 The promise of QD lasers

For efficient lasing, it's important that there exists a large density of carriers in both the electron and hole bands at energies close to the band-edge. In QD lasers, the atom-like density of states ensures that low densities of carriers are needed for population inversion. The short carrier lifetime of the ground state supports a high level of radiative optical transitions per unit volume. This mechanism contributes to a low threshold carrier density, and

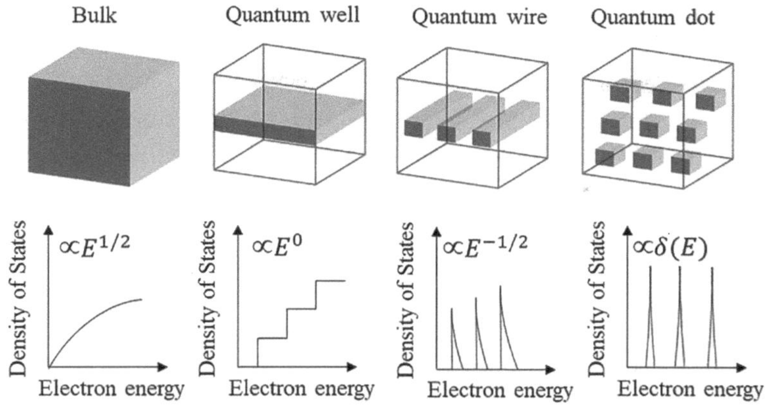

Fig. 1 Variation in density of states with degrees of quantum confinement.

subsequently low threshold current densities (Coleman et al., 2011). In bulk material, when the temperature rises, the charge carriers are distributed continuously to higher energy states. In QDs, each individual dot is separated to neighboring dots by large intersubband separations on the order of 60–85 meV and a total separation of a few hundred meV from the dot ground state. The much suppressed carrier thermalization contributes to temperature-insensitive operation (Sugawara and Usami, 2009). In addition, the gain bandwidth of QDs can be engineered from broad to narrow and to emit at a wide range of wavelengths (Feng et al., 2019). The stochastic process of dot formation results in a Gaussian gain distribution which enables ultralow linewidth enhancement factors and isolator-free stability (Otto, 2014). The carriers injected into the dots are confined by a lateral energy barrier with a few hundred meV from the dot ground state. The reduced carrier lateral diffusion from the device center to the sidewall, and to dislocations generated during epitaxial growth, is important for future ultradense photonic integration up to 450 mm diameter Si wafer.

2.3 GaAs-on-Si epitaxy

For laser growth and processing up to 450 mm Si wafers, high crystalline quality III/V-on-Si metamorphic buffers are a prerequisite, in which the main challenge relates to dislocations generated from the mismatch of lattice constant, thermal expansion coefficient, and polar/nonpolar incompatibility between the III/V and Si. Nowadays, dislocation densities in the III-V materials on Si has been reduced to the level of $10^6\,cm^{-2}$ using techniques such as two-step growth (Bolkhovityanov and Pchelyakov, 2008), strained-layer superlattices (El-Masry et al., 1988), thermal cycle annealing (Lee et al., 1987), relaxed and graded SiGe layers (Groenert et al., 2003), selective area or patterned growth (Li et al., 2016), and GaP buffer layers (Yaung et al., 2016). To maintain full compatibility with CMOS processing, microelectronics-standard nominal (001) Si substrates are preferred. While conventional heteroepitaxy on planar Si wafers needs a 4°–6° miscut angle to prevent APB formation, recent breakthroughs with no miscut angle have demonstrated comparable or even better performance (Wan et al., 2017c). These approaches range from patterned growth on exposed {111} V-groove Si facets (Wan et al., 2015), 0.15° misorientation in the [110] direction (Alcotte et al., 2016), direct nucleation of a GaAs film with special Si wafer preparation (Chen et al., 2017), growing an $Al_{0.3}Ga_{0.7}As$ seed layer (Kwoen et al., 2018), the use of a thin GaP buffer layer (Jung et al., 2017a), and U-shaped patterned Si (001) substrates with {111}-faceted-sawtooth Si hollow structures (Wei et al., 2018). A comprehensive discussion of the physics of the growth mechanisms can be found in Fang et al. (1990). An overview of defect control in III/V-on-Si heteroepitaxy can be found in Kunert et al. (2018) and Li and Lau (2017).

2.4 Current status of QD lasers on Si

Despite the impressive progress of heteroepitaxy of III/V materials on Si, defect-free material or even material comparable to the state-of-the-art III-V wafers with defect level at 10^3–$10^4\,cm^{-2}$ remains elusive. This, to a large extent, deteriorates the reliability of conventional III-V devices on Si, particularly lasers. In fact, the most prolonged lifetime reported among GaAs-based QW lasers on Si after more than two decades of research is merely ~200h (Kazi et al., 2001). On the contrary, for QD devices, with typical dislocation densities of ~10^6–$10^7\,cm^{-2}$ in optimized III/V-on-Si buffers and typical dot densities of ~$6 \times 10^{10}\,cm^{-2}$, the chance of charge carriers encountering a defect is far less than that of finding a dot and recombining radiatively. In addition, the strain field in QDs restricts the in-plane movement of dislocations. Evolution of dark line defects, which is the most critical mechanism responsible for rapid device failure (Ueda and Pearton, 2012), is thus effectively restrained. While the first QD laser grown on Si could only operate at 80K under pulse biased conditions with a large threshold current density of $3.85\,kA/cm^2$ and a center wavelength at 1.038 μm (Linder et al., 1999), the performance has progressed rapidly in a short time frame. In 2011, 1.3 μm operation wavelength was reported (Wang et al., 2011). A threshold current density of $725\,A/cm^2$ with an output power of ~26 mW was achieved in broad-area lasers (3 mm × 50 μm) at room temperature under pulsed injection. In this case, current was injected through the defective GaAs/Si interface with bottom contact placed on the Si substrate. Three years later, the pulsed lasing threshold for a 3 mm × 25 μm broad-area laser was reduced to 200 A/cm^2 (Chen et al., 2014). This was accomplished with an improved dislocation filtering by replacing $In_{0.15}Ga_{0.85}As$/GaAs with $In_{0.15}Al_{0.85}As$/GaAs strained layer superlattices, and an improved current injection scheme with top-top contact geometry to avoid current flow via the defective GaAs/Si interface. The maximum temperature under pulse injection was increased to be 111 °C, with more than 100 mW of output power from a single cleaved facet. Using an intermediate Ge buffer nearly lattice matched to GaAs, the first CW lasing from InAs/GaAs QDs was reported with a maximum output power of 3.7 mW from both facets, a maximum operation temperature of 30 °C, and a low threshold current density of 163 A/cm^2 (Lee et al., 2012). Using a higher number of QD layers, narrow ridge waveguide structures, and high-reflection coatings, lasers grown on a similar Ge/Si substrates were demonstrated with low thresholds (16 mA), high output power (176 mW), high-temperature lasing (up to 119 °C), and high characteristic temperature T_0 (>200 K) (Liu et al., 2014). For the first time, over 2700 h of CW operation with no catastrophic failures at 30 °C was reported, with extrapolated mean time to failures up to 4600 h (Liu et al., 2015a). This is much longer than any previous lifetime test of III-V lasers epitaxially grown on a Si substrate. Starting from 2016, efforts have been focused on developing CMOS compatible epitaxial platform on (001) Si, where III-V growth can be scaled up to 300-mm Si wafers (Alcotte et al., 2016). Exceptional laser performance has been realized with high single-side output power of 175 mW, low threshold currents below 1 mA, and extrapolated lifetime of more than 10,000,000 h at 35 °C (Norman et al., 2018). Chronological evolution of InAs/GaAs QD lasers epitaxially grown on Si is shown in Fig. 2. Using QDs to achieve a commercially-viable monolithically integrated on-chip laser, with a push from the Si industry, is also boosted by the numerous performance advantages of QDs over QWs as described above.

3. Miniaturization of QD lasers using microcavities

Parallel to the success in engineering electronic confinement using QDs, optical confinement can be achieved with whispering-gallery-mode (WGM) resonators, enabling scalable, low-threshold, efficient light

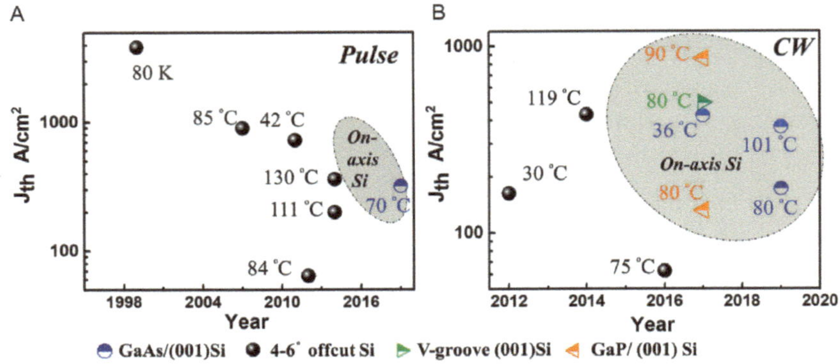

Fig. 2 Chronological evolution of InAs/GaAs QD lasers epitaxially grown on Si.

sources that occupy a small volume on chips with compact sizes and high integration densities.

3.1 Whispering gallery mode microcavity lasers

The concept of WGMs was first proposed by Rayleigh in the study of the acoustic properties in St Paul's Cathedral, where reflection of sound from walls forms high-quality resonators for acoustic waves to circulate around the wall (Rayleigh, 1910). In 1961, an optical analogue of the above architecture was realized by Garrett et al. (1961), where stimulated emission was achieved in a samarium-doped CaF_2 sphere. Since then, WGMs have been extensively and intensively investigated in a range of micro-sized systems used in a range of applications from sensing, frequency combs, optomechanics, quantum optics, to microcavity lasers (Nosich et al., 2007). For microcavity lasers, lasing can be achieved by the intrinsic nonlinearities of the resonator, or introducing active materials to the resonator (He et al., 2013). Resonators made from various liquid and solid materials such as water droplets, silica, two-dimensional materials, bulk semiconductors, organic semiconductors, epitaxial QDs, colloidal QDs, polymers, etc., have been reported. Various geometries including microdroplets, microspheres, photonic crystals, microtoroids, microdisks, microrings, microtubes, and microcylinders have been demonstrated (Song, 2019). In what follows, we will be interested primarily in QD microdisks and microrings where the resonator is formed by the total internal reflection along the circular perimeter, allowing both optical pumping and electrical injection, as well as output coupling to a waveguide.

3.2 Microdisks and microrings: Major figure of merits

Microdisks and microrings are relatively thin (<1 μm) waveguide structures with a suspended disk or a surface ring. The refractive index contrast between the resonator material and the surrounding material contributes to strongly confined modes. The mode circulates the cavity in a closed-loop path, eliminating the necessity of mirror facets, gratings, or reflective coatings for optical feedback, which is beneficial for in-plane photonic integration. The two designs differ from one and another primarily in the schemes for vertical optical confinement. For microdisks with a suspended cavity, a relatively thin post was formed to connect the disk to the substrate by selectively removing the buffer layer consisting of InP in InGaAsP-based structures or of AlGaAs with high-Al compositions in GaAs-based structures. At least a half-wavelength of air gap was formed between the disk and substrate to avoid mode leakage into the substrate. For microrings with a circular mesa structure, the vertical optical confinement is obtained via a lower index semiconductor cladding at the bottom and a semiconductor-air refractive index contrast on the top surface. Generally, the suspended disks provide better optical confinement, and are more commonly adopted for optically pumped devices, while the microrings enable smaller thermal resistance and are more commonly used for electrically injected devices. In the following, we first define several important and key parameters of cavity-based light sources such as quality factor, spontaneous emission factor, threshold current, etc., and explain how critically the cavity design and carrier pumping/injection schemes are related.

3.2.1 Quality factor

Lasing threshold, wavelength, and spectral linewidth are the primary parameters to characterize a microcavity laser, and are all related to the volume (V) of the resonant mode and the quality factor (Q) of the laser cavity. While the V determines the level of light field confinement, Q defines the loss of the cavity. By examining the relation between the cavity loss and the power storage, the cavity Q can be deduced as the ratio of peak wavelength (λ_c) to the peak width $\Delta\lambda$:

$$Q = \frac{\lambda_c}{\Delta\lambda}$$

In conventional "macroscale" optical resonators consisting of two or more mirrors, for example, a *F-P* cavity, the resonant condition implies that there is an integer number of the wavelength of light in a round-trip optical path. Since the modes are reflected at near normal incidence, the Q factor is generally below 100, and is given by

$$Q=\frac{2\pi nL}{\lambda(1-R)}$$

where L is the cavity length, R is the reflectance, n is the refractive index. From the equation, the Q factor is inversely correlated to the mirror reflectivity, and proportional to the cavity length. To achieve a high light intensity confined in the resonator, high-reflectivity mirrors and small cavity size are required to reduce the optical loss as well as to form strong light confinement. This is challenging for mirror-based cavity constructions from both a technical and economical standpoint. On the contrary, microcavities can confine light to small volumes via resonant recirculation, with reflections at small grazing angles. While the Q-factor of microcavities is determined by a number of complicated mechanisms, including intrinsic loss from tunnel coupling of the confined mode to a continuum of propagating modes, and optical loss caused by absorption of the active layers and scattering mechanisms. This value is determined to a significant degree by the latter, and can be approximately as

$$Q=\frac{2\pi n}{\lambda\alpha}$$

where α is the optical loss (Kryzhanovskaya et al., 2014). In this case, by addressing the initial material quality and the microresonator fabrication process, cavities with $Q>10^4$ can be fabricated using GaAs/AlGaAs materials (Kryzhanovskaya et al., 2015b). Characterizing the cavity Q-factor requires mode linewidth measured at transparency when stimulated emission exactly compensates the absorption. As an example, Fig. 3A shows emission spectra from a 4-μm diameter microdisk grown on Si with increasing pump power from a 532 nm CW diode laser (Wan et al., 2016b). At a low pump intensity of 165 μW, a weak cavity mode at 1308 nm can be observed with a broad background emission. This sub-threshold emission can be fitted with bi-Lorentzian curves to extract the narrow cavity emission (Fig. 3B). The extracted cavity emission has a small full-width at half-maximum (FWHM) of 0.27 nm (Fig. 3C), and Q can be estimated to be $\lambda/\Delta\lambda=4844$. It should be noted that this Q is limited by the resolution limit of the spectrometer (0.07 nm). Subject to a similar resolution limit of the instrument, a Q factor of 3700 is obtained in a 1.2-μm diameter

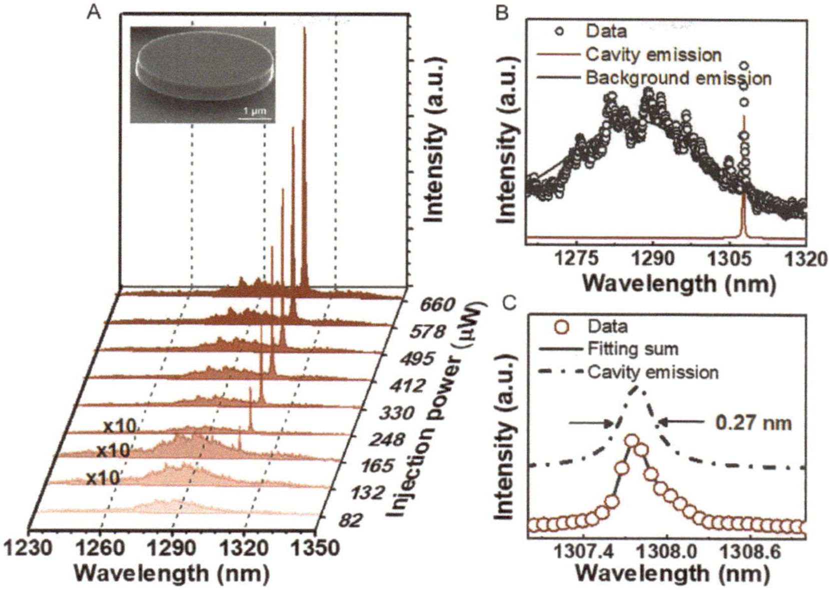

Fig. 3 Lasing characteristics of a 4-μm diameter QD microdisk grown on Si. (A) Spectra taken at progressively higher pump powers. (B) Sub-threshold spectrum taken at a pump power of 165 μW. The black circles are measured data, the blue line is a fit to the broad photoluminescence background of the QDs, the red line is a fit to the narrow cavity mode emission. (C) High-resolution spectrum taken at a pump power of 165 μW. The circles are measured data, the black line is a fitting sum of the measured data using bi-Lorentzian curves, and the dashed black line is a fit to the narrow cavity mode emission (Wan et al., 2016b).

GaN/InGaN microdisk (Tamboli et al., 2007). By conducting the measurement with a high spectral resolution of 30 pm in a cryostat environment at 78 K, a much higher value of ~20,000 is obtained in a 7-μm diameter InAs QD mcirodisk (Kryzhanovskaya et al., 2015b).

3.2.2 Characteristic curves

To characterize the onset of lasing, L-L curve (laser output power versus pump power) and a L-I-V (light-current-voltage) plots are typically used, in optically pumped and electrically injected devices, respectively. Fig. 4 shows two typical L-L curves from a 4-μm diameter microdisk grown on Si (Wan et al., 2016c): a linear plot in (a) and a log-log plot in (b). Three regimes of lasing operation can be observed. The first regime of operation is dominated by spontaneous emission, in which the gain medium is excited well below the lasing threshold. In this region, the characteristic curve in the

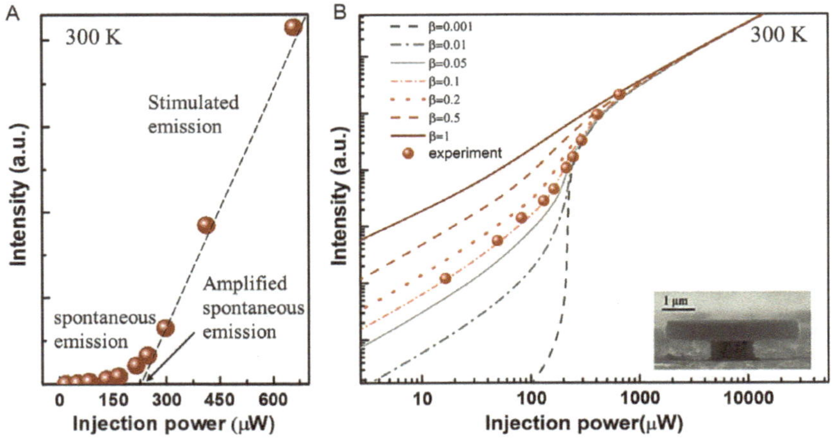

Fig. 4 A typical (A) linear and (B) log-log lasing curve plotting output power against pump power for a 4-μm diameter microdisk grown on Si. In both plots, three regimes of lasing operation are clearly identified, spontaneous emission, amplified spontaneous emission, and stimulated emission (Wan et al., 2016c).

linear plot doesn't differ from that of a light emitting diode, and has a constant slope in the log-log plot. With an increase in pump power, the rate of stimulated emission becomes comparable to that of the spontaneous emission. This point denotes the transition to the second regime of operation, named the amplified spontaneous emission regime. In this region, the overall photons emitted are incoherent, with random phases and a significant spread in wavelengths due to the large number of spontaneous emission events involved. Both plots have an exponential increase in photon population, denoting a kink in the linear plot and a turning point of a "S" shape in the log-log plot. As the pump power is further increased beyond the threshold, the rate of the stimulated emission dominates over the spontaneous emission, with photons emitted coherently in phase and wavelength. Both plots show a linear dependence between the output power and pump power. Lasing threshold can then be determined by either extracting the kink in the linear plot, or the turning point of the "S" shape characteristics in the log-log plot, which is 175 μW in this case.

3.2.3 Spontaneous emission factor

At this point, it's important to introduce the spontaneous emission factor (β), which characterizes the amount of spontaneous emission into a single mode (Coldren et al., 2012). β can be extracted by calculating theoretical values and fitting them to experimental data. In the log-log plot of L-L curve in Fig. 4B, the rate equation model solutions for various values of β are calculated by solving equations using a coupled rate equation model (Wang et al., 2015) for the carrier density N and the photon density P:

$$\begin{cases} \dfrac{dN}{dt} = \eta \cdot \dfrac{P_{in}}{\hbar\omega \cdot V} - V_g \cdot g \cdot P - B \cdot N^2 - C \cdot N^3 \\ \dfrac{dP}{dt} = \Gamma \cdot V_g \cdot g \cdot P + \Gamma \cdot \beta \cdot B \cdot N^2 - \dfrac{P}{\tau_p} \end{cases}$$

Here, η is the internal quantum efficiency, P_{in} is the pump power, V_g is the group velocity (7.8×10^9 cm/s), B is the bimolecular recombination coefficient (2×10^{-10} cm^3/s), C is the Auger coefficient (8×10^{-17} cm^6/s), $\hbar\omega$ is the pump photon energy, V is the volume of the active region, Γ is the optical confinement factor (0.21), and τ_p is the photon lifetime (4.1 ps). Assuming a gain function of $g = g_0 \bullet \frac{N - N_t}{1 + \varepsilon \bullet P}$, where g_0 is the differential gain (3×10^{-16} cm^2), N_t is the transparency carrier density (1.23×10^{18} cm^{-3}), ε is the gain compression factor (1×10^{-17} cm^3), β was extracted to be 0.1.

In conventional semiconductor *F-P* lasers or for any gas or solid-state laser, typical values of β are in the range of 10^{-5} to 10^{-4}. In such lasers, there is a distinct increase in output intensity with the onset of stimulated emission, similar to the scenario described above. It's noted that β is inversely proportional to the laser cavity size, and can be theoretically close to 1 when the cavity volume becomes so small, such that all the spontaneous emission occurs in a single optical mode. This case is termed a zero-threshold laser, where the L-L curve is linear without a clear distinction between the spontaneous and the stimulated emission region. As a matter of fact, for ultrasmall microdisk lasers where the diameter of the disk is smaller than the lasing wavelength (termed subwavelength), the lasing kink is gradually replaced with an increasingly smoother feature when β increases. In Fig. 5A and B, laser output power and mode linewidth are plotted as a function of pump power for a subwavelength microdisk with a diameter of 1 μm grown on Si (Wan et al., 2016a). From the emission spectra (Fig. 5C), a weak cavity mode with a broad background emission appears at a low pump intensity of 16.5 μW (Fig. 5D). Upon surpassing the threshold, the cavity mode increases sharply in intensity, with a clear saturation of the background QD emission and a measurably larger signal, compared to the background optical emission. In Fig. 5E, emission below threshold (8.25 μW) was amplified by 100 times and overlaid with the emission well above threshold (660 μW),

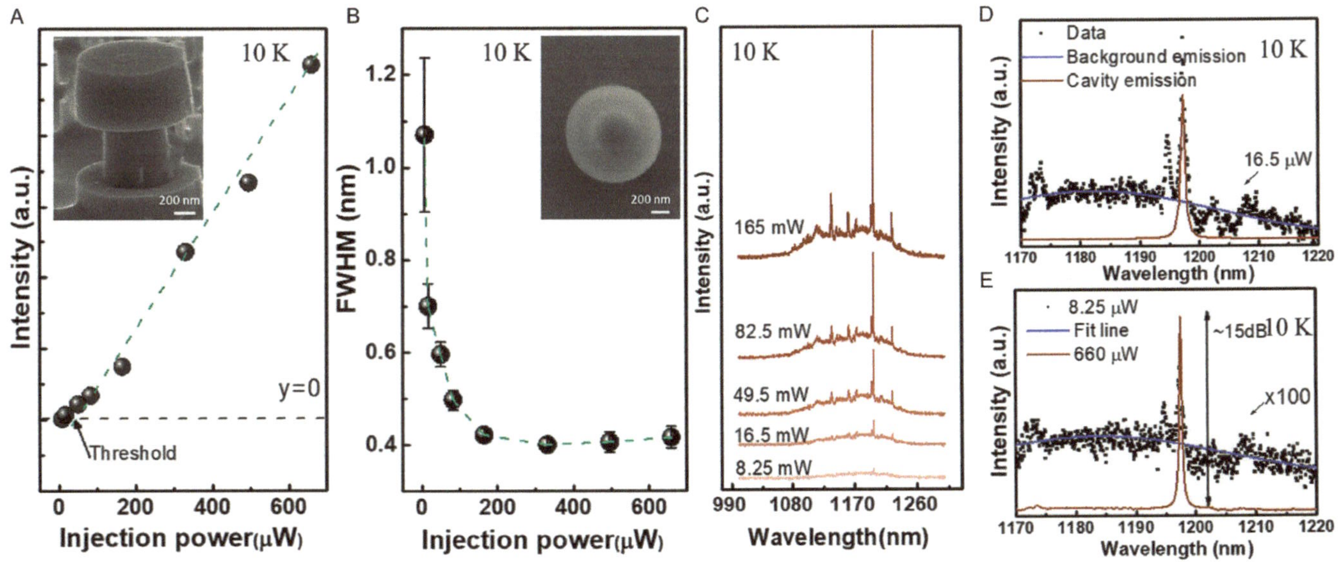

Fig. 5 Laser characteristics for a 1-μm diameter microdisk grown on silicon: (A) output power and (B) mode linewidth against pump power; (C) Power dependent emission spectra; (D) emission spectrum below threshold (taken at a pump power of 16.5 μW). The black dots are measured data, the blue line is a fit to the broad photoluminescence background of the QDs; (E) Emission spectrum measured below (light blue, 8.25 μW) and above (red, 660 μW) the threshold. Black dots are measured data and blue lines are Gaussian fits. The spectrum below the threshold (light blue, 8.25 μW) was amplified 100 times to be visible (Wan et al., 2016a).

to highlight the optical transition positions in the gain medium. While lasing is evidenced with the pronounced narrowing of the lasing mode and the strong saturation of spontaneous emission with pumping, no well-defined transition can be observed from the L-L curve (Fig. 5A).

The well-established laser rate equations used for the 4-μm diameter microdisk described above in Fig. 4 is not sufficient to determine device behavior here (Strauf et al., 2006). Instead, β of the subwavelength microdisk laser can be extracted using a semiconductor cavity quantum electrodynamics (CQED) model (Gies et al., 2007). In Fig. 6A, L-L curve were fitted with the computed input/output curves for various values of β that bound the experimental data in the log-log scale (Wan et al., 2016a). In the model, for a given carrier injection rate into the QD states, equations of motion for carrier, photon populations, and photon correlations are solved numerically using the light-matter coupling coefficient and the dephasing rate as inputs to the model. The light-matter coupling coefficient can be computed from the active region and the optical resonator configuration, represented by a Rabi frequency of $6 \times 10^{10}\,\text{s}^{-1}$. The dephasing rate was assumed with a value of 1 ps^{-1} at 10 K. By analyzing both the onset of linewidth reduction and the bends in the "S" shape of the input/output intensity in the log-log plot (Fig. 6), β was extracted to be 0.3. This high β contributes to a low lasing threshold of 35 μW determined by a combination of characteristic L-L curve, linewidth narrowing profile, and strong saturation of spontaneous emission. While a truly zero-threshold laser hasn't been demonstrated due to a number of practical problems not including fabrication issues, high-β can be obtained synthesizing a small modal

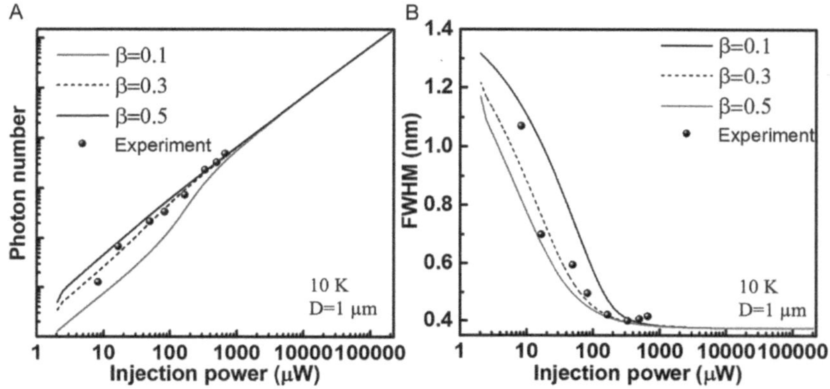

Fig. 6 Calculation and experiment for (A) input/output intensity and (B) the linewidth versus pump intensity log-log plot (Wan et al., 2016a).

volume, a high-quality factor of the mode, and an enlarged overlap of the single mode with the gain medium (Chow et al., 2014). To date, a record high β of 0.95 has been reported in nanoscale coaxial cavities (Khajavikhan et al., 2012). A value of 0.85 has been reported on photonic crystal lasers sustained by only 2–4 QDs as the active gain material (Strauf et al., 2006). More information of the recent experimental progress in the control of β by manipulating optical modes with nanocavities can be found in Noda et al. (2007).

3.2.4 Modal field distribution

WGMs are the eigenstates of a microcavity system and can be solved through the Maxwell equations for a given field $\{E, H\}$ and a wave vector k, given below:

$$\begin{aligned}
&\nabla^2 E + k^2 E = 0\\
&\nabla^2 H + k^2 H = 0\\
&\nabla \cdot E = 0\\
&\nabla \cdot H = 0
\end{aligned}$$

Using the cylindrical coordinate system (ρ, ϕ, z) and arbitrarily choosing the longitudinal component E_Z and H_Z as the generating field, the wave function can be written as:

$$\psi(\rho, \phi, z) = P(\rho)\Theta(\phi)Z(z)$$

where $P(\rho)$ is the m_{th} order Bessel function, $\Theta(\phi)$ is a Harmonic function of ϕ, $Z(z)$ is an exponential decay outside the cavity. The wave function can thus be separated into three equations:

$$\begin{aligned}
&\frac{1}{\rho}\frac{d}{d\rho}\left(\rho\frac{d}{d\rho}P\right) + \left((k_\rho \rho)^2 - m^2\right)P = 0\\
&\frac{1}{\Theta}\frac{d^2\Theta}{d\phi^2} = -m^2\\
&\frac{1}{Z}\frac{d^2 Z}{dz^2} = {\beta_\perp}^2\\
&{k_\rho}^2 = k^2 + {\beta_\perp}^2
\end{aligned}$$

where m and $\beta_\perp$ (mode index and propagation constant) are constants to solve the respective differential equations. By considering the boundary condition where E_Z and H_Z are finite at the origin and periodic with a period of 2π with respect to ϕ inside the cavity, E_Z and H_Z can be written in the following form inside the cavity

$$E_{Z1} = AJ_m(k_{\rho 1}\rho)\begin{pmatrix} \cos(m\phi) \\ \sin(m\phi) \end{pmatrix} e^{\beta z}$$
$$H_{Z1} = BJ_m(k_{\rho 1}\rho)\begin{pmatrix} \sin(m\phi) \\ \cos(m\phi) \end{pmatrix} e^{\beta z}$$
$$k_{\rho 1}{}^2 = k^2 + \beta^2$$

and can be written as the following form outside the cavity

$$E_{Z2} = CH_m(k_{\rho 2}\rho)\begin{pmatrix} \cos(m\phi) \\ \sin(m\phi) \end{pmatrix} e^{\beta z}$$
$$H_{Z2} = DH_m(k_{\rho 2}\rho)\begin{pmatrix} \sin(m\phi) \\ \cos(m\phi) \end{pmatrix} e^{\beta z}$$
$$-k_{\rho 2}{}^2 = k^2 + \beta^2$$

In this way, the field is related to the azimuth angle ϕ by $\exp(\pm im\phi)$. The radial distribution within the cavity takes on the form of a Bessel function (J_m of the first kind and the m_{th} order), while the evanescent radial field outside the disk is described by the Hankel function (H_m of the first kind and the m_{th} order). WGMs can then be characterized by (n, m) pairs where n and m are the radial and azimuthal mode numbers that describe the spatial field distribution in the radial direction and the azimuthal direction. While there are no analytical solutions to determine the eigenvalues of the mode, finite difference time domain method can be used to numerically solve the equations and study the WGM emission patterns. In a microdisk cavity, high-radial order modes generally have low quality factors and are easily lost via the underlying pedestal. Since gain material gives more TE gain, only the transverse electric (TE)-polarized modes with the first radial mode number will be considered. A computed 1st order WGM mode profile of the 1-μm diameter microdisk discussed above is shown in Fig. 7. Only one maximum in the radial direction is observed. Here, The $TE_{1,5}$ mode has been identified with a resonant wavelength at 1.194 μm, which matches well with the measured lasing wavelength at 1.197 μm from the spectrum in Fig. 5C.

3.2.5 Free spectral range (FSR)

The azimuthal solution discussed above also offers a method to calculate the FSR of the WGM modes. To form a standing wave in the azimuthal direction, the wave must repeat itself upon one round trip around the circle. In this regard, the resonant condition in a disk of radius R can be

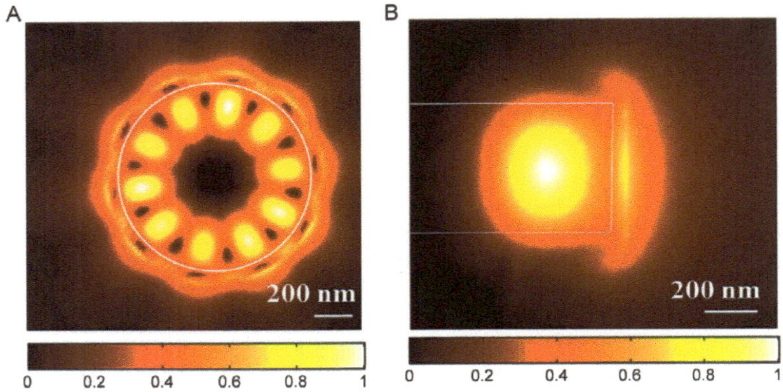

Fig. 7 (A) Top down view and (B) Side view of the 1st order WGM of a 1-μm diameter (Wan et al., 2016a).

approximated when the field equals to zero on the disk surface (Slusher et al., 1993), given by

$$2\pi R n_{eff} = m\lambda$$

The resonant condition is achieved when there is an integer number m of waves with wavelength of λ/n_{eff} along the circumference of a waveguide with a cavity length of $2\pi R$. This can be used to derive FSR as a function of the circumference of the cavity, as

$$FSR = \frac{\lambda^2}{2\pi R n_{eff}}$$

FSR increases with the decrease of R. This value can reach several tens of nanometers in micron-sized resonators. This is important for low-threshold lasers as reduced number of modes increases the spontaneous emission factor. Besides the cavity size, the FSR also decreases with the decrease of the mode wavelength, and an increase of n_{eff}. The former implies more internal bounces within the cavity, and the latter is mostly influenced by the variation of temperature and pumping power for a given material and structure. In Fig. 8A and B, emission spectra from a 4-μm diameter microdisk were measured at both 10 and 300 K (Wan et al., 2016c). To denote the optical transition positions in the gain medium, emission below threshold (black dots) was magnified 100 × and plotted together with that well above the threshold (red lines). At 10 K, four lasing peaks were observed. The sub-peak (1188 nm) close to the main peak (1191 nm) comes from

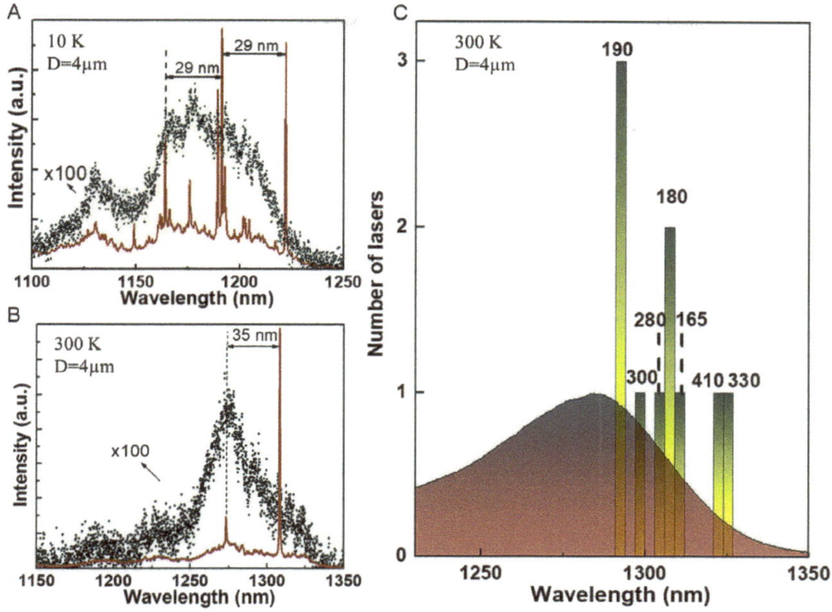

Fig. 8 Laser emission spectra measured from a 4-μm diameter microdisk grown on Si at below (black, 8.25 μW) and above (red, 495 μW) threshold in (A) 10 K and (B) 300 K. Spectrum taken at 8.25 μW was amplified 100 times to be visible; (C) Histogram of the lasing wavelength from a series of 4-μm diameter microdisks on Si. The average lasing threshold of the devices within each histogram are listed at the top of the bar. The normalized photoluminescence spectrum of the as-grown sample is denoted in red (Wan et al., 2016c).

out-of-plane polarization while the other three well-separated modes (1162, 1191, 1220 nm) are WGMs with different azimuthal orders. At 10 K, FSR is extracted to be around 29 nm, corresponding to a n_{eff} of 3.95. At room temperature, FSR increases to 35 nm, corresponding to a lower n_{eff} of 3.84. The large FSR (compared to the FWHM of the gain media) results in well-separated neighboring resonances that favor single mode operation, which is the case at room temperature in Fig. 8B. At lower temperature in Fig. 8A, WGMs of various radial orders form their own sequences of resonances, which may overlap and complicate the optical spectrum with simultaneous emission of multiple modes. Let's consider the modes as photon states: If the modes have high Q, then once a photon occupies a particular mode, it has a reasonably long life-time in that state. The gain spectrum emitted by QDs has a relatively broad luminescence peak because of the variation in size of the dots. Suppose the gain material emits a photon at the exact energy where there is a mode: the higher the Q and the longer the lifetime of the mode, the more likely the photons will be seen at the mode energy when optically interrogating the microdisk. The role of the cavity (and its modes) is thus to selectively filter out photons of a certain energy. That is an important first step to lasing: using the cavity to selectively filter out a certain optical transition, and avoiding pumping energy into all other (ultimately non-lasing) transitions. In both cases of high injection and low temperature, there is more efficient production of photons which can interact with the higher order modes, but because these modes have very low Q, with a relatively high photon loss, the effect is still minimal on the spectrum and tends to disappear at room temperature. At room temperature, a histogram of the dominant lasing mode for a series of 4-μm diameter microdisks on Si is presented in Fig. 8C. The histograms are plotted together with a normalized photoluminescence spectrum from the as-grown material. The broad photoluminescence spectrum corresponds to optical emissions from the entire range of QDs, and the dominant modes denote optical transitions that are most efficiently coupled to the cavity, due to advantages in spatial overlap with a mode, radiative emission rate, or resonance frequency.

3.2.6 Characteristic temperature

Two figures of merit are primarily used to describe the laser performance at elevated temperatures: the maximum temperature, and the characteristic temperature, T_0. T_0 measures how much a laser's threshold power (P_{th}) or threshold current (I_{th}), changes with temperature, and is defined by $P_{th} = P_0 e^{T/T_0}$ and $I_{th} = I_0 e^{T/T_0}$ for an optically pumped device and an electrically injected device, respectively. While temperature invariant operation ($T_0 = \infty$) in a range of 5–70 °C (Fathpour et al., 2004) and maximum operating temperature up to 220 °C (Kageyama et al., 2011) have been demonstrated for QD lasers with "macroscale" resonators on native substrates, the high aspect ratio of sidewall/active region volume deteriorates the temperature performance of microcavity lasers to some extent. Despite the fact that thermal conductance and injection pumping are difficult to be achieved in suspended disks, it is in this configuration that QD WGMs were first observed (Gayral et al., 1999). Among the limited research carried out on the temperature characteristics of InAs QD microdisk lasers on a native GaAs substrate, Ide et al. reported CW pumped micordisk lasers with a T_0 of 64 K in the temperature range of 130–230 K and of 36 K above 230 K (Ide et al., 2005), while Yang et al. demonstrated a T_0 of 31 K above room temperature under pulsed excitation (Yang et al., 2007). A record-high T_0 of 105 K was reported in a 4-μm diameter microdisk epitaxially grown on Si through continuous optical pumping (Wan et al., 2016c). This was achieved by optimizing the lateral undercut for high light confinement within the disk while maintaining sufficient heat sinking via the underlying pedestal. Normalized lasing spectra of this microdisk laser were compared from 10 to 300 K in Fig. 9A, together with the corresponding L-L curves in Fig. 9B. In the spectra, narrow lines of high-Q WGMs shift to longer wavelengths with the increase of temperature, with a small temperature coefficient of less than 0.04 nm/K from 10 to 80 K, and from 220 to 280 K (Fig. 9C). The lasing line is also determined by the temperature shrinkage of an active region bandgap. Mode hopping toward longer wavelengths at 80 and 280 K was observed due to thermal redshifting of the gain spectrum. The overall trend follows the theoretical shrinkage line of InAs band-gap, given below (Heitz et al., 1999).

$$\Delta E = A \cdot \frac{T^2}{T + B}$$

assuming $A = 0.00042\,eV/K^2$ and $B = 199\,K$. In Fig. 9D, the threshold pump power increases by a factor of ~2 as temperature increases from 10 to 300 K, giving a T_0 of 105 K.

Ring resonators provide much improved thermal conductivity and are better suited to electrical injection than microdisks. A similar structure using ring geometry achieved a record-high lasing temperature of 107 °C and a T_0 of 77 K under continuous optical pumping with a 6-μm outer diameter and a 2-μm inner diameter (Kryzhanovskaya et al., 2014). To achieve electrical injection in a QD microlaser, a microdisk was the first configuration where WGMs were observed at a temperature of 5 K, with a complicated air bridge structure (Zhang and Hu, 2003). However, the highest temperature to date was 300 K, achieved by burying the disk in benzocyclobutene (BCB) cladding to alleviate the thermal and fragility problem (Mao et al., 2011). On the contrary, microring QD lasers show much superior temperature performance, with the highest CW temperature up to 100 °C, and T_0 of 197 K (Fig. 10) (Wan et al., 2017a).

In addition, the temperature characteristics have been improved to a great deal, attributed to special techniques such as tunnel injection of electrons and modulation *p*-doping of the active region (Bhattacharya and Mi, 2007). In tunnel injection, cold carriers (electrons) are tunneling injected into the QD lasing states so that other carriers are not heated. This leads to an enhanced T_0 by minimizing carrier occupation in the states from wetting layer and barrier (Bhattacharya et al., 2003). *P*-modulation doping of the QDs is more generally utilized to reduce the sensitivity to temperature.

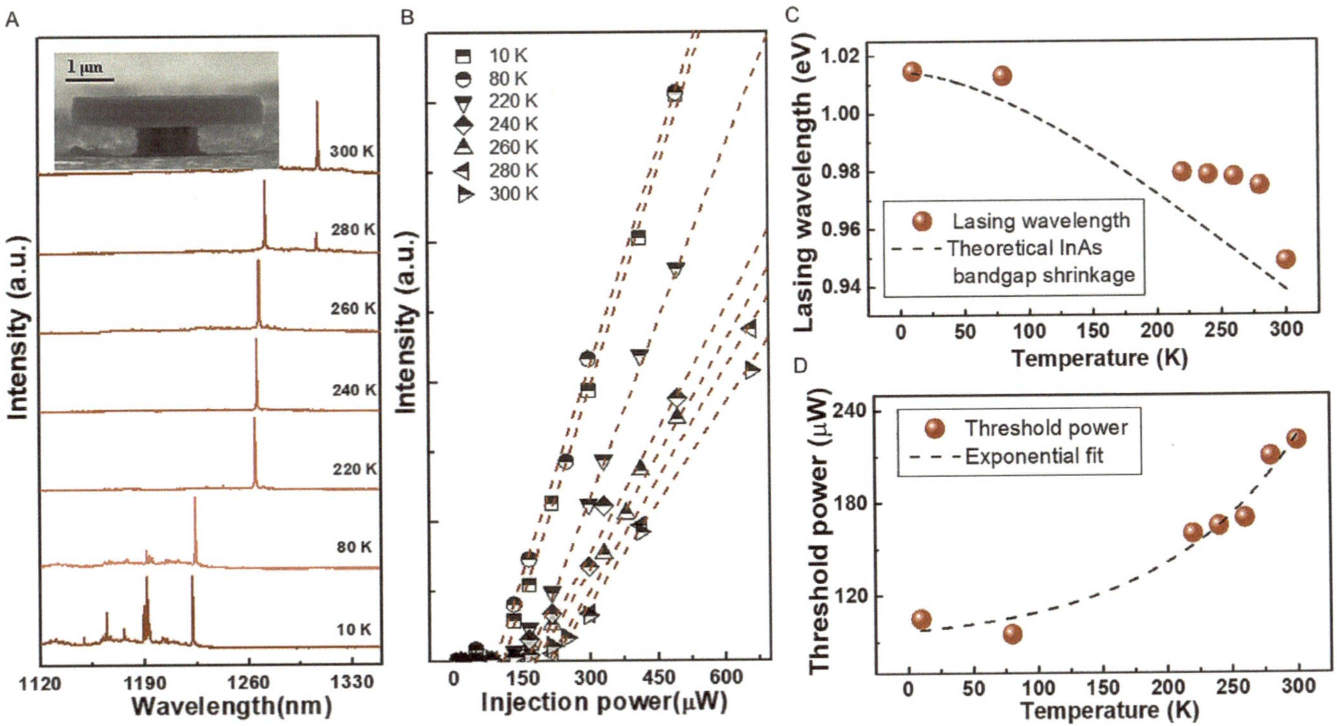

Fig. 9 Temperature dependent characteristics of a 4-μm diameter microdisk epitaxially grown on Si: (A) normalized lasing spectra from 10 to 300 K taken at three times the threshold. Inset: SEM image of the disk; (B) L-L curves of the lasing peak from 10 to 300 K. (C) Lasing wavelength versus temperature and theoretical InAs band-gap shrinkage; (D) Threshold power versus temperature (Wan et al., 2016c).

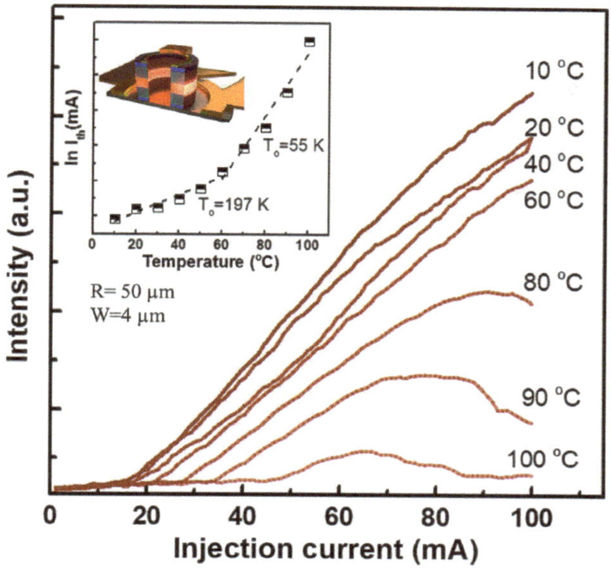

Fig. 10 Measured temperature dependent L-I curves of an electrically injected microring laser with a 50-μm radius of and a 4-μm width. Inset: threshold current versus temperature (Wan et al., 2017a).

Typically, the separation between the ground state and first excited state band is up to 80 meV in the conduction, but is only around 10 meV in the valence band (Kageyama et al., 2011). At room temperature, while the electrons can be well confined in the dots, the holes can easily thermalize and escape, as the valence band offset is well below kT. This effect of thermalization can be suppressed by adding extra holes via *p*-type modulation doping, leading to increased temperature stability. As a direct comparison, two microring lasers were tested at various heatsink temperatures and their *L-I* characteristics were analyzed in Fig. 11A and B, respectively (Wan et al., 2018a). For the laser in Fig. 11B, the GaAs barriers separating the QDs was modulation *p*-doped using beryllium with a hole concentration of $5 \times 10^{17}\,cm^{-3}$, otherwise the two structures are nominally the same. In Fig. 11A, CW lasing was sustained up to 40 °C for the undoped laser with a $T_0 \sim 22$ K. On the contrary, for the same laser structure grown on a separate wafer but with modulation *p*-doped active regions, CW lasing temperature was elevated up to 80 °C, with a T_0 of ~103 K near room temperature (20–40 °C) (Fig. 11B).

3.2.7 3dB bandwidth

The two techniques described above to improve temperature characteristics also increased the differential gain, leading to an enhanced modulation

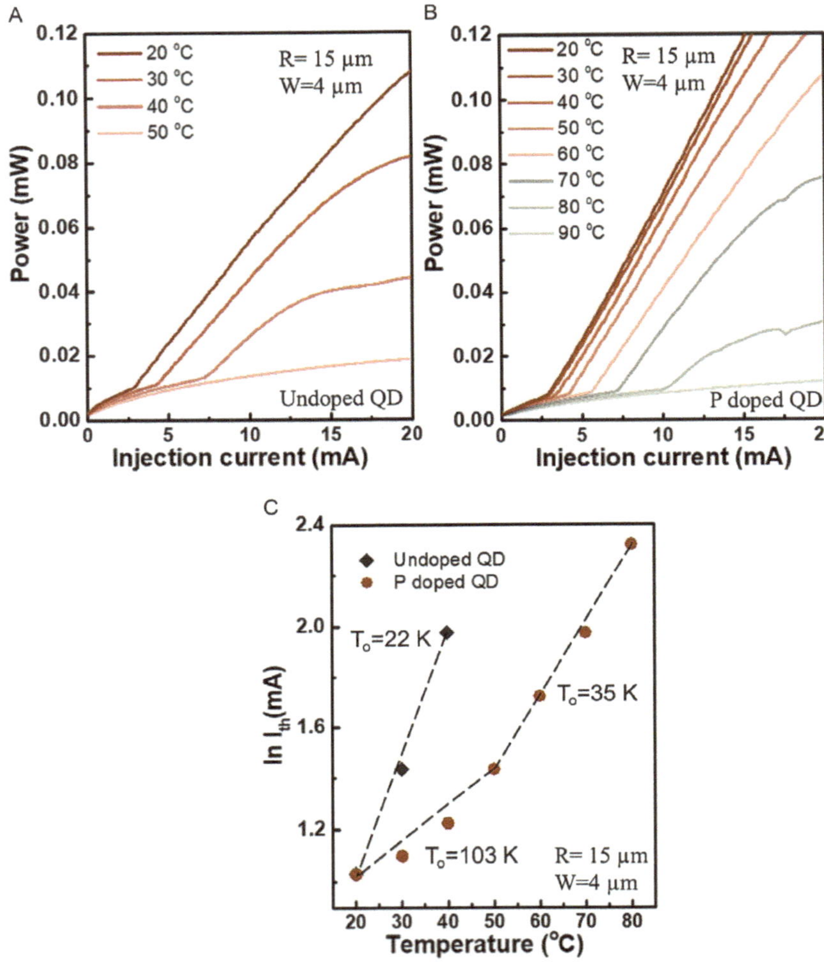

Fig. 11 Temperature characteristics of two QD microring lasers grown on Si with an outer ring radius of 15 μm and ring waveguide width of 4 μm: L-I curves with varied heatsink temperature for (A) QDs with an intrinsic active region and (B) QDs with a modulation *p*-doped active region. (C) Threshold current versus heatsink temperature (Wan et al., 2018a).

bandwidth. The small signal modulation bandwidth describes the laser frequency response and is proportional to the square root of the differential gain. The small-signal response, S_{21} of ring lasers can be measured by directly probing the device using a signal/ground (SG) RF probe. The experimental set up is schematically shown in Fig. 12A.

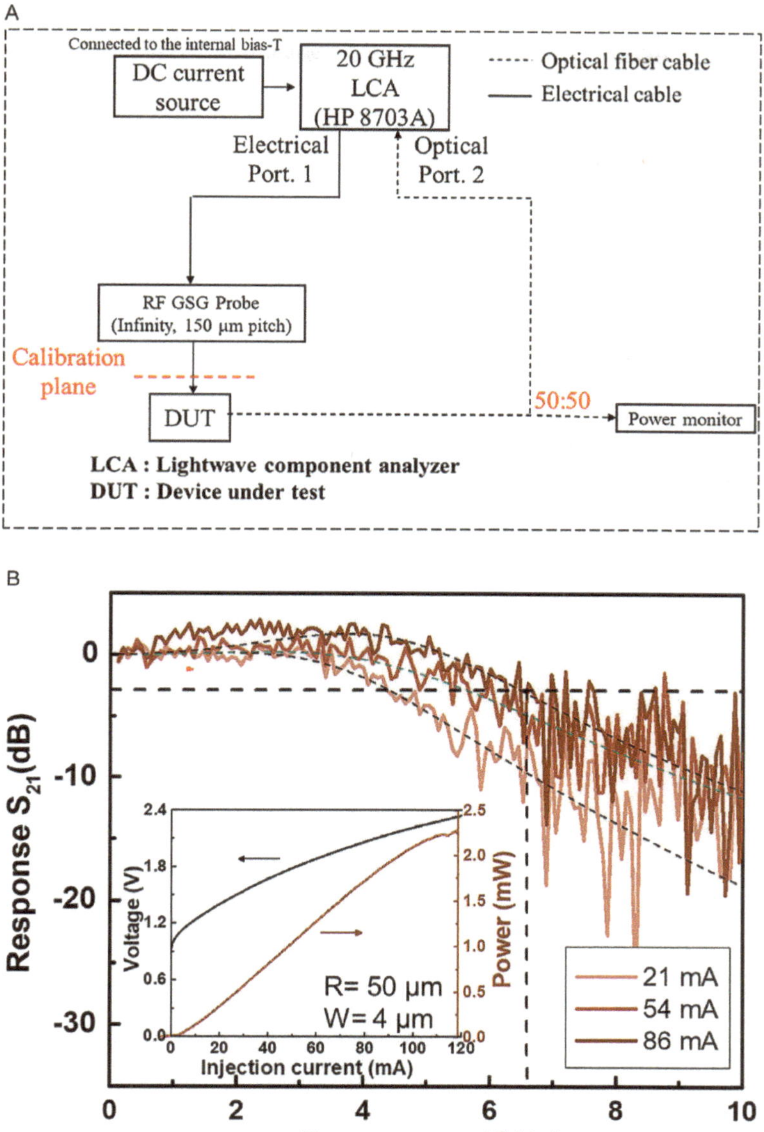

Fig. 12 (A) Measurement setup for small-signal modulation. (B) Small-signal modulation responses of the QD ring laser with an outer ring radius of 50 μm and ring waveguide width of 4 μm. Inset: *L-I-V* characteristics (Wan et al., 2018a).

Fig. 12B shows the measured S_{21} of a microring laser with an outer ring radius of 50 μm and ring waveguide width of 4 μm, under three injection currents (21, 54, and 86 mA). A 3dB bandwidth of 6.5 GHz can be attained at a bias current of 86 mA. Generally, a three-pole fitting function H(*f*) is used to fit the S_{21} curves to extract the damping rate γ and relaxation oscillation frequency f_r (Nagarajan et al., 1992).

$$H(f) = \frac{1}{\left(1 + \left(2\pi f \tau_p\right)^2\right)} \frac{f_r^4}{\left(f_r^2 - f\right)^2 + \left(\frac{\gamma f}{2\pi}\right)^2}$$

where τ_p is the RC or carrier transport delay. The extracted f_r, together with the measured f_{3dB} can be used to extract the modulation efficiencies by plotting the resonance frequency versus $(I_b - I_{th})^{1/2}$. In Fig. 12B, modulation efficiencies of 0.38 GHz/mA$^{1/2}$ for f_{3dB} and 0.34 GHz/mA$^{1/2}$ for f_r were extracted by linear fitting, using the data points below $(I_b - I_{th})^{1/2} = 8$ mA$^{1/2}$ (Wan et al., 2018a). An analysis of the small signal response shows that the highest modulation rate attainable at the transition of a laser diode to conditions of the pronounced damping of relaxation oscillations is defined in terms of the so-called *K* factor (Spencer et al., 1997).

$$f_{3dB,\max} = 2\sqrt{2}\pi / K$$

Theoretically, the K factor is the summation of contributions from the photon lifetime (τ_{ph}) and gain compression (ξ), and can be represented as a function of optical loss (α) (Zhukov et al., 2012).

$$K = \begin{cases} K_{ph} = 4\pi^2 \tau_{ph} \\ + \\ K_\xi = 4\pi^2 \frac{\xi}{v_{gr} g} \end{cases} = \frac{4\pi^2}{v_{gr}} \left[\frac{1}{\alpha} + \frac{\xi}{g_{\max}\left(1 - \frac{\alpha}{G_{sat}}\right)} \right]$$

where v_{gr} is the group velocity of photons, $g_{\max}$ is the largest value of differential gain (at zero losses), G_{sat} is the saturated gain. Due to the unequal carrier scattering rates of electrons and holes within the QD lasers, the gain compression in QDs is $>10\times$ higher than that in QW lasers. This results in strong damping of relaxation oscillations, a high K factor value, and a limited maximum (small signal) direct modulation bandwidth of ~10–15 GHz, as is illustrated in Fig. 13.

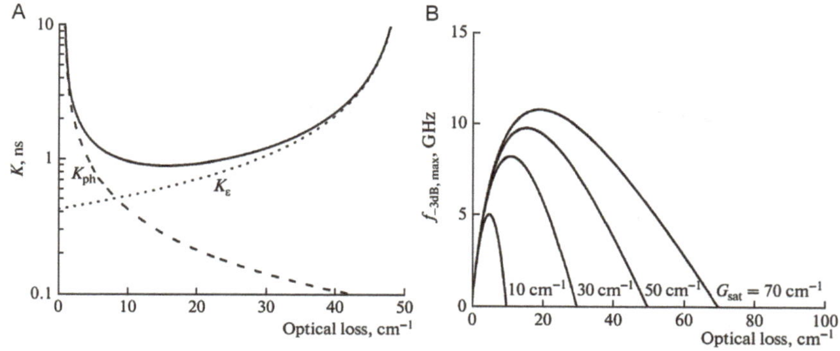

Fig. 13 (A) Dependence of the K factor (the solid line) and its components caused by the lifetime of photons (dashed line) and nonlinear gain saturation (dotted line) on losses for $Gsat = 50\,\text{cm}^{-1}$ and (B) the dependence of the highest modulation frequency on losses for various values of the saturated gain. The values $\varepsilon/g_{max} = 0.1\,\text{cm}$ and $v_{gr} = 0.9 \times 10^{10}\,\text{cm/s}$ were used in the calculations (Zhukov et al., 2012).

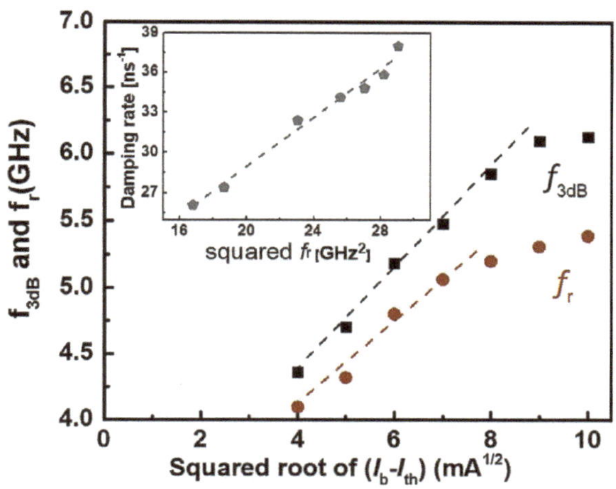

Fig. 14 3dB bandwidth f_{3dB} and relaxation oscillation frequency f_r versus square-root of the bias current above threshold. Inset: Damping rate γ versus squared relaxation oscillation frequency f_r (Wan et al., 2018a).

Experimentally, the K-factor can be derived by linear fitting the damping rate γ as a function of squared f_r using the following equation (Inoue et al., 2018a)

$$\gamma = K \cdot f_r^2 + \gamma_0$$

where γ_0 represents damping offset. In Fig. 14, the K-factor and a maximum K-factor limited $f_{3dB,\max}$ was calculated to be 0.91 ns and 9.7 GHz, respectively.

The modulation bandwidth is at the same time affected by the device design and extrinsic electrical properties. To analyze the discrepancy between the theoretical $f_{3dB,\ max}$ and the measured f_{3dB}, reflection coefficient S_{11} is generally used to extract the impedance and RC cutoff frequency. A general equivalent circuit model (Bowers et al., 1986) consists of an inductance L, a total capacitance C, a device resistance R, a voltage source V_s, and a characteristic impedance of the transmission line (50 Ohm) Z_0 (Fig. 15). The inductance L can be neglected if the electrical contact was performed directly by a RF probe without any wire bonding. The pad capacitance was calculated to be 0.33 pF by approximating the *p*-probe pad area with 80-μm width and 120-μm length over a 1-μm-thick SiO_2 layer (RF dielectric constant $\varepsilon = 3.9$). The total capacitance was extracted to be 0.74 pF from the 50-Ohm normalized Smith chart. RC cutoff frequency is calculated to be 10.3 GHz using the circuit parameters ($C = 0.74$ pF, $R = 20.9$ Ohm). The extrinsic electrical properties can be improved by reducing the pad capacitance to tens of fF using thick dielectric passivation and dielectric embedded channel ridge structure (Kobayashi et al., 2013).

While 25 Gbit/s direct modulation has been reported for high-mesa waveguide *F-P* lasers on native substrates (Tanaka et al., 2010), 12.5 Gbit/s NRZ operation and a 25 Gbit/s PAM-4 operation have been reported for a similar "macroscale" resonator epitaxially grown on a Si substrate (Inoue et al., 2018b). For microcavity lasers, the relatively low fiber-coupling output power makes it difficult to perform an eye-diagram characterization.

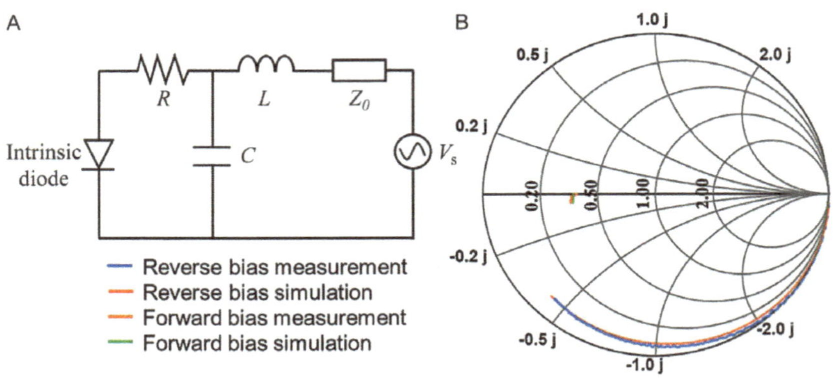

Fig. 15 (A) A general equivalent circuit model. (B) Measured and fitted curves of S_{11} under reverse (−3 V) and forward (50 mA) biased condition from 0.14 to 5 GHz (Wan et al., 2018a).

Still, with the design of an asymmetric 2-tap equalizer (Fan et al., 2019), a recent report from Hewlett Packard Labs achieved 12 Gbit/s operation from a hybrid-integrated directly-modulated QD microring laser.

4. Designs of WGM resonators based on III-V materials

4.1 Optically pumped microdisks

A typical cross-sectional transmission electron microscope (TEM) image of a microdisk laser structure is shown in Fig. 16A. In the disk, the principal emission direction is radially in-plane. To avoid mode leakage into the substrate, the gap between the microdisk and substrate should be at least half of a wavelength in air, which is formed from a 600-nm $Al_{0.7}Ga_{0.3}As$ sacrificial layer. The thickness of the disk was chosen so that it's thick enough to incorporate an adequate number of QD layers to achieve enough gain and thin enough to be below the cutoff thickness of the second order planar waveguide mode. A five-layer InAs/InGaAs dot-in-a-well (DWELL) structure was adopted as active region, giving a total thickness of 500 nm (Wan et al., 2016a). At this thickness, the gain for the modes with polarization perpendicular to the disk is highly suppressed, so that only modes with polarization parallel to the plane of the disk will lase. To acquire high density of uniform QDs (typical density $\sim 6.5 \times 10^{10}\,cm^{-2}$), the Stranski-Krastanov growth mode is the most common approach where self-organized islands

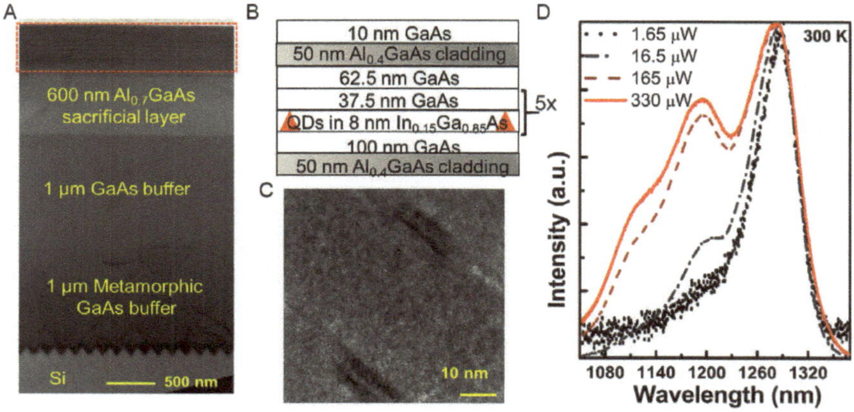

Fig. 16 (A) Cross-sectional TEM of a microdisk grown on Si; (B) Schematic of the DWELL; (C) High-resolution TEM image of the QD; (D) QD photoluminescence spectra at different pump power (Wan et al., 2016a).

are formed after a few monolayers of layer-by-layer growth (Wu and Jin, 2015). The DWELL structure (Fig. 16B), in which the dots are inserted in a QW so that the well captures a high density of carriers and keeps them localized around the dots, is widely used to promote effective capture of electrons and holes in the relatively small and spatially separated QDs (Stintz et al., 2000). For typical growth parameters used in molecular beam epitaxial (MBE) or metal–organic vapor phase epitaxy (MOVPE), the dots have lateral size of 10–40 nm and heights of 5–8 nm. An example of a TEM image of a single dot is shown in Fig. 16C, where the dot has a size of ~21 nm in diameter and ~6 nm in height. A useful feature of the QD size distribution is a reduction in the photoluminescence spectral linewidth. In the PL characterization from the as-grown QDs in Fig. 16D, the PL FWHM is as narrow as 28 meV centered at the 1.3 μm telecommunication band. Similar results have been reported by other groups, with a linewidth of 28 meV (at 300 K) in Guimard et al. (2010), a linewidth of 29 meV (300 K) in Chen et al. (2016), and a record-low linewidth of 17.5 meV (10 K) in Bhattacharya and Mi (2007), providing evidence of high-quality defect-free QDs, which are needed for the fabrication of devices.

Fig. 17A schematically shows one process for the fabrication of a QD microdisk resonator. A dilute suspension of SiO_2 spheres in isopropyl alcohol (IPA) was first dispersed onto the sample surface by pipette and quickly dried on a hot plate at 110 °C. The diameters of the SiO_2 spheres are varied from 1 to 5 μm within ±5% precision. Using the isolated sphere-masks, the cylinder mesa was etched to a depth of ~1.1 μm into the full structure using Inductively coupled plasma etching. It's essential to acquire a clean and smooth circumference sidewall surface, a vertical profile, and a low degree of plasma-induced damage. In refining the etching, dominant parameters include the gas ratio, RF power, temperature, pressure, and selection of the etching mask. Among the different plasma chemistries used to etch GaAs-based compound semiconductor films, the preference for chlorine-based plasmas to etch Ga containing III-V materials is due to the high volatility of the Ga and group-V chlorides. Mixed BCl_3 and Ar gases were used with flow rates of 20 and 6.5 sccm, respectively. The physical nature of BCl_3 plasma helps to achieve a moderate etch rate due to the generation of less reactive Cl compared to Cl_2, while Ar was added as an etching buffer which performs physical anisotropic etching to carry away the etching by-product as well as to keep the cavity plasma stable. Substrate temperature controls the volatility of the etching species, and was adjusted to be 10 °C. Low pressure causes the plasma to diffuse from the generation region and drift to the

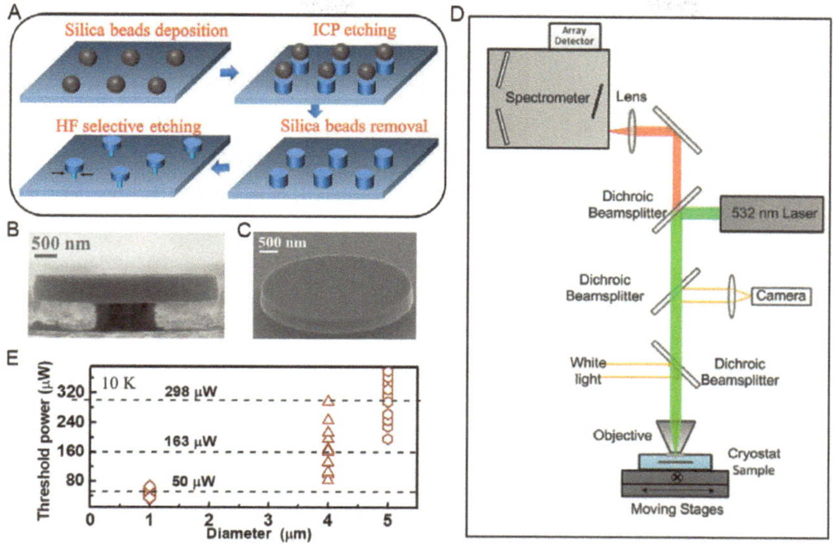

Fig. 17 (A) Schematic illustration of the fabrication procedure; (B) and (C) SEM images of a 4-μm fabricated micrdisk; (D) Measurement set up; (E) Threshold power versus disk diameter (Wan et al., 2016a).

substrate at relatively low ion energy and is therefore maintained at 5 mTorr. Coil and platen power are set as 500 and 100 W, respectively, to obtain anisotropic profiles. After the cylinder mesa formation, the remaining SiO_2 microspheres were removed using acetone in an ultrasonic bath, and the underlying pedestal was defined by a 5% HF solution, which has an extremely high selectivity in etching aluminum-rich AlGaAs in preference to the GaAs in the disk area. By controlling the time of wet etching, the lateral undercut of the disk from the outer periphery was optimized to maintain effective light confinement within the disk, as well as low mode leakage into the substrate. A fabricated SEM images of a 4-μm microdisk can be seen in Fig. 17B and C, revealing a smooth sidewall, and a circularity of the disk cavity. Devices can be measured through a micro-photoluminescence (μPL) system in a surface-normal pump/collection configuration (Fig. 17D). Excitation from a 532-nm laser spot can be focused to approximately the size of the disk by a high-NA microscope objective, and the pumping intensity can be varied using a polarizer and monitored using a power meter. The collected signal can be dispersed by a monochromator and recorded by a liquid-nitrogen-cooled InGaAs detector array. The sample stage contains a helium gas flow cryostat, which can control the temperature from 10 to 300 K. Fig. 17E summarizes the threshold powers obtained from a series of microdisk lasers on Si with different diameters. An apparent monotonic decrease of threshold power can be seen as the disk diameter is scaled from 5 to 1 μm (Wan et al., 2016a). A similar result of microdisk lasers grown on native substrate indicated that the threshold power density remains nearly constant when decreasing the diameter of the disk down to 1.7 μm (Kryzhanovskaya et al., 2015a).

4.2 Electrically injected microrings

While optical pumping is a critical step for demonstrating lasing and is a useful method for studying cavity and material properties, electrical injection is a necessity for commercial applications, especially for integrated photonics. The main fabrication challenge associated with electrically injected microcavities is the requirement for metal contacts, which, if in direct contact with the cavity, represent a major source of optical loss. To address this issue, it's important to acquire small metal contacts that maintain a low resistance and thermal impedance while not significantly perturbing the optical cavity. Contrary to the case of micropillars where vertical confinement requires complicated distributed Bragg reflectors (Feng et al., 2019), or the case of suspended microdisks where an airbridge contact contains a "double-disk" structure that separates the contact layer and the active layer (Zhang and Hu, 2003), the epitaxial structure of electrically injected QD microrings presented in the following sections do not differ from a typical design of $GaAs/Al_xGa_{1-x}As$ graded-index separate-confinement heterostructure lasers.

Fig. 18 shows the critical processing steps in fabricating electrically injected microrings (Wan et al., 2017a). Using oxide hard mask, the ring structures were patterned and etched all the way to the *n*-contact layer. The sidewall of the ring was covered with 12-nm conformal Al_2O_3 by atomic-layer deposition, which provides excellent surface passivation

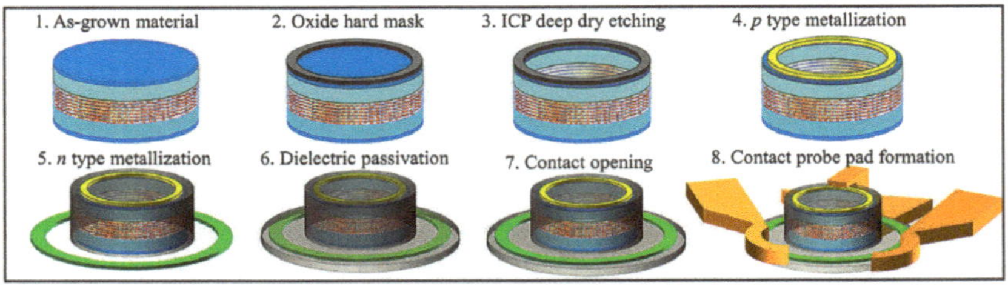

Fig. 18 Electrically injected microrings: typical process flow (Wan et al., 2017a).

(Zhou et al., 2018), widely used to improve photoluminescence efficiency (Black et al., 2017) and reduce scattering loss (Ikku et al., 2012). *p*- and *n*-contacts were formed using Pd/Ti/Pd/Au and Pd/Ge/Au layer stacks, respectively. A 1 μm-thick SiO_2 layer was sputtered to further passivate the sidewall and to isolate the optical modes from the metal contacts. After via opening, *n*-contact pads were displaced laterally around the ring mesa and *p*-contact were arranged on top of the ring mesa for efficient current injection and spreading.

4.2.1 Typical characteristics

Typical L-I-V characteristics and spectra of a QD microring laser under CW injetion are shown in Fig. 19. The cavity has a radius of 15 μm and a mesa width of 4 μm. The lasing threshold is around 3 mA. The primary lasing mode centered at ∼1.3 μm, with an extinction ratio over 30 dB and a side-mode suppression ratio of 16 dB in the injection range of 20–25 mA (Wan et al., 2018a). Despite the high ratio of sidewall to active region volume of the ring structure, a record low threshold current density of 306 A/cm^2 has been reported with a radius of 80 μm and a mesa width of 4 μm (Wan et al., 2018b). The optical power of the microring is generally

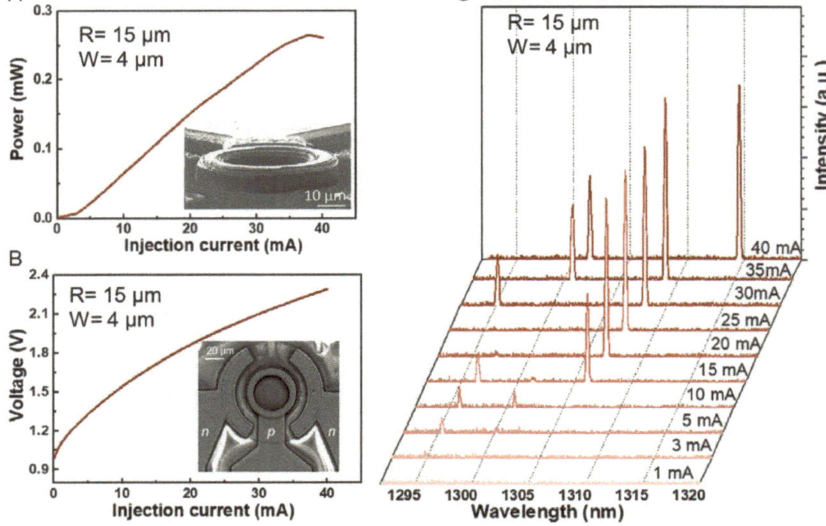

Fig. 19 (A) L-I, (B) I-V characteristics, and (C) emission spectra for a microring laser with a 15-μm radius and a 4-μm ring width (Wan et al., 2018a).

collected by capturing the radiation out-coupling at one side. While the angular directivity pattern of radiation brings about only a fraction of total power that can be measured, no output saturation was observed when the output power reached 1.2 mW at an injection current of 100 mA with a radius of 50 μm and a mesa width of 4 μm (Wan et al., 2017a). This power level is large enough for optical interconnects where sensitive photodetectors exist. With the same size QD microring laser, a maximum 3 dB bandwidth of 6.5 GHz has been reported with a modulation efficiencies of 0.38 GHz/mA$^{1/2}$ (Wan et al., 2018a). A maximum CW temperature of 100 °C has been achieved, with T_0 of 197 K between 10 °C and 50 °C and of 55 K between 60 °C and 100 °C (Wan et al., 2017a). Probably the most important advantage of using epitaxially grown QDs over heterogeneous QWs in a microcavity laser is their high temperature stability. Si is one of the most thermally conductive semiconductors, which is beneficial for dissipating heating. However, in the hybrid Si platform, a 7.5-μm diameter microdisk was produced from an InP-based QW epitaxial structure and then molecularly bonded to the SOI waveguide wafer. Fast thermal rollover is obvious even at a small injection current of 1.5 mA (Van Campenhout et al., 2007). An InP-based QW microring that sits on a silicon microdisk with the same diameter was reported through a low temperature O_2 plasma-assisted wafer bonding process (Liang et al., 2009). While the hybrid ring has a similar size (a radius of 50 μm and a mesa width of 4 μm) as the epitaxial QD microring described above (Wan et al., 2017a), lasing can only sustain up to 65 °C, and CW L-I measurement typically stop at 25 mA in order to avoid excessive device heating. The primary cause of serious device heating and inefficient heat dissipation in QW hybrid microcavity lasers is due to the thick buried oxide layer (typically 1–3 μm, with a 100 times smaller thermal conductivity compared to Si), and the inferior thermal tolerance of QWs, as compared to their QD counterparts.

Another distinguishing characteristic of QDs used by microcavity lasers is the reduced sensitivity to recombination at device sidewalls that enables ultra-dense photonic integration. Fig. 20 plot the threshold currents of the microring lasers as a function of the outer-ring radii. The threshold current decreases monotonically with the reduction of the ring radius. For the smallest dimension with a radius of 5 μm and a ring width of 3 μm, an ultra-low threshold of ~0.5 mA is obtained with single mode operation centered at ~1.29 μm.

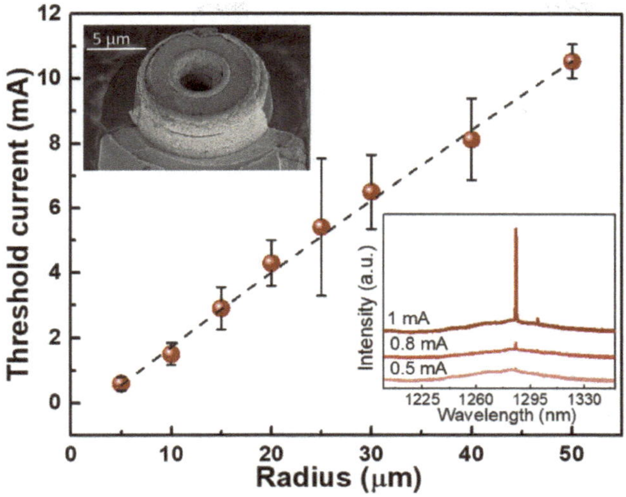

Fig. 20 Threshold currents of microring lasers with different outer-ring radii and a constant ring waveguide width of 4 µm. Inset: Emission spectra of the sub-milliamp threshold microring laser (Wan et al., 2017a).

5. Future outlook

The field of epitaxially integrated QD microcavity lasers is evolving quickly. However, compared to conventional linear cavity lasers, the total output power of the ring cavity is relatively low. Much research is needed to achieve single mode operation, wavelength-division multiplexing (WDM) sources, and other emission wavelengths. A systematic study of reliability and reproducibility at the device level is required to warrant further large-scale on-chip integration. In terms of integration via epitaxy, majority of research still focus on individual devices, particularly lasers. The performance of amplifiers, photodetectors, and modulators using the QD materials need to be demonstrated. Low-loss active-passive coupling schemes need to be developed to integrate these QD-based components with waveguides, preferably, in Si photonics. With that achieved, one can envisage the leverage of photonics for post-Moore performance scaling of electronic systems, and the intimate integration of photonics with electronics into a package. In the sections below, some possible avenues of future research to further advance the field will be outlined.

5.1 Increasing the output power

Generally, the output power of microcavity lasers is isotropic and can't be extracted efficiently due to their rotational symmetry. These challenges are less acute for applications requiring low output powers. For example, in Stock et al. (2013), on-chip quantum optics experiments can be conducted using an electrically pumped micropost WGM microlaser to optically excite the nearby microresonators, despite a low lateral free-space outcoupling efficiency. In Wan et al. (2017d), proof-of-principle demonstrations of on-chip data transmission has been realized from a microring laser to an on-chip photodetector sharing the same InAs/GaAs QD active region. In Munnelly et al. (2017), tunable emission energy is achieved on a QD micropillar-based single-photon source triggered by a monolithically integrated on-chip QD WGM microlaser. However, for more efficient data transmission, coupling to external waveguides with directional couplers or multimode interferometers (MMI) is necessary. Advanced device implementation methods need to be utilized, such as vertical coupling into a silicon-on-insulator waveguide via bonding (Van Campenhout et al., 2007), deformed microcavities (Harayama and Shinohara, 2011), lateral evanescent coupling to a straight waveguide (Lv et al., 2013), waveguide-coupled microspirals (Yang et al., 2014), and nano-antennas (Moiseev et al., 2017).

5.2 Single mode operation

In addition to output power, the resonant wavelength of a microcavity is sensitive to ambient temperature changes and fabrication variations. While shrinking the cavity size to a microscale volume helps to get single mode operation as the resonant frequencies are spaced by more than the gain bandwidth. Mode competition can be complicated at increased temperature or pumping, with simultaneous emission from multiple modes, as described previously. Actively tuning and locking the ring resonances to the desired wavelengths is important, and can be acquired with harnessing notions from parity-time symmetry (Hodaei et al., 2014), integrated second order gratings on the ring waveguide (Arbabi et al., 2015), and light confinement in subwavelength structures using metallic cavities (Feng et al., 2014).

5.3 WDM source

Microring cavities can also be very versatile because they can both serve as on-chip light sources with active regions based on QDs, and as wavelength-selective modulators of the QD laser output based on passive silicon microrings. It's widely recognized that WDM may be the only practical choice for very short reach inter-processor links in the future, and a single broad-spectrum light emitter replacing a group of DFB lasers appears to be the most competitive solution. The ultrafast gain recovery and reduced feedback sensitivity of QD-based gain media promise a lower jitter than expected compared to their QW counterparts (Rafailov et al., 2007). Broad, flat gain spectrum can be obtained via the inhomogeneous broadening from different size of QDs, and the deliberate broadening due to the use of several layers of QDs differing in center wavelengths. These features of QDs make them naturally suited for broadband emission sources in WDM applications. One example is QD mode lock lasers, where 4.1 Tb/s transmission using a single mode lock laser grown on Si was demonstrated (Liu et al., 2019). Alternatively, WDM transmitters can be built by cascading a simple QD *F-P* laser diode with microring modulator arrays, which simultaneously provide multiple functionalities including wavelength multiplexing/de-multiplexing and modulation. Experimentally, a DWDM transmitter based on a single QD comb laser and an array of microring modulators has been reported (Chen et al., 2015; Wojcik et al., 2009).

5.4 Emission at 1.55 μm

While the emission wavelengths of most of the QD lasers grown on Si have been limited to ~1.3 μm, developing InP-based QD lasers on Si with 1550 nm telecom wavelength is of immense technological interest. 1.55 μm emission from InAs quantum dashes (QDashes) grown on InP substrates can be achieved with InAs/InP material systems (Khan et al., 2014). Since the first demonstration of InAs QDash lasers grown on a InP substrate (001) (Wang et al., 2001), device characteristics have been quickly improved (Sadeev et al., 2016), with lowest threshold current density being 730 A/cm^2 (46 A/cm^2 per layer) (Hein et al., 2009). By exploiting ultra-broad gain profile of QDashes, various broadband optical components have been made (Ooi et al., 2008), such as optical amplifiers, mode-locked lasers, and superluminescent diodes (Khan et al., 2014). On native substrates, large ring lasers with millimeter-length have been reported with a threshold current density of 6.4 kA/cm^2 at near room temperature (283 K) (Hill et al., 2008), CW lasing up to 55 °C has been reported with a threshold current density of 528 A/cm^2 in a 15-μm radius ring (Wan et al., 2019). Epitaxially grown on Si substrates, only optically pumped QD microdisk lasers have been demonstrated with a threshold power of 1.6 mW (Shi et al., 2017b), and no electrical injected embodiment has been reported yet. While Shockley-Read-Hall recombination in InP related compounds is less severe than in GaAs so that even longer operating lifetimes are expected for InP-based QD lasers on Si, significant research is still needed to catch up with the state-of-the-art GaAs-based devices epitaxially grown on Si.

5.5 Other components employing ensembles of QDs

From the perspective of using QDs for improving other devices: QD solar cells offer high power conversion efficiency, where the partially occupied intermediate band of the confined QD levels results in additional sub-bandgap absorption in a single-junction cell (Popescu et al., 2008). QD modulators are shown to have enhanced electro-optic coefficients and steep absorption edges due to the quantum confined Stark effect (Akca et al., 2008). Experimentally, InAs QD cores in GaAs/AlGaAs epitaxial layers grown on Si have demonstrated 1.76 V-cm modulation efficiency, 2.2 V V_{π}, and less than 3 dB/cm propagation loss in a Mach-Zehnder modulator with 8 mm long electrodes (Bhasker et al., 2018). QD semiconductor optical amplifiers (SOAs) possess distinct superiorities compared to the QW counterparts at higher current levels due to the fast carrier dynamics and the decoupling of gain and phase dynamics. This property is manifested through comparative measurements with a commercially available QW SOA (Meuer et al., 2009), where an almost threefold increase of cross-gain modulation bandwidth is measured in QD SOAs (which can be tuned well beyond 40 GHz). The QD heterostructure can be engineered for the detection of photons in the midinfrared, far-infrared, and terahertz frequencies, with low dark current, intrinsic sensitivity to normal incidence light, and much longer lifetime of photoexcited electrons due to the reduced electron–phonon scattering (Bhattacharya et al., 2007). Sharing the same material with the QD-based lasers epitaxially grown on Si, monolithically integrated waveguide photodetectors (Fig. 21) operating in the O-band have been reported with an internal responsivity of 0.9 A/W, 10 Gbit/s eye opening with non-return-to-zero signals, and most notably, an ultra-low dark current density of 0.13 mA/cm^2 (Inoue et al., 2018c), which is over two orders of magnitude better than the state-of-art dark Ge-on-Si detectors (Michel et al., 2010). The waveguide QD photodetector can detect free-space coupled light from a nearby microring laser, both heteroepitaxially integrated on the same chip (Wan et al., 2018c). To more efficiently extract

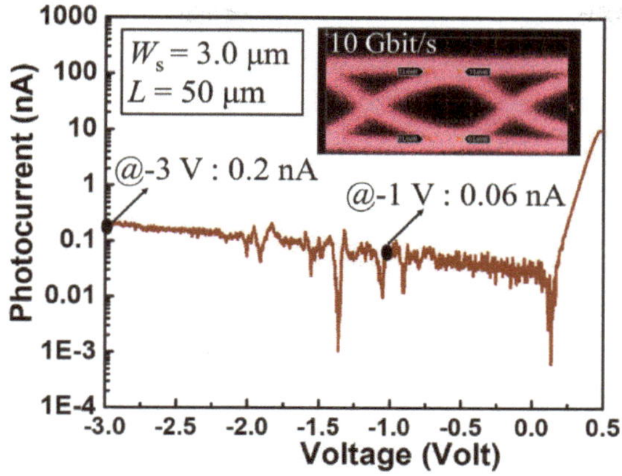

Fig. 21 Current-voltage characteristics of $3.0 \times 50\,\mu m^2$ waveguide photodetector. Inset: Measured eye diagrams for data-rates at 10 Gbit/s with bias voltage of −5 V (Inoue et al., 2018c).

the laser emission, in-plane coupling waveguide structures were designed, where the light can be evanescently coupled to the bus waveguide, and detected at the integrated photodiodes at both ends (Shang et al., 2018).

5.6 On-chip integration with epitaxial III-V on Si

Epitaxial integration can be achieved by coupling light generated in the QD section to either the Si waveguide in an SOI substrate, deposited foreign waveguides, or within the epitaxially deposited III-V layers.

5.6.1 Couple light to Si waveguide of the SOI substrate

From the perspective of co-integration of QD lasers with other Si photonic components that are defined in the top Si layer of a SOI platform: Liu et al. proposed that the optical mode can be butt-coupled to a Si rib waveguide by selective area growth of III-V cavity onto a prepatterned SOI substrate, and the optical coupling can be maximized by aligning the height of the active QD region with the Si waveguide layer (Liu et al., 2015b). In this way, the Si is used as both a substrate carrier and a low-loss wave-guiding material. However, the height alignment is challenging with a thick buffer, efficient electrical injection is also difficult as the material volume obtained is very limited. Buffer-less selective growth of III-V on Si is expected to simplify the coupling as the III-V active region is close to the underlying Si waveguide (Wang et al., 2015). Furthermore, the shape of the material growing out of the trench can be engineered with a box shape, where the lateral width of the box is on the order of a micrometer, comparable to the typical width of a "classical" III-V laser (Shi et al., 2017a). Electrical injection is more feasible as the material volume obtained is larger. Using the integration approach discussed above, QW nano-ridge lasers have been demonstrated under pulsed optical pumping, covering the emission wavelengths of 1050 nm (Shi et al., 2017a) and telecom bands (Han et al., 2019). Incorporating QDs inside the cavity can further widen the range of achievable emission wavelengths, as well as improve the thermal performance to demonstrate CW lasing. However, for both selective area growth approaches discussed above, either through butt coupling of the QD layer to a SOI waveguide, or buffer-less approach where the active layers are within reasonable coupling distances to the underlying Si, the key obstacle is electrical injection. This requires doping of both the Si and GaAs layers, which need to go through highly defective III-V/Si heterointerfaces that are of large resistance. Furthermore, selective area growth relies on MOCVD, which has lagged substantially behind MBE in QD laser performance due to the inherent temperature limitations in the growth technique. To reduce the electrical and coupling issues, the III-V layers can be either directly grown on SOI wafers, or be bonded to SOI wafers after its growth on Si. Light coupling from the QD region to the Si part in SOI wafers can be achieved by either tapers, or photonic wire bonding. In the former, emitted light from a laser at 54.6° with a low divergence angle of 1.7° have been effectively directed onto a vertical coupler in the Si substrate for waveguide integration (Zhang et al., 2018b). In the latter, insertion loss down to 0.4 dB has been obtained between the InP-based horizontal-cavity surface-emitting lasers and the passive silicon photonic circuits with a direct-write two-photon lithography (Billah et al., 2018).

5.6.2 Couple light to deposited foreign waveguides

The light coupling scheme described above, using either butt coupling (Liu et al., 2015b), inclined coupling (Zhang et al., 2018b), or photonic wire coupling (Billah et al., 2018), can be used to couple light from the QD region to foreign waveguides, like crystalline Si, amorphous Si (*a*:Si-H), and SiN. It has been demonstrated that crystalline Si membranes, with reasonably low loss of $\leq$0.5 dB/cm, can be transferred onto foreign substrates using a liftoff process (Yuan and Ma, 2006), and be readily extended for the monolithic integration of single crystalline Si waveguides with III-V lasers and modulators (Mi et al., 2009). Alternatively, *a*:Si-H waveguides can be deposited by plasma-enhanced chemical-vapor deposition with low temperature processing, tunable refractive index (De Dood et al., 2002), and relatively low losses of 2.0 and 5 dB/cm at 1.55 and 1.3 μm, respectively (Harke et al., 2005). With *a*:Si-H waveguides, on-chip integration of edge-emitting QD lasers has been outlined by evanescent coupling (Wan et al., 2017e) and has been experimentally demonstrated by butt coupling (Yang and Bhattacharya, 2008). Compared to Si, SiN has even lower propagation loss and temperature sensitivity that enables advanced wavelength multiplexing functionality through the high waveguide thickness and the low index contrast. Monolithic co-integration of QD-based InP on SiN has been proposed using interposer-based coupling scheme (Vyrsokinos et al., 2019). Using a novel Si-Rich Nitride (SRN) material with low stress and high refractive index ($n > 3.16$), light from the InP-based emitter can be transferred to the SRN intermediate waveguide and then guided to the Si_3N_4 photonic circuitry.

5.6.3 Couple light to epitaxially grown III-V waveguides

A final approach is to follow the established methods in monolithic integration, similar to photonic integration on native III-V substrate (Coldren et al., 2011), but with a major difference being that the III-V layers are grown on a Si substrate. While the relatively small index contrast provided by III-V waveguides compared to that of the SOI material system reduces optical confinement and dense integration prospects to some extent, just the application of Si as a cheap, robust, and large substrate is very advantageous, as the fabrication could benefit from a well-developed massive manufacturing ecosystem that is capable of producing tens of millions of wafers per year at the 300-mm wafer scale with near perfect yield in all fabricated components.

5.6.4 CMOS compatibility

From the perspective of future co-integration of the epitaxially grown QD lasers with Si CMOS: the III-V thermal budget is a challenge. One approach is to begin with the front-end-of-line of Si CMOS part, then cover the Si CMOS part and expose the area where III-V needs to be processed. Both the mask sets and the process window need to be well designed. Recently, there is a lot of effort from both electronic and photonic communities for III-V integration on Si. IMEC is investing heavily to III-V high mobility logic transistors on Si using aspect ratio trapping process, where the process modules are developed in a Si-III/V hybrid 300 mm R&D pilot line, compatible for future CMOS high-volume manufacturing. Intel is currently using a 300 mm heterogeneous integration Si photonics foundry line for volume production of optical transceivers that include III-V semiconductor gain regions bonded onto a single large Si wafer (Jones et al., 2019). A similar optimization has already been carried out when introducing Ge-detectors in a standard CMOS-flow, as was demonstrated in industry (Schow et al., 2007). Introducing epitaxially grown III-V on Si into the CMOS line should be of a similar exercise in engineering, and some of the lessons there can be leveraged. As the field matures and prices come down, on-chip integration with epitaxial III-V on Si may become ubiquitous and affordable for mass-production.

6. Conclusions

The field of QD microresonators is advancing rapidly and promises to complement the rise of Si photonics by populating these chips with ultralow-threshold and small-footprint light sources. As the field matures and prices reduce, future epitaxial integration of these devices into optoelectronic circuits may become ubiquitous and affordable for mass-production.

Acknowledgements

The authors thank Advanced Research Projects Agency-Energy (ARPA-E) (DE-AR0000672), American Institute for Manufacturing (AIM) Integrated Photonics, Research Grants Council of Hong Kong (RGC) (No 16212115), Innovation Technology Fund of Hong Kong (ITS/320/14), for funding this research. Important contributors to the epitaxial growth on Si research described here include Arthur C. Gossard, Roger Helkey, Evelyn L. Hu, Kei May Lau, Alan Y. Liu, Daehwan Jung, Qiang Li, Chen Shang, Daisuke Inoue, M. J. Kennedy, Di Liang, Chong Zhang, Songtao Liu, Paolo Pintus, Zeyu Zhang, Kaiyin Feng, Noelle Collins, and Mario Dumont.

References

Akca, I.B., Dana, A., Aydinli, A., Rossetti, M., Li, L., Fiore, A., Dagli, N., 2008. Electro-optic and electro-absorption characterization of InAs quantum dot waveguides. Opt. Express 16 (5), 3439. Optical Society of America.

Alcotte, R., Martin, M., Moeyaert, J., Cipro, R., David, S., Bassani, F., Ducroquet, F., Bogumilowicz, Y., Sanchez, E., Ye, Z., Bao, X.Y., 2016. Epitaxial growth of antiphase boundary free GaAs layer on 300 mm Si (001) substrate by metalorganic chemical vapour deposition with high mobility. APL Mater. 4 (4), 46101. AIP Publishing.

Arakawa, Y., Sakaki, H., 1982. Multidimensional quantum well laser and temperature dependence of its threshold current. Appl. Phys. Lett. 40, 939.

Arbabi, A., Kamali, S.M., Arbabi, E., Griffin, B.G., Goddard, L.L., 2015. Grating integrated single mode microring laser. Opt. Express 23 (4), 5335.

Bhasker, P., Norman, J., Bowers, J., Dagli, N., 2018. Intensity and phase modulators at 1.55 μm with InAs/InGaAs quantum dots epitaxially grown on silicon. In: CLEO Pacific Rim Conference. OSA, Washington, D.C., p. Th2J.3.

Bhattacharya, P., Mi, Z., 2007. Quantum-dot optoelectronic devices. Proc. IEEE 95 (9), 1723–1740.
Bhattacharya, P., Ghosh, S., Pradhan, S., Singh, J., Zong-Kwei, W., Urayama, J., Kim, K., Norris, T.B., 2003. Carrier dynamics and high-speed modulation properties of tunnel injection InGaAs-GaAs quantum-dot lasers. IEEE J. Quantum Electron. 39 (8), 952–962.
Bhattacharya, P., Su, X., Ariyawansa, G., Perera, A.U., 2007. High-temperature tunneling quantum-dot intersublevel detectors for mid-infrared to terahertz frequencies. Proc. IEEE 95 (9), 1828–1837.
Billah, M.R., Blaicher, M., Hoose, T., Dietrich, P.I., Marin-Palomo, P., Lindenmann, N., Nesic, A., Hofmann, A., Troppenz, U., Moehrle, M., Randel, S., 2018. Hybrid integration of silicon photonics circuits and InP lasers by photonic wire bonding. Optica 5 (7), 876–883.
Bimberg, D., Pohl, U.W., 2011. Quantum dots: promises and accomplishments. Mater. Today 14 (9), 388–397. Elsevier.
Bimberg, D., Grundmann, M., Heinrichsdorff, F., Ledentsov, N.N., Ustinov, V.M., Zhukov, A.E., Kovsh, A.R., Maximov, M.V., Shernyakov, Y.M., Volovik, B.V., Tsatsul'nikov, A.F., 2000. Quantum dot lasers: breakthrough in optoelectronics. Thin Solid Films 367 (1–2), 235–249.
Black, L.E., Cavalli, A., Verheijen, M.A., Haverkort, J.E.M., Bakkers, E.P.A.M., Kessels, W.M.M., 2017. Effective surface passivation of InP nanowires by atomic-layer-deposited Al2O3 with PO x interlayer. Nano Lett. 17 (10), 6287–6294.
Bolkhovityanov, Y.B., Pchelyakov, O.P., 2008. GaAs epitaxy on Si substrates: modern status of research and engineering. Physics-Uspekhi 51 (5), 437. IOP Publishing.
Bowers, J.E., Hemenway, B., Gnauck, A., Wilt, D., 1986. High-speed InGaAsP constricted-mesa lasers. IEEE J. Quantum Electron. 22 (6), 833–844.
Bowers, J.E., Huang, D., Jung, D., Norman, J., Tran, M.A., Wan, Y., Xie, W., Zhang, Z., 2019. Realities and challenges of III-V/Si integration technologies. In: Optical Fiber Communication Conference. Optical Society of America, p. Tu3E-1. March.
Chen, S., Tang, M.C., Wu, J., Jiang, Q., Dorogan, V.G., Benamara, M., Mazur, Y.I., Salamo, G.J., Seeds, A.J., Liu, H., 2014. 1.3 μm InAs/GaAs quantum-dot laser monolithically grown on Si substrates operating over 100°. Electron. Lett. 50 (20), 1467–1468.
Chen, C.H., Seyedi, M.A., Fiorentino, M., Livshits, D., Gubenko, A., Mikhrin, S., Mikhrin, V., Beausoleil, R.G., 2015. A comb laser-driven DWDM silicon photonic transmitter based on microring modulators. Opt. Express 23 (16), 21541–21548.
Chen, S., Li, W., Wu, J., Jiang, Q., Tang, M., Shutts, S., Elliott, S.N., Sobiesierski, A., Seeds, A.J., Ross, I., Smowton, P.M., 2016. Electrically pumped continuous-wave III–V quantum dot lasers on silicon. Nat. Photonics 10 (5), 307.
Chen, S., Liao, M., Tang, M., Wu, J., Martin, M., Baron, T., Seeds, A., Liu, H., 2017. Electrically pumped continuous-wave 1.3 μm InAs/GaAs quantum dot lasers monolithically grown on on-axis Si (001) substrates. Opt. Express 25 (5), 4632–4639.
Cheng, Q., Bahadori, M., Glick, M., Rumley, S., Bergman, K., 2018. Recent advances in optical technologies for data centers: a review. Optica 5 (11), 1354–1370.
Chow, W.W., Jahnke, F., Gies, C., 2014. Emission properties of nanolasers during the transition to lasing. Light-Sci. Appl. 3 (8), e201.
Coldren, L.A., Nicholes, S.C., Johansson, L., Ristic, S., Guzzon, R.S., Norberg, E.J., Krishnamachari, U., 2011. High performance InP-based photonic ICs—a tutorial. J. Lightwave Technol. 29 (4), 554–570.
Coldren, L.A., Corzine, S.W., Mashanovitch, M.L., 2012. Diode Lasers and Photonic Integrated Circuits. vol. 218. John Wiley & Sons.
Coleman, J.J., Young, J.D., Garg, A., 2011. Semiconductor quantum dot lasers: a tutorial. J. Lightwave Technol. 29 (4), 499–510.

De Dood, M.J.A., Polman, A., Zijlstra, T., Van der Drift, E.W.J.M., 2002. Amorphous silicon waveguides for microphotonics. J. Appl. Phys. 92 (2), 649–653.
De La Rue, R., Lourtioz, J.-M., Yu, S., 2014. Compact Semiconductor Lasers. John Wiley & Sons.
Dingle, R., Henry, C.H., 1976. Quantum Effects in Heterostructure Lasers. U.S. Patent and Trademark Office, Washington, DC. U.S. Patent 3,982,207, issued September 21, 1976.
Duan, J., Huang, H., Dong, B., Jung, D., Norman, J.C., Bowers, J.E., Grillot, F., 2019. 1.3-μm reflection insensitive InAs/GaAs quantum dot lasers directly grown on silicon. IEEE Photon. Technol. Lett. 31 (5), 345–348.
El-Masry, N.A., Tarn, J.C., Karam, N.H., 1988. Interactions of dislocations in GaAs grown on Si substrates with InGaAs-GaAsP strained layered superlattices. J. Appl. Phys. 64 (7), 3672–3677. AIP.
Fan, Y.H., Liang, D., Roshan-Zamir, A., Zhang, C., Wang, B., Fiorentino, M., Beausoleil, R., Palermo, S., 2019. A directly modulated quantum dot microring laser transmitter with integrated CMOS driver. March, In: Optical Fiber Communication Conference. Optical Society of America, p. W3E-5.
Fang, S.F., Adomi, K., Iyer, S., Morkoc, H., Zabel, H., Choi, C., Otsuka, N., 1990. Gallium arsenide and other compound semiconductors on silicon. J. Appl. Phys. 68 (7), R31–R58.
Fathpour, S., Mi, Z., Bhattacharya, P., Kovsh, A.R., Mikhrin, S.S., Krestnikov, I.L., Kozhukhov, A.V., Ledentsov, N.N., 2004. The role of auger recombination in the temperature-dependent output characteristics ($T_0=\infty$) of p-doped 1.3 μm quantum dot lasers. Appl. Phys. Lett. 85 (22), 5164–5166.
Feng, L., Wong, Z.J., Ma, R.M., Wang, Y., Zhang, X., 2014. Single-mode laser by parity-time symmetry breaking. Science 346 (6212), 972–975.
Feng, Q., Wei, W., Zhang, B., Wang, H., Wang, J., Cong, H., Wang, T., Zhang, J., 2019. O-band and C/L-band III-V quantum dot lasers monolithically grown on Ge and Si substrate. Appl. Sci. 9 (3), 385.
Garrett, C.G.B., Kaiser, W., Bond, W.L., 1961. Stimulated emission into optical whispering modes of spheres. Phys. Rev. 124 (6), 1807.
Gayral, B., Gérard, J.M., Lemaıtre, A., Dupuis, C., Manin, L., Pelouard, J.L., 1999. High-Q wet-etched GaAs microdisks containing InAs quantum boxes. Appl. Phys. Lett. 75 (13), 1908–1910.
Gies, C., Wiersig, J., Lorke, M., Jahnke, F., 2007. Semiconductor model for quantum-dot-based microcavity lasers. Phys. Rev. A 75 (1), 013803.
Groenert, M.E., Leitz, C.W., Pitera, A.J., Yang, V., Lee, H., Ram, R.J., Fitzgerald, E.A., 2003. Monolithic integration of room-temperature cw GaAs/AlGaAs lasers on Si substrates via relaxed graded GeSi buffer layers. J. Appl. Phys. 93 (1), 362–367.
Guimard, D., Ishida, M., Bordel, D., Li, L., Nishioka, M., Tanaka, Y., Ekawa, M., Sudo, H., Yamamoto, T., Kondo, H., Sugawara, M., 2010. Ground state lasing at 1.30 μm from InAs/GaAs quantum dot lasers grown by metal–organic chemical vapor deposition. Nanotechnology 21 (10), 105604.
Han, Y., Ng, W.K., Xue, Y., Li, Q., Wong, K.S., Lau, K.M., 2019. Telecom InP/InGaAs nanolaser array directly grown on (001) silicon-on-insulator. Opt. Lett. 44 (4), 767–770.
Harayama, T., Shinohara, S., 2011. Two-dimensional microcavity lasers. Laser Photonics Rev. 5 (2), 247–271.
Harke, A., Krause, M., Mueller, J., 2005. Low-loss singlemode amorphous silicon waveguides. Electron. Lett. 41 (25), 1377–1379.
He, L., Özdemir, Ş.K., Yang, L., 2013. Whispering gallery microcavity lasers. Laser Photonics Rev. 7 (1), 60–82.
Hein, S., Podemski, P., Sęk, G., Misiewicz, J., Ridha, P., Fiore, A., Patriarche, G., Höfling, S., Forchel, A., 2009. Orientation dependent emission properties of columnar quantum dash laser structures. Appl. Phys. Lett. 94 (24), 241113.

Heitz, R., Mukhametzhanov, I., Madhukar, A., Hoffmann, A., Bimberg, D., 1999. Temperature dependent optical properties of self-organized InAs/GaAs quantum dots. J. Electron. Mater. 28 (5), 520–527.
Hill, M.T., Anantathanasarn, S., Zhu, Y., Oei, Y.S., Van Veldhoven, P.J., Smit, M.K., Notzel, R., 2008. InAs–InP (1.55-μm region) quantum-dot microring lasers. IEEE Photon. Technol. Lett. 20 (6), 446–448.
Hodaei, H., Miri, M.A., Heinrich, M., Christodoulides, D.N., Khajavikhan, M., 2014. Parity-time–symmetric microring lasers. Science 346 (6212), 975–978.
Ide, T., Baba, T., Tatebayashi, J., Iwamoto, S., Nakaoka, T., Arakawa, Y., 2005. Room temperature continuous wave lasing in InAs quantum-dot microdisks with air cladding. Opt. Express 13 (5), 1615–1620.
Ikku, Y., Yokoyama, M., Takenaka, M., Takagi, S., Ichikawa, O., Hata, M., 2012. Propagation-loss reduction in InGaAsP photonic-wire waveguides by InP and Al_2O_3 passivation layers. In: ECIO 2012, 25–28.
Inoue, D., Jung, D., Norman, J., Wan, Y., Nishiyama, N., Arai, S., Gossard, A.C., Bowers, J.E., 2018a. Directly modulated 1.3 μm quantum dot lasers epitaxially grown on silicon. Opt. Express 26 (6), 7022–7033.
Inoue, D., Wan, Y., Jung, D., Norman, J., Shang, C., Nishiyama, N., Arai, S., Gossard, A.C., Bowers, J.E., 2018b. Low-dark current 10 Gbit/s operation of InAs/InGaAs quantum dot pin photodiode grown on on-axis (001) GaP/Si. Appl. Phys. Lett. 113 (9), 093506.
Inoue, D., Jung, D., Norman, J., Wan, Y., Nishiyama, N., Arai, S., Gossard, A.C., Bowers, J.E., 2018c. NRZ and PAM-4 direct modulation of 1.3 μm quantum dot lasers grown directly on on-axis (001) Si. In: 2018 European Conference on Optical Communication (ECOC). IEEE, pp. 1–3. September.
Jones, R., Doussiere, P., Driscoll, J.B., Lin, W., Yu, H., Akulova, Y., Komljenovic, T., Bowers, J.E., 2019. Heterogeneously integrated InP\/silicon photonics: fabricating fully functional transceivers. IEEE Nanotechnol. Mag. 13 (2), 17–26.
Jung, D., Norman, J., Kennedy, M.J., Shang, C., Shin, B., Wan, Y., Gossard, A.C., Bowers, J.E., 2017a. High efficiency low threshold current 1.3 μm InAs quantum dot lasers on on-axis (001) GaP/Si. Appl. Phys. Lett. 111 (12), 122107.
Jung, D., Zhang, Z., Norman, J., Herrick, R., Kennedy, M.J., Patel, P., Turnlund, K., Jan, C., Wan, Y., Gossard, A.C., Bowers, J.E., 2017b. Highly reliable low-threshold InAs quantum dot lasers on on-axis (001) Si with 87% injection efficiency. ACS Photonics 5 (3), 1094–1100.
Kageyama, T., Nishi, K., Yamaguchi, M., Mochida, R., Maeda, Y., Takemasa, K., Tanaka, Y., Yamamoto, T., Sugawara, M., Arakawa, Y., 2011. Extremely high temperature (220° C) continuous-wave operation of 1300-nm-range quantum-dot lasers. In: The European Conference on Lasers and Electro-Optics. Optical Society of America, p. PDA_1. May.
Kazi, Z.I., Thilakan, P., Egawa, T., Umeno, M., Jimbo, T., 2001. Realization of GaAs/AlGaAs lasers on Si substrates using epitaxial lateral overgrowth by metalorganic chemical vapor deposition. Jpn. J. Appl. Phys. 40 (8R), 4903.
Khajavikhan, M., Simic, A., Katz, M., Lee, J.H., Slutsky, B., Mizrahi, A., Lomakin, V., Fainman, Y., 2012. Thresholdless nanoscale coaxial lasers. Nature 482 (7384), 204.
Khan, M.Z.M., Ng, T.K., Ooi, B.S., 2014. Self-assembled InAs/InP quantum dots and quantum dashes: material structures and devices. Prog. Quantum Electron. 38 (6), 237–313. Pergamon.
Kobayashi, W., Ito, T., Yamanaka, T., Fujisawa, T., Shibata, Y., Kurosaki, T., Kohtoku, M., Tadokoro, T., Sanjoh, H., 2013. 50-Gb/s direct modulation of a 1.3-μm InGaAlAs-based DFB laser with a ridge waveguide structure. IEEE J. Sel. Top. Quantum Electron. 19 (4), 1500908.
Kryzhanovskaya, N.V., Maximov, M.V., Zhukov, A.E., 2014. Whispering-gallery mode microcavity quantum-dot lasers. Quantum Electron. 44 (3), 189. IOP Publishing.

Kryzhanovskaya, N.V., Zhukov, A.E., Maximov, M.V., Moiseev, E.I., Shostak, I.I., Nadtochiy, A.M., Kudashova, Y.V., Lipovskii, A.A., Kulagina, M.M., Troshkov, S.I., 2015a. Room temperature lasing in 1-μm microdisk quantum dot lasers. IEEE J. Sel. Top. Quantum Electron. 21 (6), 709–713.

Kryzhanovskaya, N.V., Maximov, M.V., Zhukov, A.E., Nadtochiy, A.M., Moiseev, E.I., Shostak, I.I., Kulagina, M.M., Vashanova, K.A., Zadiranov, Y.M., Troshkov, S.I., Nevedomsky, V.V., 2015b. Single-mode emission from 4–9-μm microdisk lasers with dense array of InGaAs quantum dots. J. Lightwave Technol. 33 (1), 171–175.

Kunert, B., Mols, Y., Baryshniskova, M., Waldron, N., Schulze, A., Langer, R., 2018. How to control defect formation in monolithic III/V hetero-epitaxy on (100) Si? A critical review on current approaches. Semicond. Sci. Technol. 33 (9), 093002.

Kwoen, J., Jang, B., Lee, J., Kageyama, T., Watanabe, K., Arakawa, Y., 2018. All MBE grown InAs/GaAs quantum dot lasers on on-axis Si (001). Opt. Express 26 (9), 11568–11576.

Lee, J.W., Shichijo, H., Tsai, H.L., Matyi, R.J., 1987. Defect reduction by thermal annealing of GaAs layers grown by molecular beam epitaxy on Si substrates. Appl. Phys. Lett. 50 (1), 31–33.

Lee, A., Jiang, Q., Tang, M., Seeds, A., Liu, H., 2012. Continuous-wave InAs/GaAs quantum-dot laser diodes monolithically grown on Si substrate with low threshold current densities. Opt. Express 20 (20), 22181–22187.

Li, Q., Lau, K.M., 2017. Epitaxial growth of highly mismatched III-V materials on (001) silicon for electronics and optoelectronics. Prog. Cryst. Growth Charact. Mater. 63 (4), 105–120. Elsevier.

Li, Q., Wan, Y., Liu, A.Y., Gossard, A.C., Bowers, J.E., Hu, E.L., Lau, K.M., 2016. 1.3-μm InAs quantum-dot micro-disk lasers on V-groove patterned and unpatterned (001) silicon. Opt. Express 24 (18), 21038–21045.

Liang, D., Fiorentino, M., Okumura, T., Chang, H.H., Spencer, D.T., Kuo, Y.H., Fang, A.W., Dai, D., Beausoleil, R.G., Bowers, J.E., 2009. Electrically-pumped compact hybrid silicon microring lasers for optical interconnects. Opt. Express 17 (22), 20355–20364.

Linder, K.K., Phillips, J., Qasaimeh, O., Liu, X.F., Krishna, S., Bhattacharya, P., Jiang, J.C., 1999. Self-organized $In_{0.4}Ga_{0.6}As$ quantum-dot lasers grown on Si substrates. Appl. Phys. Lett. 74 (10), 1355–1357.

Liu, A.Y., Bowers, J., 2018. Photonic integration with epitaxial III–V on silicon. IEEE J. Sel. Top. Quantum Electron. 24 (6), 1–12.

Liu, A.Y., Zhang, C., Norman, J., Snyder, A., Lubyshev, D., Fastenau, J.M., Liu, A.W., Gossard, A.C., Bowers, J.E., 2014. High performance continuous wave 1.3 μm quantum dot lasers on silicon. Appl. Phys. Lett. 104 (4), 041104.

Liu, A.Y., Srinivasan, S., Norman, J., Gossard, A.C., Bowers, J.E., 2015a. Quantum dot lasers for silicon photonics. Photonics Res. 3 (5), B1–B9.

Liu, A.Y., Herrick, R.W., Ueda, O., Petroff, P.M., Gossard, A.C., Bowers, J.E., 2015b. Reliability of InAs/GaAs quantum dot lasers epitaxially grown on silicon. IEEE J. Sel. Top. Quantum Electron. 21 (6), 690–697.

Liu, S., Wu, X., Jung, D., Norman, J.C., Kennedy, M.J., Tsang, H.K., Gossard, A.C., Bowers, J.E., 2019. High-channel-count 20 GHz passively mode-locked quantum dot laser directly grown on Si with 4.1 Tbit/s transmission capacity. Optica 6 (2), 128–134.

Lv, X.M., Huang, Y.Z., Zou, L.X., Long, H., Du, Y., 2013. Optimization of direct modulation rate for circular microlasers by adjusting mode Q factor. Laser Photonics Rev. 7 (5), 818–829.

Mao, M.H., Chien, H.C., Hong, J.Z., Cheng, C.Y., 2011. Room-temperature low-threshold current-injection InGaAs quantum-dot microdisk lasers with single-mode emission. Opt. Express 19 (15), 14145–14151.

Meuer, C., Kim, J., Laemmlin, M., Liebich, S., Eisenstein, G., Bonk, R., Vallaitis, T., Leuthold, J., Kovsh, A., Krestnikov, I., Bimberg, D., 2009. High-speed small-signal cross-gain modulation in quantum-dot semiconductor optical amplifiers at 1.3 μm. IEEE J. Sel. Top. Quantum Electron. 15 (3), 749–756.
Mi, Z., Yang, J., Bhattacharya, P., Qin, G., Ma, Z., 2009. High-performance quantum dot lasers and integrated optoelectronics on Si. Proc. IEEE 97 (7), 1239–1249.
Michel, J., Liu, J., Kimerling, L.C., 2010. High-performance Ge-on-Si photodetectors. Nat. Photonics 4 (8), 527.
Moiseev, E.I., Kryzhanovskaya, N., Polubavkina, Y.S., Maximov, M.V., Kulagina, M.M., Zadiranov, Y.M., Lipovskii, A.A., Mukhin, I.S., Mozharov, A.M., Komissarenko, F.E., Sadrieva, Z.F., 2017. Light outcoupling from quantum dot-based microdisk laser via plasmonic nanoantenna. ACS Photonics 4 (2), 275–281.
Moore, S.A., O'Faolain, L., Cataluna, M.A., Flynn, M.B., Kotlyar, M.V., Krauss, T.F., 2006. Reduced surface sidewall recombination and diffusion in quantum-dot lasers. IEEE Photon. Technol. Lett. 18 (17), 1861–1863.
Munnelly, P., Heindel, T., Thoma, A., Kamp, M., Höfling, S., Schneider, C., Reitzenstein, S., 2017. Electrically tunable single-photon source triggered by a monolithically integrated quantum dot microlaser. ACS Photonics 4 (4), 790–794.
Nagarajan, R., Ishikawa, M., Fukushima, T., Geels, R.S., Bowers, J.E., 1992. High speed quantum-well lasers and carrier transport effects. IEEE J. Sel. Top. Quantum Electron. 28 (10), 1990–2008.
No, Y.S., 2019. Electrically driven micro-and nano-scale semiconductor light sources. Appl. Sci. 9 (4), 802.
Noda, S., Fujita, M., Asano, T., 2007. Spontaneous-emission control by photonic crystals and nanocavities. Nat. Photonics 1, 449.
Norman, J.C., Jung, D., Wan, Y., Bowers, J.E., 2018. Perspective: the future of quantum dot photonic integrated circuits. APL Photonics 3 (3), 030901.
Norman, J.C., Jung, D., Zhang, Z., Wan, Y., Liu, S., Shang, C., Herrick, R.W., Chow, W.W., Gossard, A.C., Bowers, J.E., 2019. A review of high-performance quantum dot lasers on silicon. IEEE J. Quantum Electron. 55 (2), 1–11.
Nosich, A.I., Smotrova, E.I., Boriskina, S.V., Benson, T.M., Sewell, P., 2007. Trends in microdisk laser research and linear optical modelling. Opt. Quant. Electron. 39 (15), 1253–1272.
Ooi, B.S., Djie, H.S., Wang, Y., Tan, C.L., Hwang, J.C., Fang, X.M., Fastenau, J.M., Liu, A.W., Dang, G.T., Chang, W.H., 2008. Quantum dashes on InP substrate for broadband emitter applications. IEEE J. Sel. Top. Quantum Electron. 14 (4), 1230–1238.
Otto, C., 2014. Dynamics of Quantum Dot Lasers: Effects of Optical Feedback and External Optical Injection. Springer Science & Business Media.
Popescu, V., Bester, G., Hanna, M.C., Norman, A.G., Zunger, A., 2008. Theoretical and experimental examination of the intermediate-band concept for strain-balanced (In, Ga) As/Ga (As, P) quantum dot solar cells. Phys. Rev. B 78 (20), 205321.
Rafailov, E.U., Cataluna, M.A., Sibbett, W., 2007. Mode-locked quantum-dot lasers. Nat. Photonics 1 (7), 395.
Rayleigh, L., 1910. CXII. The problem of the whispering gallery. Lond. Edinb. Dublin Philos. Mag. 20 (120), 1001–1004.
Sadeev, T., Arsenijević, D., Bimberg, D., 2016. Comparison of dynamic properties of InP/InAs quantum-dot and quantum-dash lasers. Appl. Phys. Lett. 109 (16), 161104.
Schow, C.L., Koester, S.J., Schares, L., Dehlinger, G., John, R.A., 2007. High-speed, low-voltage optical receivers consisting of Ge-on-SOI photodiodes paired with CMOS ICs. Silicon Photonics II, vol. 6477. International Society for Optics and Photonics, 647705.

Scifres, D.R., Burnham, R.D., Streifer, W., 1976. Grating-coupled GaAs single heterostructure ring laser. Appl. Phys. Lett. 28 (11), 681–683.
Shang, C., Wan, Y., Jung, D., Norman, J., Kennedy, M.J., Liang, D., Zhang, C., Gossard, A.C., Bowers, J.E., 2018. Quantum dot micro-lasers integrated with photodetectors and optical amplifiers on (001) Si via waveguide coupling. MayIn: 2018 Conference on Lasers and Electro-Optics (CLEO). IEEE, pp. 1–2.
Shi, Y., Wang, Z., Van Campenhout, J., Pantouvaki, M., Guo, W., Kunert, B., Van Thourhout, D., 2017a. Optical pumped InGaAs/GaAs nano-ridge laser epitaxially grown on a standard 300-mm Si wafer. Optica 4 (12), 1468–1473.
Shi, B., Zhu, S., Li, Q., Wan, Y., Hu, E.L., Lau, K.M., 2017b. Continuous-wave optically pumped 1.55 μm InAs/InAlGaAs quantum dot microdisk lasers epitaxially grown on silicon. ACS Photonics 4 (2), 204–210.
Slusher, R.E., Levi, A.F.J., Mohideen, U., McCall, S.L., Pearton, S.J., Logan, R.A., 1993. Threshold characteristics of semiconductor microdisk lasers. Appl. Phys. Lett. 63 (10), 1310–1312.
Song, Q., 2019. Emerging opportunities for ultra-high Q whispering gallery mode microcavities. Sci. China Phys. Mech. Astron. 62 (7), 074231.
Spencer, R.M., Greenberg, J., Eastman, L.F., Tsai, C.Y., O'Keefe, S.S., 1997. High-speed direct modulation of semiconductor lasers. Int. J. High Speed Electron. Syst. 8 (03), 417–456.
Stintz, A., Liu, G.T., Li, H., Lester, L.F., Malloy, K.J., 2000. Low-threshold current density 1.3-μm InAs quantum-dot lasers with the dots-in-a-well (DWELL) structure. IEEE Photon. Technol. Lett. 12 (6), 591–593.
Stock, E., Albert, F., Hopfmann, C., Lermer, M., Schneider, C., Höfling, S., Forchel, A., Kamp, M., Reitzenstein, S., 2013. On-chip quantum optics with quantum dot microcavities. Adv. Mater. 25 (5), 707–710.
Strauf, S., Hennessy, K., Rakher, M.T., Choi, Y.S., Badolato, A., Andreani, L.C., Hu, E.L., Petroff, P.M., Bouwmeester, D., 2006. Self-tuned quantum dot gain in photonic crystal lasers. Phys. Rev. Lett. 96 (12), 127404.
Sugawara, M., Usami, M., 2009. Quantum dot devices: handling the heat. Nat. Photonics 3 (1), 30.
Tamboli, A.C., Haberer, E.D., Sharma, R., Lee, K.H., Nakamura, S., Hu, E.L., 2007. Room-temperature continuous-wave lasing in GaN/InGaN microdisks. Nat. Photonics 1 (1), 61.
Tanaka, Y., Ishida, M., Takada, K., Yamamoto, T., Song, H.Z., Nakata, Y., Yamaguchi, M., Nishi, K., Sugawara, M., Arakawa, Y., 2010. May. 25 Gbps direct modulation in 1.3-μm InAs/GaAs high-density quantum dot lasers. In: CLEO/QELS: 2010 Laser Science to Photonic Applications. IEEE, pp. 1–2.
Ueda, O., Pearton, S.J. (Eds.), 2012. Materials and Reliability Handbook for Semiconductor Optical and Electron Devices. Springer Science & Business Media.
Van Campenhout, J., Rojo-Romeo, P., Regreny, P., Seassal, C., Van Thourhout, D., Verstuyft, S., Di Cioccio, L., Fedeli, J.M., Lagahe, C., Baets, R., 2007. Electrically pumped InP-based microdisk lasers integrated with a nanophotonic silicon-on-insulator waveguide circuit. Opt. Express 15 (11), 6744–6749.
Vyrsokinos, K., Chatzitheocharis, D., Papadovasilakis, M., Ketzaki, D., Calo, C., Caillaud, C., Sacchetto, D., Zervas, M., Reithmaier, J.P., Sichkovskyi, V., Eisenstein, G., 2019. MOICANA: monolithic cointegration of QD-based InP on SiN as a versatile platform for the demonstration of high-performance and low-cost PIC transmitters. March, In: Optical Interconnects XIX. vol. 10924. International Society for Optics and Photonics, p. 1092410.
Wan, Y., Li, Q., Geng, Y., Shi, B., Lau, K.M., 2015. InAs/GaAs quantum dots on GaAs-on-V-grooved-Si substrate with high optical quality in the 1.3 μm band. Appl. Phys. Lett. 107 (8), 081106.

Wan, Y., Li, Q., Liu, A.Y., Gossard, A.C., Bowers, J.E., Hu, E.L., Lau, K.M., 2016a. Optically pumped 1.3 μm room-temperature InAs quantum-dot micro-disk lasers directly grown on (001) silicon. Opt. Lett. 41 (7), 1664–1667.

Wan, Y., Li, Q., Liu, A.Y., Chow, W.W., Gossard, A.C., Bowers, J.E., Hu, E.L., Lau, K.M., 2016b. Sub-wavelength InAs quantum dot micro-disk lasers epitaxially grown on exact Si (001) substrates. Appl. Phys. Lett. 108 (22), 221101.

Wan, Y., Li, Q., Liu, A.Y., Gossard, A.C., Bowers, J.E., Hu, E.L., Lau, K.M., 2016c. Temperature characteristics of epitaxially grown InAs quantum dot micro-disk lasers on silicon for on-chip light sources. Appl. Phys. Lett. 109 (1), 011104.

Wan, Y., Norman, J., Li, Q., Kennedy, M.J., Liang, D., Zhang, C., Huang, D., Zhang, Z., Liu, A.Y., Torres, A., Jung, D., Gossard, A.C., Hu, E.L., Lau, K.M., Bowers, J.E., 2017a. 1.3 μm submilliamp threshold quantum dot micro-lasers on Si. Optica 4 (8), 940–944.

Wan, Y., Zhang, Z., Chao, R., Norman, J., Jung, D., Shang, C., Li, Q., Kennedy, M.J., Liang, D., Zhang, C., Shi, J.W., Gossard, A.C., Lau, K.M., Bowers, J.E., 2017b. Monolithically integrated InAs/InGaAs quantum dot photodetectors on silicon substrates. Opt. Express 25 (22), 27715–27723.

Wan, Y., Jung, D., Norman, J., Shang, C., MacFarlane, I., Li, Q., Kennedy, M.J., Gossard, A.C., Lau, K.M., Bowers, J.E., 2017c. O-band electrically injected quantum dot micro-ring lasers on on-axis (001) GaP/Si and V-groove Si. Opt. Express 25 (22), 26853–26860.

Wan, Y., Li, Q., Liu, A.Y., Geng, Y., Norman, J., Chow, W.W., Gossard, A.C., Bowers, J.E., Hu, E.L., Lau, K.M., 2017d. Quantum dot lasers grown on (001) Si substrate for integration with amorphous Si waveguides. March, In: 2017 Optical Fiber Communications Conference and Exhibition (OFC). IEEE, pp. 1–3.

Wan, Y., Norman, J., Li, Q., Kennedy, M.J., Liang, D., Zhang, C., Huang, D., Liu, A.Y., Torres, A., Jung, D., Gossard, A.C., Hu, E.L., Lau, K.M., Bowers, J.E., 2017e. Sub-mA threshold 1.3 μm CW lasing from electrically pumped micro-rings grown on (001) Si. May, In: 2017 Conference on Lasers and Electro-Optics (CLEO). IEEE, pp. 1–2.

Wan, Y., Inoue, D., Jung, D., Norman, J.C., Shang, C., Gossard, A.C., Bowers, J.E., 2018a. Directly modulated quantum dot lasers on silicon with a milliampere threshold and high temperature stability. Photonics Res. 6 (8), 776–781.

Wan, Y., Jung, D., Inoue, D., Norman, J.C., Shang, C., Cossard, A.C., Bowers, J.E., 2018b. On-chip detection from directly modulated quantum dot microring lasers on Si. August, In: 2018 Progress in Electromagnetics Research Symposium (PIERS-Toyama). IEEE, pp. 249–253.

Wan, Y., Jung, D., Norman, J., Feng, K., Dagli, A., Gossard, A.C., Bowers, J.E., 2018c. Quadruple reduction of threshold current density for micro-ring quantum dot lasers epitaxially grown on (001) Si. May, In: CLEO: Science and Innovations. Optical Society of America, p. SW3Q-3.

Wan, Y., Jung, D., Shang, C., Collins, N., MacFarlane, I., Norman, J., Dumont, M., Gossard, A.C., Bowers, J.E., 2019. Low-threshold continuous-wave operation of electrically pumped 1.55 μm InAs quantum dash microring lasers. ACS Photonics 6 (2), 279–285.

Wang, R.H., Stintz, A., Varangis, P.M., Newell, T.C., Li, H., Malloy, K.J., Lester, L.F., 2001. Room-temperature operation of InAs quantum-dash lasers on InP [001]. IEEE Photon. Technol. Lett. 13 (8), 767–769.

Wang, T., Liu, H., Lee, A., Pozzi, F., Seeds, A., 2011. 1.3-μm InAs/GaAs quantum-dot lasers monolithically grown on Si substrates. Opt. Express 19 (12), 11381–11386.

Wang, Z., Tian, B., Pantouvaki, M., Guo, W., Absil, P., Van Campenhout, J., Merckling, C., Van Thourhout, D., 2015. Room-temperature InP distributed feedback laser array directly grown on silicon. Nat. Photonics 9 (12), 837.

Wang, Z., Abbasi, A., Dave, U., De Groote, A., Kumari, S., Kunert, B., Merckling, C., Pantouvaki, M., Shi, Y., Tian, B., Van Gasse, K., 2017. Novel light source integration approaches for silicon photonics. Laser Photonics Rev. 11 (4), 1700063.

Wei, W.Q., Wang, J.H., Zhang, B., Zhang, J.Y., Wang, H.L., Feng, Q., Xu, H.X., Wang, T., Zhang, J.J., 2018. InAs QDs on (111)-faceted Si (001) hollow substrates with strong emission at 1300 nm and 1550 nm. Appl. Phys. Lett. 113 (5), 053107.

Wojcik, G.L., Yin, D., Kovsh, A.R., Gubenko, A.E., Krestnikov, I.L., Mikhrin, S.S., Livshits, D.A., Fattal, D.A., Fiorentino, M., Beausoleil, R.G., 2009. A single comb laser source for short reach WDM interconnects. February, In: Novel in-Plane Semiconductor Lasers VIII. vol. 7230. International Society for Optics and Photonics, p. 72300M.

Wu, J., Jin, P., 2015. Self-assembly of InAs quantum dots on GaAs (001) by molecular beam epitaxy. Front. Phys. 10 (1), 7–58.

Yang, J., Bhattacharya, P., 2008. Integration of epitaxially-grown InGaAs GaAs quantum dot lasers with hydrogenated amorphous silicon waveguides on silicon. Opt. Express 16 (7), 5136–5140.

Yang, T., Mock, A., O'Brien, J.D., Lipson, S., Deppe, D.G., 2007. Lasing characteristics of InAs quantum dot microcavity lasers as a function of temperature and wavelength. Opt. Express 15 (12), 7281–7289.

Yang, Y.D., Zhang, Y., Huang, Y.Z., Poon, A.W., 2014. Direct-modulated waveguide-coupled microspiral disk lasers with spatially selective injection for on-chip optical interconnects. Opt. Express 22 (1), 824–838.

Yao, Z., Wu, K., Tan, B.X., Wang, J., Li, Y., Zhang, Y., Poon, A.W., 2018. Integrated silicon photonic microresonators: emerging technologies. IEEE J. Sel. Top. Quantum Electron. 24 (6), 1–24.

Yaung, K.N., Vaisman, M., Lang, J., Lee, M.L., 2016. GaAsP solar cells on GaP/Si with low threading dislocation density. Appl. Phys. Lett. 109 (3), 032107.

Yuan, H.-C., Ma, Z., 2006. Microwave thin-film transistors using Si nanomembranes on flexible polymer substrate. Appl. Phys. Lett. 89 (21), 212105.

Zhang, L., Hu, E., 2003. Lasing from InGaAs quantum dots in an injection microdisk. Appl. Phys. Lett. 82 (3), 319–321.

Zhang, Z., Jung, D., Norman, J.C., Patel, P., Chow, W.W., Bowers, J.E., 2018a. Effects of modulation p doping in InAs quantum dot lasers on silicon. Appl. Phys. Lett. 113 (6), 061105.

Zhang, Y., Su, Y., Bi, Y., Pan, J., Yu, H., Zhang, Y., Sun, J., Sun, X., Chong, M., 2018b. Inclined emitting slotted single-mode laser with 1.7° vertical divergence angle for PIC applications. Opt. Lett. 43 (1), 86–89.

Zhou, L., Bo, B., Yan, X., Wang, C., Chi, Y., Yang, X., 2018. Brief review of surface passivation on III-V semiconductor. Crystals 8 (5), 226.

Zhukov, A.E., Maksimov, M.V., Kovsh, A.R., 2012. Device characteristics of long-wavelength lasers based on self-organized quantum dots. Sem. Ther. 46 (10), 1225–1250.

Further reading

Koester, S.J., Schow, C.L., Schares, L., Dehlinger, G., Schaub, J.D., Doany, F.E., John, R.A., 2007. Ge-on-SOI-detector/Si-CMOS-amplifier receivers for high-performance optical-communication applications. J. Lightwave Technol. 25 (1), 46–57.

Kryzhanovskaya, N.V., Zhukov, A.E., Nadtochy, A.M., Slovinsky, I.A., Maximov, M.V., Kulagina, M.M., Savelev, A.V., Arakcheeva, E.M., Zadiranov, Y.M., Troshkov, S.I., Lipovskii, A.A., 2012. High-temperature lasing in a microring laser with an active region based on InAs/InGaAs quantum dots. Sem. Ther. 46 (8), 1040–1043.

Index

Note: Page numbers followed by "*f*" indicate figures.

R

S

U

V

FSC
www.fsc.org
MIX
Papier | Fördert gute Waldnutzung
FSC® C083411

Kolibri 360 GmbH
Ferdinand-Jühlke-Straße 7
99095 Erfurt, Deutschland
produktsicherheit@kolibri360.de